OPERATIONAL TOOLS IN THE MANAGEMENT OF FINANCIAL RISKS

OPERATIONAL TOOLS IN THE MANAGEMENT OF FINANCIAL RISKS

edited by
Constantin Zopounidis

Technical University of Crete
Dept. of Production Engineering and Management
Decision Support Systems Laboratory
University Campus
73100 Chania, Greece

Kluwer Academic Publishers

Distributors for North America:
Kluwer Academic Publishers
101 Philip Drive
Assinippi Park
Norwell, Massachusetts 02061 USA

Distributors for all other countries:
Kluwer Academic Publishers Group
Distribution Centre
Post Office Box 322
3300 AH Dordrecht, THE NETHERLANDS

Library of Congress Cataloging-in-Publication Data

Operational tools in the management of financial risks / edited by
Constantin Zopounidis.
p. cm.
Includes bibliographical references and index.
ISBN 0-7923-8055-X
1. Venture capital--Mathematical models. 2. Portfolio management--Mathematical models. 3. Risk management--Mathematical models. 4. Financial futures--Mathematical models. I. Zopounidis, Constantin.
HG4751.064 1997 97-35193
658.15'5--dc21 CIP

Printed on acid-free paper.

Printed in the United States of America

In the memory of my father Dimitris

Contents

III. Linear and Stochastic Programming in Portfolio Management

IV. Fuzzy Sets and Artificial Intelligence Techniques in Financial Decisions

V. Multicriteria Analysis in Country Risk Evaluation

Author Index

Editorial

The management of financial risks has become a very important task for every organization (i.e. firms, banks, insurance companies, etc.) in the 1990s. The importance of financial risks in organizations has been shown very recently by the published works of several authors such as Mulvey et al. (1997), Thomas (1992), Williams (1995), Zenios (1993), Ziemba and Mulvey (1996). In their work, first the financial risks are determined and second the scientific tools are developed to assess and to manage these risks. For example, Thomas (1992) suggests for the portfolio analysis problem the classical Markowitz model (Mean-Variance model), while for the credit scoring problem he cites some techniques such as discriminant analysis, logistic regression, mathematical programming and recursive partitioning. The use of optimization models in several fields of financial modeling has been explored in the work by Zenios (1993). Mulvey et al. (1997) in their invited review in the European Journal of Operational Research, propose the asset/liability management (ALM) via multi-stage stochastic optimization. According to the authors, "the ALM is an important dimension of risk management in which the exposure to various risks is minimized by holding the appropriate combination of assets and liabilities so as to meet the firms objectives". The application of multi-stage stochastic programming for managing asset-liability risk over extended time periods can be found in the book by Ziemba and Mulvey (1996). Williams (1995) provides a classified bibliography of recent work related to project risk management.

In parallel with the above work, some other new operational tools started to be applied in the assessment and management of financial risks coming from multicriteria analysis (an advanced field of operations research), decision support systems, chaos theory, fuzzy sets, artificial intelligence, etc. For example, the use of multicriteria analysis in the modeling of financial problems has been studied for three important reasons (cf. Zopounidis, 1997):

(1) Formulating the problem in terms of seeking the optimum, financial decision makers (i.e. financial analysts, portfolio managers, investors,, etc.) get involved in a very narrow problematic, often irrelevant to the real decision problem.

(2) The different financial decisions are taken by the people (i.e. financial managers) and not by the models; the decision makers get more and more involved in the decision making process and, in order to solve problems, it becomes necessary to take into consideration their preferences, their experiences and their knowledge.

(3) For financial decision problems such as the choice of investment projects, the portfolio selection, the evaluation of business failure risk, etc., it seems illusory to speak of optimality since multiple criteria must be taken into consideration.

The solution of some financial problems (i.e. venture capital investment, business failure risk, bond rating, country risk, choice of investment projects, portfolio management, financial planning, etc.) based on the logic of multiple criteria (i.e. multicriteria paradigm, cf. Roy, 1988), must take into account the following elements:

- multiple criteria;
- conflict situation among the criteria;
- complex evaluation process, subjective and ill-structured;
- introduction of financial decision makers in the evaluation process.

In the same way, tools coming from artificial intelligence (i.e. expert support systems, neural nets) and decision support systems contribute in an original way, in the solution of some financial decision problems (cf. Klein and Methlie, 1995). Furthermore, the combination of the above methodologies (multicriteria analysis, decision support systems and artificial intelligence) gives more powerful tools for the analysis and assessment of financial risks (i.e. the systems INVEST, CGX, CREDEX, FINEVA, cf. Heuer et al., 1988; Scrinivasan and Ruparel, 1990; Pinson, 1992; Zopounidis et al., 1996).

On the basis of the above remarks, the basic aim of this book is to present a set of new operational tools coming from multivariate statistical analysis, multicriteria analysis, mathematical programming, fuzzy sets and artificial intelligence for the assessment and the management of some financial risks in several organizations. In some papers in this volume, the authors proceed to the combination of classical methods and new ones in order to create methodological tools which are more powerful and suitable for the solution of the financial risk problem.

The present volume is divided in five chapters.

The first chapter involves three papers and refers to the application of the multivariate data analysis and multicriteria analysis in the classical problem of the portfolio selection. Two of the three papers combine classical methods (i.e. discriminant analysis and arbitrage pricing theory) with new ones (i.e. chaos theory and multicriteria analysis) in the decision process of portfolio selection (cf. papers of Karapistolis et al., and Hurson and Ricci-Xella).

The seven papers of the second chapter deal, also, with the application of the multivariate data analysis and the multicriteria analysis in the related fields of business failure, corporate performance and viability and bank bankruptcy. Some innovative ideas are proposed in this chapter, for example, the application of the multi-factor model in the analysis of corporate failure (paper of Vermeulen et al.); the ELECTRE TRI method for the analysis and prediction of business failure in Greece (paper of Zopounidis et al.); a new rough set approach for the evaluation of bankruptcy risk by Greco et al. and, finally, a new multicriteria decision support system for financial classification problems based on the preference disaggregation method (paper of Zopoundis and Doumpos).

The third chapter includes four papers which examine the contribution of several techniques of mathematical programming such as linear, dynamic, stochastic, to the problem of portfolio management. The last paper of this chapter by Clewlow et al. presents a good review of these techniques in the risk management of derivative securities.

The fourth chapter studies the introduction of fuzzy sets and artificial intelligence techniques in some financial decisions. Gil Aluja, using fuzzy sets analyzes the problem of financial risk in the investment decision. Lopez-Gonzalez et al. apply a fuzzy genetic algorithm (the POFUGENA model) in the portfolio selection problem, while Politof and Ulmer use artificial neural networks for the forecasting of interest rates.

Finally, the fifth chapter examines the contribution of several multicriteria decision aid methods in the assessment of country risk.

Sincere thanks must be expressed to the authors whose contribution have been essential in creating this volume. I owe a great debt to those who worked long and hard to review the contributions and advised the high standard of this book.

Finally, I would also like to thank Michael Doumpos, Thelma Mavridou and Konstantina Pentaraki for their assistance in my contacts with the authors and for helping me in the material collection and management.

Constantin Zopounidis
Technical University of Crete
Dept. of Production Engineering and
Management
Decision Support Systems Laboratory
University Campus
73100 Chania, Greece.

References

Heuer, S., U. Koch and C. Cryer (1988) INVEST: An expert system for financial investments, *IEEE Expert*, Summer, 60-68.

Klein, M. and L.B. Methlie (1995) *Expert Systems: A Decision Support Approach with Applications in Management and Finance*, Addison-Wesley, Wokingham.

Mulvey, J.M., D.P. Rosenbaum and B. Shetty (1997) Strategic financial risk management and operations research, *European Journal of Operational Research* 97, 1-16.

Pinson, S. (1992) A multi-expert architecture for credit risk assessment: The CREDEX system, in: O' Leary, D.E. and Watkins P.R. (eds.) *Expert Systems in Finance*, Elsevier Science Publishers, 27-64.

Roy, B. (1988) Des critères multiples en rechrerche opérationnelle: Pourquoi? In: Rand G.K. (ed.), *Operational Research '87*, Elsevier Science Publishers, North Holland, Amsterdam, 829-842.

Scrinivasan, V. and B. Ruparel (1990) CGX: An expert support system for credit granting, *European Journal of Operational Research* 45, 293-308.

Thomas, L.C. (1992) Financial risk management models, in: Ansell, J. and Wharton, F. (eds.), *Risk: Analysis, Assessment and Management*, John Wiley and Sons, Chichester, 55-70.

Williams, T. (1995) A classified bibliography of recent research relating to project risk management, *European Journal of Operational Research* 85, 18-38.

Zenios, S.A. (1993) *Financial Optimization*, Cambridge University Press, Cambridge.

Ziemba, W.T. and J.M. Mulvey (1996) *World Wide Asset and Liability Modeling*, Cambridge University Press, Cambridge.

Zopounidis, C., N.F. Matsatsinis and M. Doumpos (1996) Developing a multicriteria knowledge-based decision support system for the assessment of corporate performance and viability: The FINEVA system, *Fuzzy Economic Review* 1 / 2, 35-53.

Zopounidis, C. (1997) Multicriteria decision aid in financial management, in: Barcelo, J. (ed.), *Plenaries and Tutorials of EURO XV-INFORMS XXXIV Joint International Meeting*, 7-31.

Beginning from youth we must keep learning

Protagoras

I. MULTIVARIATE DATA ANALYSIS AND MULTICRITERIA ANALYSIS IN PORTFOLIO SELECTION

PROPOSAL FOR THE COMPOSITION OF A SOLVENT PORTFOLIO WITH CHAOS THEORY AND DATA ANALYSIS

Dimitris Karapistolis [1], Costas Siriopoulos [2], Iannis Papadimitriou [3], Raphael Markellos[4]

[1] Technological Educational Institute of Thessaloniki,
Thessaloniki GR 54101, Greece

[2] University of Macedonia, Dept. of Economics,
156 Egnatia str., P.O.Box 1591, Thessaloniki GR 540 06, Greece

[3] University of Macedonia, Dept. of Informatics,
156 Egnatia str., P.O.Box 1591, Thessaloniki GR 540 06, Greece

[4] Loughborough University, Dept. of Economics,
Loughborough, Leics LE11 3TU, UK

Abstract: This paper deals with the structure and dynamics of the Athens Stock Exchange (ASE) in Greece. Chaos Theory and Data Analysis methods are applied and produce evidence of a reasonably low-dimensional system, in both the phase and data space domain respectively. Based on the determined the concept of the solvent firm is identified and the solvent portfolio is constructed and traded according to a passive and active strategy. While the solvent portfolio returns of ASE for the period 1/1/1993 - 31/12/1993, clearly outperforms the market return, it is shown that it should be used as an investment tool, rather than for speculation.

Key words: Greek Stock Market, Chaos Theory, Data Analysis, Portfolio Management

1. Introduction

One of the fundamental issues of empirical finance in the study of stock markets is: given knowledge about the system and its past behaviour, what can be said about its future evolution. As shown in Figure 1. two basic approaches exist and may be classified into the econometric or model driven approach [1] and the non parametric or data driven approach.

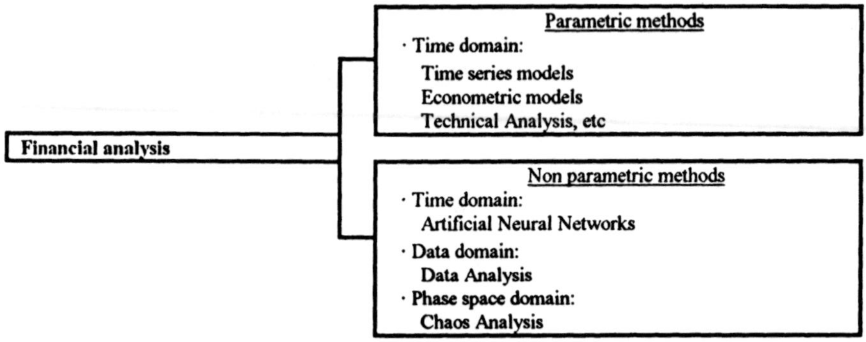

Figure 1: Approaches in Financial Analysis

The first approach attempts to analyse the sequence of observations produced by the underlying mechanism directly. From the statistics obtained from the observation sequence one hopes to be able to infer some knowledge about the future observation of the observation sequence. The strict statistical restrictions imposed by parametric model driven methods have often proven themselves to be unrealistic since properties such as noise, non stationary, nonlinearies and non-normality have been found to dominate stock market returns [2], [3].

The second approach postulates that no a priori assumption can be made about the structure of the stock market and the interaction of its components and that a data driven methodology should be adopted in order to estimate both interactions and components. Such methodology includes Data analysis, Artificial Neural Networks, Chaos analysis etc [21]. In this paper the perspective of interest is based on nonparametric methods of Chaos theory and Data Analysis. According to Chaos theory fluctuations are endogenous to the system and reflect the presence of important nonlinearities in the behavioural relationships of the system.

We applied chaos analysis to calculate Hurst's exponent in an attempt to determine whether our system is of long term memory. Next, we estimated the system's fractal dimension - which is a number that quantitatively describes how the system fills its space - and that lead us to the number of components least needed to represent the stock market [18]

Since we cannot determine those components through chaos analysis, we apply data analysis (DA) in the study of the system's behavioural structures and relationships and the isolation of critical properties by specifying its components. The isolated properties and relationships resulting from DA, are used by the critical construction of the "solvent portfolio" [4]. The solvent portfolio is based on the complex interaction of many qualitative and quantitative criteria it is a concept much wider than the efficient portfolio. The empirical findings of Data analysis indicate that it can be defined in terms of stock Corporate validity, Acceptability and Economic vigour. It is found that the ASE does not confirm to the strict assumptions made by traditional portfolio management and parametric modelling techniques, it of a low dimensional-system characterised by complex nonlinear regularities. The proposed 3-aspect solvent portfolio, produced by DA, is empirically justified and validated as a powerful investment tool and a satisfactory alternative to other methods [17]. Additionally, the solvent portfolio performance is consistent to Rossenberg's portfolio theory of extra market covariance, while its theoretical construction is based on an extension of Larrain's nonlinear model [8] It must be noted that the philosophy of Chaos and Data analysis is very similar: in Data analysis we reduce the degrees of freedom of a system described by a large number of variables, while in Chaos analysis we assess the degrees of freedom that are needed to reconstruct the system, but in a topological sense.

The paper is organised as follows: in the next section, a statistical description of the data is given. In section 3, the basic concepts of Chaos theory are introduced along with empirical evidence on the dimensionality and properties of the ASE returns. In section 4, Data analysis is applied in order to decrease the degrees of freedom, in a quantitative and qualitative data space of criteria concerning stocks listed in the ASE. The concept of the solvent portfolio and firm is then defined and determined via Data analysis. In section 5, the effectiveness of the proposed solvent portfolio is examined and in section 6, the major findings are presented along with a route for future research.

2. The data

The data analysed in section 3 consist of closing prices for the Athens Stock Exchange General Index (GIASE) for the period October 1986 to February 1994, a total of 1810 daily observations. To reduce the effects of non-stationarity and serial correlation, the raw prices are transformed to logarithmic returns. In figure 2 the descriptive statistics of the data are presented. The kurtosis and skewness measures indicate that the distribution of the GIASE returns has heavy tails and is skewed towards the right while the Jarque-Bera test strongly rejects the assumption of normality. The ARCH heteroskedasticity and McLeod - Li tests detect substantial non-linearities in the variance and mean respectively.

Average	0.0014
StDev	0.0221
Skewness	0.4
Kurtosis	17.6
Jarque Berra test	14315.5
ARCH test	176.4
McLeod-Li test	541.3

Figure 2: Descriptive statistics of the GIASE returns

The data used in section 4 of the paper, consists of 15 items for 240 stocks listed in the ASE for the year 1992. These items are: Company size, Stock market value, Capitalisation ratio, Financial position progress Marketability, Traded shares per day, Transaction value, Flow ratio, capital gain, dividend profits, Dept to equity, Equity to assets, Current ratio, P/E and Equity earnings. As shown in figure 3, these 15 aspects represent the 3 aspects of solvency: Corporate validity, Acceptability and Economic vigour.

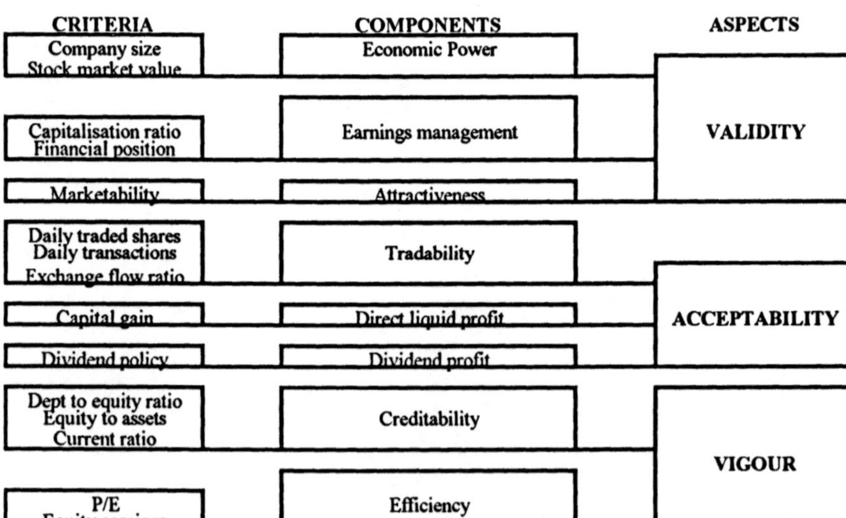

Figure 3: Aspects of corporate solvency

3. Chaos Analysis

Rather than assuming that the observation sequence may be considered as one specific realisation of a random process - where randomness arises from the many independent degrees of freedom interacting linearly - an emerging view exists in finance that postulates that apparently random behaviour may be generated in the long term by chaotic deterministic systems with only a few degrees of freedom that interact nonlinearly. A necessary, though not sufficient, condition for the occurrence of chaos is the presence of non-linearity. Chaotic systems are described by fractal dimensions and have strange attractors. In a nonlinear dynamic series, an attractor, is a definition of the equilibrium level of the system. Also, chaotic systems have some very interesting characteristics: due to their sensible "dependence on initial conditions", it is possible to make only very short-term predictions.

By plotting one variable - in our case stock returns - with different lags in different embedding dimensions (m), one can represent a system in the so-called phase space domain and treat it as a geometric object with invariant properties. A phase space is a graph that allows all possible states of a system. In this graph, the value of a variable is plotted against possible values of the other variables at the same time. Due to a theorem by Takens [5] one can fully reconstruct the original, unknown phase space with only one dynamic observal variable and obtain the attractor of the system, using the so called time delay method. Takens showed that a topologically equivalent picture of the attractor in phase space can be constructed by the time delay method, which consists of choosing a proper delay time and reconstructing a set of n-dimensional vectors, where n is non known a priori. The reconstructed phase space refers to the true dynamical system that generated the series and gives us information on the possibilities of the system.

Using this reconstructed phase space we can calculate the fractal dimension which measures how much m-dimensional space is occupied by and object. The most commonly used method to estimate the fractal dimension is the Grassberger Procaccia method [6], that uses the correlation dimension (CD). The Grassberger-Procaccia method offers a reliable, relatively simple method for estimating the fractal dimension when only one dynamical observal variable is known, as it is in our case. The CD measures the probability that two points chosen at random in phase space, will be within a certain distance of each other and examines how this probability changes as the distance is increased. The CD can be interpreted as a lower bound to the significant degrees of freedom of a dynamical system. Although only logarithmic returns are analysed it must be clear that according to the Takens theorem the estimated degrees of freedom refer to the stock market system as a whole and not to the return series alone. The CD has been also used to differentiate between deterministic, stochastic and chaotic systems. If chaos is present, a strange attractor can be identified that only occupies a small fraction of the available phase space. The computation of the CD allows us to find the dimension of this attractor. If the value of the CD does not change further with embeddings, it is assumed that the CD has converged to its correct value. That is, if chaos is present in the data the correlation dimension saturates with increasing embedding dimensions of the phase space. If this stabilisation does not occur, the system is considered high-dimensional or stochastic. The data are generated from some deterministic process when the correlation dimension remains well below the value of the embedding dimension and keeps increasing with increasing embedding without ever saturating. The estimated CD, for embedding dimensions from 2 to 10 and initial distance 0.0405 increased 10% each time, are presented in figure 4. The CD shows a strong saturating tendency for increasing number of dimensions in about 2,35. We can presume that the ASE dynamics are possibly chaotic and that at least 3 variables are needed to represent the system [17].

Embedding dimension	Correlation dimension
2	0.61917
3	0.95229
4	1.33443
5	1.73523
6	2.17384
7	2.33678*
8	2.44838
9	2.74870
10	2.21977

Figure 4: Correlation dimension of the GIASE returns

Another method of analysis in the phase space domain involves the calculation of the Largest Lyapunov exponents (LLE). These exponents show the average* exponential rates of divergence or convergence of nearby points in the phase space. The LLE can be interpreted as the average loss of predictive power in terms of bits of information. By applying the Wolf method [7], the average loss of information was estimated to be 0.0013, that is 0.0013 bits of information lost per day. So, the whole information set is lost in 1/0.0013 days which is about 37 months. A bit is a measure of information.

While various theoretical market models have been proposed by those who study chaos, their common point is that nonlinearities and determinism results from the interaction of long term fundamental and short term technical analysis or sentimental factors. Fundamental factors help to determine overall trends while technical factors help to determine near-term volatility. Such a model has been proposed by Larrain [8] and is called the K-Z map. Larrain shows that since fundamental factors (Z-map) alone cannot produce a true distribution of prices the nonlinear dynamics are produced by the effect of sentimental factors and short-term speculators (K-map) that use technical analysis. Erratic stock market behaviour occurs when the K-map overpowers the Z map. Larrain also argues that financial markets do not respond instantaneously to fundamental events as a result of the time and costs involved in acquiring, processing and interpreting information and the availability of such information.

From fractal geometry and topology we know that although a chaotic system is totally unpredictable in the long term it is bounded in a certain area of the attractor. In the case of the stock exchange these bounds are set by fundamental information and the structure of the market: speculation, sentimental and technical factors can move the price of a stock in a nonlinear fashion but this movement is restricted by the non-systematic fundamental boundaries of the stock, in the long term. An attractor exists for every stock and although we can assume that the structure of these attractors will be in general similar to the overall attractor of the market, their fundamental boundaries differ substantially.

The results of Chaos analysis indicate that predicting the ASE is essentially impossible in the long term. The evidence of nonlinearity and nonnormality of returns do not confirm to the strict statistical assumptions of the EMH and parametric modelling and to mean/variance portfolio optimisations. In the next section it is attempted to extract, discriminate and extrapolate the fundamental boundaries of each stock in order to exploit their differences in the determination of the solvent portfolio.

4. Data Analysis

Many studies that make use of Data analysis have been reported in the financial literature [19]. Recurrence plot analysis (RPA) [9] uses the same techniques as in Chaos analysis, particularly in the reconstruction of the phase space. With RPA we try to find the characteristics of a time series in terms of geometric criteria. The idea is to identify the similarities of the behaviour of points in time. Similarities between Chaos and Data analysis methods go beyond surface since they both are nonparametric methods and attempt to reduce the dimensionality of a system outside the time series domain.

The statistical properties of the returns justify the adoption of a nonparametric methodology such as Data analysis, since no a priori hypotheses are needed. The time series of returns has been found extremely difficult to model, thus one should focus on other fundamental qualitative and quantitative information about the stocks.

The concept of portfolio solvency receives special attention in this study. The solvent portfolio of firms is fundamentally different from the efficient portfolio but does confirm to Rossenberg's [10] portfolio theory. Rossenberg reformed Merkowitz's and Sharpe's ideas in an enriched and more applicable form. He introduced the concept of extra market covariance, which means that many stocks move together independently of what the market does as a whole. For example, stocks of companies in the same industry, stocks that are small in size or, as in our case, solvent stocks may move independently of the market. The term corporate solvency is defined as the competence of a firm listed in the stock exchange to fulfil its obligations in the long term. Thus solvency is strongly related to the reliability of a firm.

The selection of the fifteen criteria analysed in this section, was done in order to succeed the best possible representation of a firm's fundamental status and obtain the maximum quantity of information with minimum covariation of items. It is assumed that this data set contains sufficient information to define the boundaries of each stock. An additional assumption made is that the fundamental information contained in the criteria is not absolved immediately and that influences the market until it is drastically altered, which is consistent with the long term information decay of 37 months found by the LLE [17].

By applying methods of Data analysis the original 15 criteria are organised in three groups of five criteria each, as shown in figure 3.

The Validity of a firm is determined by the following 5 criteria: Company size, Stock market value, Capitalisation ratio, Financial position progress and Marketability. The first two criteria form the component of Economic power while the next two form the component of Earnings management. The last criteria form the attractiveness component of corporate validity. In order to decrease the large variations observed in the values of the above five criteria, they were divided in quartiles. As a result the initial quantitative criteria were transformed in qualitative binary values. The above transformation justified the use of Correspondence analysis [11] in the investigation of the data.

The Acceptability aspect is determined by the following 5 criteria: Traded shares per day, Transaction value, Exchange Flow ratio, Capital gain, and Dividend profits. The first three criteria form the component of Tradability while the Capital gain and the Dividend profits criteria form the Direct liquid profit and Dividend policy components respectively. Since the above 5 criteria concern ratios and we are interested in both their relations and factors, we apply Component analysis [12].

Finally, the aspect of Economic vigour is determined by the following 5 criteria: Dept to equity, Equity to assets, Current ratio, P/E and Equity earnings. The first three criteria form the Creditability component while the last two form the Efficiency component. Since Economic vigour is determined according to the classification of firms according to the above 5 criteria, the Range Analysis is used.

After performing the above analysis and classifying the stocks, it is possible to mark each one of the three aspects of a firm's solvency in a discrete scale of 1 to 5 according to each stock's integrity at the respective aspect. We then form a (240x3) table containing marks for each solvency aspect of 240 stocks. By summing the 3 marks of a firm we can obtain an overall measure which constitutes the solvency mark of the respective stock.

Mark	Group
[12,15]	Solvent portfolio
[9,11]	Potential alternatives
[3,8]	Uninteresting firms

Figure 5: Ranks of Corporate Solvency

As shown in figure 5, the 240 stocks can then be ranked in three groups, according to their solvency mark. Descriminant Analysis is then applied on the 240x3 table and after the possible recontructing emerging from analysis, the first group constitutes the solvent portfolio. The stocks that belong to the first group determine the solvent portfolio. The stocks ranked in the second group are acceptable in terms of solvency and can be used if needed for diversification. The stocks that form the third group are characterised for the time being as not interesting and will be evaluated in the next inflow of fundamental information. After applying the above analysis it was found that on 1/1/1993 the solvent portfolio of the ASE was constituted by nine stocks [20].

It is very interesting that the dimension estimated by Data analysis in the data domain we used is three, since Chaos analysis found that at least three components are required in order to model the Athens Stock Exchange in the phase space of the GIASE returns.

5. Evaluating the solvent portfolio *via* technical analysis indicators.

In this section we test the performance of the proposed solvent portfolio using an active strategy based on a mechanical trading rule, and a passive buy and hold strategy. The use of mechanical trading rules has been recently proposed by different authors in testing market efficiency or portfolio returns. We use the Stochastic Momentum Index (SMI) developed by Blau [13], since it is better adopted by ASE structures and magnitudes formulated as follows:

$$SMI(q,r,s) = 100\, E(s)\, \{ E(r)\, [SM(q)] \} \,/\, 0.5\, \{ E(s)\, [E(r)\, [HH{:}\, q - LL{:}q]\,] \}$$

where:
$SM(q)$ = closing price - 0.5 (HH: q + LL:q)
$E(x)$ is an exponential x-day moving average
HH:q is the highest high value over q periods
LL: q is the lowest low value over q periods
and $ASM(q) = E(s)\,\{E(r)\,[SM(q)\,]\,\}$ is the exponential moving average of period s of the exponential moving average of period r of the quantity SM(q),$s>r$. That is the average stochastic momentum, i.e. a double smoothing of stochastic momentum which is a known technical indicator. In our case q is chosen to be equal to 5 (a week), and after optimisation with MetaStock v.5.1, $r = 5$ and $s = 20$. We would choose other values for the parameters q,s and r but the results will be quite similar.
This technical indicator is based on the location of the closing price between the highest high and the lowest low value over q periods (days, weeks,...). The idea is that averaging this formula produces a relatively smooth indicator with a fast response giving buy and sell signals (trades). The SMI index is one of the followed technical indicators by portfolio managers for active trading (speculation) in stock markets. Other technical indicators are discussed in [14] and [16].

We take a long (L) and a short (S)position, respectively, if:
L: $Close(t-1) < SMI(t-1)$ and $Close(t) > SMI(t)$
S: $Close(t-1) > SMI(t-1)$ and $Close(t) < SMI(t)$
Where close(t) denotes the closing price at time t.

	S_1	S_2	S_3	S_4	S_5	S_6	S_7	S_8	S_9	Total
LT	10	8	8	7	10	11	8	7	10	
ST	11	9	8	8	11	11	7	7	10	
%LT	60	63	75	57	20	28	63	57	40	
%ST	28	11	25	25	37	37	43	43	30	
%Total	80	75	46	35	38	7	72	25	2	47
%B&H of the solvent portfolio	220	210	120	70	-5	-6	110	70	30	89
%Market										40

Si , I=1,...,9 are the nine solvent stocks, LT: # of long trades, ST: # of short trades, and %LT, %ST: profitable long and short trades respectively.

Figure 6: Solvent portfolio returns for the period 1/1/93 - 31/12/93.

The results of this strategy are shown in figure 6 for the equally weighted (11.1%) solvent portfolio of 9 stocks. These results are lower than a simple buy and hold strategy (47% vs. 89%), but greater than the return of the market portfolio (40%), for the same period. It is obvious that the proposed Solvent portfolio serves as an long term investment tool rather than for speculation since the passive buy and hold strategy clearly outperforms the speculative strategy, using technical analysis tools such as the SMI trading indicator. This is true, since the Return of the Solvent Portfolio (SP%) over the one year period is greater than the Return of a speculative strategy (SMI%) and even greater than the Return of the Market (M%): SP% > SMI% > M%.

6. Conclusions

In this paper two distinct nonparametric approaches were adopted in the study of the ASE in Greece. The statistical properties of the GIASE index returns for the period October 1986 to February 1994, were found to violate the assumptions made by traditional parametric modelling and portfolio management.

The application of Chaos analysis methods produced evidence that the ASE dynamics are not purely stochastic and can be modelled with at least three variables. Based on the 37-month average decay of information period found by Chaos analysis and on a model by Larrain Data analysis is applied on 15 criteria derived from fundamental analysis, concerning 240 stocks listed in the ASE. The results of Data analysis confirm to those of Chaos analysis showing that three significant factors can be extracted from the data. These factors are translatable in economic terms and can be used in the determination of a portfolio of stocks, that is defined as solvent. The estimated solvent portfolio return of the ASE for the period 1/1/1993 to 31/12/1993 outperformed the market return. It was also found that the it performed better when used as a long term investment tool rather than for speculation.

Future research is concentrating on the technology of Artificial Neural Networks (ANN's) and Expert systems [14] where the ratings of solvency aspects can be used as input variables. Solvency combines a wide set of information in only three aspects or measures and can provide ANN's a condensed set of three inputs that contain the maximum possible information. This is essential to the training process of ANN's since a small number of input variables results in acceleration of the training process and better generalisation without limiting the width of information to be considered. The basic problem of finding the optimal architecture of an Artificial Neural Network can be solved based on the findings of Chaos and Data analysis [15]. The estimated dimension of three can serve as a lower bound to the number of distinct features (nodes in the so-called hidden layer) that the network must recognise. Previous research has shown empirically that the ASE is best described by ANN's with 3 nodes in the hidden layer [16].

References

[1] Mills TC, *The Econometric Modelling of Financial Time Series*, Cambridge University Press, 1993.

[2] Hsieh DA, Chaos and Nonlinear Dynamics: Applications to Financial Markets, Journal of Finance, 1991;5:pp. 1839-1877.

[3] Peters, EE, *Fractal Market Analysis*, New York: John Wiley, 1994.

[4] Karapistolis D, Papadimitriou I, Construction of a Solvent Portfolio. Proceedings of 4th International Conference of the Thessaloniki Economic Society, 1994 : Thessaloniki, Greece, Vassiliadis S (ed), University Studio Press, 1995: 269-296.

[5] Takens F, "*Detecting Strange Attractors in turbulence.*" In Dynamical Systems and Turbulence, Rand D, and Young LS, eds. Berlin: Spinger-Verlag, 1980.

[6] Grassberger P, Procaccia I, Characterisation of strange attractors, Physics Review Letters, 1983;50: 3465-3490.

[7] Wolf A, Swift JB, Swinney HL, Vastano J, Determining Lyapunov Exponents From a Time Series. Physica D, 1985;16: 285-317.

[8] Larrain M, Testing Chaos and Non-Linearities in T-Bill Rates. Financial Analysts Journal, 1991;Sept.-Oct: 51-62.

[9] Eckmann JP, Kamphrost O, Ruelle D, Recurrence plot of dynamical systems. Europhysics Letters, 1986: 973-977.

[10] Deboeck GL,"The Impact of Technology on Financial Markets" in *Trading on the Edge*, ed. Deboeck, GL, 1994.

[11] Benzecri JP, *Correspondence Analysis Handbook*, New York: Marcel Dekker, 1985.

[12] Lebart L, Morineau A, Fenelon JP, *Traitement des donees Statistiques*, Dunod, Paris, 1979.

[13] Blau W, Stochastic Momentum. Technical Analysis of Stocks and Commodities, 1990;11 (1):26-32.

[14] Siriopoulos C, Doukidis G, Karakoulos G, Perantonis S, Varoufakis S, Applications of Neural Networks and Knowledge Based Systems is Stock Investment Management: A comparison of performances. Journal of Neural Network World, 1992; 6:785-795.

[15] Markellos RN, Siriopoulos C, Sirlantzis K, Testing Non- linearities and Chaos in Emerging Stock Markets: Implications for Financial Management. Proceedings (forthcoming) of the 4th Annual Meeting of the European Financial Management Association, June 1992, London, UK.

[16] Siriopoulos C, Markellos RN, Sirlantzis K, Applications of Artificial Neural Networks in Emerging Financial Markets. Proceedings of the 3rd International Conference of Neural Networks in Capital Markets, (ed) World Scientific Publishing (forthcoming), October 1995; London, UK.

[17] Sirlantzis C, Siriopoulos C, Deterministic chaos in stock market: Empirical Results from monthly returns. Journal of Neutral Network World, 1993; vol. 3, 6: 855-864.

[18] Brock W, Sayers C, Is the business cycle characterised by deterministic chaos? Journal of Monetary Economics, 1988; 22: 71-90.

[19] Benzecri JP, “L’analyse des donnees. Tome 1: La Taxinomie. Tome 2: L’analyse des correspondances, Dunod, Paris, 1973.

[20] Karapistolis D, Katos A, Papadimitriou I, Proposal for solvent portfolio selection with data analysis methods, Proceedings of the 7th National Conference of the Greek Statistical Institute, 1994 May 5-7: Nicosia. Cyprus,(ed) Hellenic Statistical Institute 1995: 90-98.

[21] Siriopoulos C, Markellos M, Semi-chaotic financial time series and neural network forecasting: evidence from the European stock markets. Yugoslav Journal of Operations Research (forthcoming), vol. 6, 2.

AN ENTROPY RISK AVERSION IN PORTFOLIO SELECTION

Antonino Scarelli
University of Tuscia
01100 Viterbo (ITALY)

Abstract: The involved risk in choosing investments, in the most of cases, is characterized and composed in manifold factors. The global risk is broken up in different attributes and it is decomposed in hierachical levels. Then, by concept of attractiveness, for each action, an evaluation is made, taking into account the weights assigned to each attribute, considered through its related partial risk. With respect to an ideal point, representing the solution risk-free, by the concept of entropy, an index is computed. In such a way it is possible to carry out an absolute degree of risk and to achieve the comparisons between the different alternatives and to single out their ranking.

Keywords: risk, anti-entropy, hierarchy, portfolio selection, ranking.

INTRODUCTION

The solution of classic decisional problems involving risk, sometimes presents relevant difficulties. For example, the problems based on the evaluation of investments, adopt procedures strongly criticized either because of the reductive factors unable to reassume entirely some quantitative information, or because of the rigid schemes on which they confine the solutions.

A first aim of the paper is to propose a procedure which tries to overcome some of the difficulties mentioned above, by an algorithm which omogeneously binds the qualitative and quantitative evaluations and looses the least information. We start from the conception that in a system "a given information" is equivalent to "a taken away uncertainty"; we try to provide an exact measure to the vague notion of "quantity of uncertainty" that is to say "quantity of information".

A second aim is to measure the acquired risk information in the decisional process as an isomorphous quantity of the negative entropy (anti-entropy) of

Physics. The measure of the information is similar to that of the anti-entropy, and as the entropy is considered to be a measure of the disorder, then the anti-entropy is a measure of the order of the system, of its organization, which, if compared with a random distribution constitutes some unlikely state. A piece of information can be transformed into a permanent anti-entropy and, vice-versa, each experience represents an anti-entropy transformation into information.

The treatment of risky investments is strictly related to the concept of organized system which has essential features such as, differentiation, hierarchical order, control, the same of the decisional process conceived as a model addressed towards a final and characteristic objective. An organized structure is a system for which we have much information, thus its anti-entropy is greater than that of a disorganised system. A decisional process involving risk, is in the category of organized structures or phenomena which respond to three principles, clearly contrary to the ones of entropy phenomena, namely the *principles of finality, non-repeatibility and differentiation or organisation.* These phenomena have been catalogued as anti-entropy [4], responding to the "*Loi de la complexification*" [12], tending towards states of greater differentiation, and finalistic structures of our perceptible universe which, together with casual structures, determine the direction of movement of phenomena.

The decison maker (D.M.) first organises knowledge acquired on different alternatives of investments that we call briefly actions, and then defines the fundamental risky attributes with respect to which he expresses and operates specific differentiations. He subsequently formulates weights to be associated with the single risky attributes, following in part, the procedure suggested by certain authors [1], based on the concept of attractiveness and on the hypothesis that a decision maker is capable of distinguishing first the less risky element, whether actions or attributes, and then the preference for other elements compared with respect to that.

BASIC DEFINITIONS

Let us consider two discrete spaces of attributes F and S, and represent by f_i and s_p two particular points of the F and S spaces, respectively. The F set is generated by defining an evaluation distribution V(f) on the F space, which assigns an evaluation $V(f_i)$ to each particular point f_i of that space. Then a F×S set is generated by assigning a joint evaluation distribution V(f,s) to the product space F× S. Provided that V(f)≠0, given the conditional evaluation distribution V(s/f), the joint evaluation distribution V(f,s) are defined in terms of V(f/s) by

$$V(f,s)=V(f)\ V(s/f).$$

Higher-order product spaces and the evaluations associated with them can be defined in a similar manner. For instance, let us consider a third discrete space T of which t_h is a particular point. Then, relating to the product space F×S×T, a joint evaluation distribution V(f,s,t) is equal to the evaluation V(f,s) multiplied by the conditional evaluation V(t/fs):

$$V(t,s,f)=V(f,s)\ V(t/fs).$$

A MEASURE OF INFORMATION

An evaluation distribution V(f,s) is given on the product space F×S and we want to define a measure of the information provided by f_i about s_p. Let us think f_i, as representing the input to the box of fig. 1 and s_p the corresponding output.

Taking the point of view of an external observer, knowing both input and output, we would like to define a measure of the extent to which f_i specifies s_p: that is, a measure of the amount of evaluation communicated through the box.

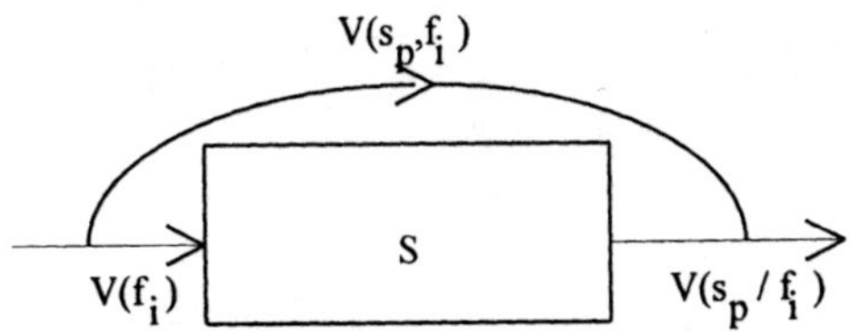

Fig. 1 _ Input and output of the information system S

The information provided by f_i about s_p consists in changing the evaluation of f_i from the priori value $V(f_i)$ to the posteriori joint evaluation $V(f_i,s_p)$. The measure of this change of evaluation, which proves to be convenient for our purposes, is the logarithm of the ratio between the joint evaluation and the self evaluation. Thus, we make the following general definition:

The amount of joint information $I(s_p;f_i)$ provided by the evaluation of the attribute represented by s_p about the evaluation of the attribute represented by f_i, is defined as

(1a) $I(s_p;f_i) = \log\left[\dfrac{V(s_p,f_i)}{V(f_i)}\right] = \log V(s_p / f_i)$; that is to say

(1b) $I(s_p;f_i) = \log V(s_p,f_i) - \log V(f_i) = \log V(s_p,f_i) + I(f_i)$.

In words, the information provided by the pair of attributes (s_p,f_i) about s_p is equal to the difference between the logarithm of the joint information provided by s_p and f_i, and the logarithm of the self evaluation f_i. This result matches perfectly with the definition provided by Shannon [10]: "the information provided by y about x is equal to the difference between the amounts of information required to specify x, before and after y becomes known".

The base of the logarithm used in this definition fixes the magnitude of the unit of information. The most commonly base used is e, the natural one because of its mathematical convenience. In such a case a unit of information is provided about s_p when the evaluation is increased by a factor to e. In our case, the measure just defined has not the property of being symmetrical with respect to s_p and f_i, because of the hierarchical levels of the attributes, being in the hierarchy the attribute f higher than the attribute s. For this reason, we will refer to a measure just defined as the "hierarchical information" between s_p and f_i.

The righthand side of equation (1a) suggests the interpretation of the hierarchical information defined as a measure of the statistical constraint between f_i and s_p. In fact, the measure is equal to zero when the self evaluation in question is statistically independent from the joint evaluation, i.e. $V(s_p,f_i)=V(f_i)$. This is the case on which the only sub-attribute or both attributes are at the worst level . Moreover, the measure is not null as soon as the level of the two attributes has a better evaluation respect to the worst one.

Let us consider now the product set F×S×T and represent a particular point of this set by the triplet (f_i, s_p, t_h) having an evaluation $V(f_i,s_p,t_h)$. The joint information between t_h for a given f_i,s_p is defined consistently with formula (1a):

$$(2)\qquad I(t_h,s_p;f_i)=\log\left[\frac{V(t_h,s_p,f_i)}{V(f_i)}\right].$$

The joint information is defined just as in formula (1a) except for the fact that the joint evaluation $V(t_h,s_pf_i)$ has been introduced. This definition generalizes the situation in which the joint information is conditioned by more levels.

$$(3a)\qquad I(t_h,s_p;f_i)=\log\left[\frac{V(t_h,s_p,f_i)}{V(f_i)}\right]=\log\left[\frac{V(t_h,s_p,f_i)}{V(s_p,f_i)}\frac{V(s_p,f_i)}{V(f_i)}\right]=$$

$$(3b)\qquad =\log\left[\frac{V(t_h,s_p,f_i)}{V(s_p,f_i)}\right]+\log\left[\frac{V(s_p,f_i)}{V(f_i)}\right]=I(t_h;s_p,f_i)+I(s_p;f_i).$$

An additional property of the mentioned measure is very important. The expressions given by equations (2a),(3a) allow us to expand in successive steps any mutual information between members of subsets of an arbitrary product ensemble into a sum of mutual information between elements of the elementary sets costituting the product ensemble. Thus, for instance, we have for the product set S× F×S'×F' with typical elements $f_is_pf_js_q$,

$$(4a)\qquad I(s_p,s_q;f_i,f_j)=\log\left[\frac{V(s_ps_qf_if_j)}{V(s_ps_q/f_if_j)}\right]=I(s_p;f_if_j)+I(s_ps_q;f_jf_i)=$$

$$(4b)\qquad =I(s_p,f_i)+I(s_pf_i;f_j)+I(s_q,s_p;f_i)+I(s_q,f_j/f_i,s_p).$$

We assumed the special case in which the product set S×F is statistically indipendent of the product S'×F': $I(s_pf_i;f_j)=I(s_q,f_j/s_p,f_i)=0$ and $I(s_q,s_p/f_j)=I(s_q,f_j)$ and $I(s_ps_q/f_if_j)=I(s_q,f_j)$. The expression provided by the equations (4) has a very simple interpretation if we regard f_i,f_j as independent inputs to two separate channels, and s_p and s_q as the corresponding outputs. Then the equation states that the information provided by the pair of outputs about the pair of inputs, is equal to the sum of the information provided separately by each output about the corresponding input.

EVALUATION OF THE RISKY ATTRIBUTES

The procedure used to determine the evaluations $V(f_i)$ ($f_i\in F$) and the conditional evaluations $V(s_p/f_i)$, $V(t_h/s_pf_i)$ of the attributes ($s_p\in S$ and $t_h\in T$) follows the same scheme. Let us consider the mono-dimensional space of the risky attributes $\mathbf{F=X^1}$. We may think its points arranged along the X axis. We want to point out how to assign an evaluation $V(f_i)$ to the particular point f_i. The numerical scale on F and the preference structure of a D.M on the attributes will be modeled by means of an ordinal function, quantifying the relative importance or attractiveness for the D.M. of the elements of F, as following:

i) The DM first fixes on the axis in the zero position, a dummy attribute f_0 for which the risk is null, and second carefully chooses in F the attributes f_1 having the least risk and positions it on the same axis, more or less near the dummy attributes. In this way two anchors have been fixed, a sort of unit of measure useful for the further evaluations.

ii) After positioning the former elements we ask the DM to put into the semi-axis the other risky attributes $f_i \in F$, i=2,3...m, spacing them out and taking into account the riskiness of each one with respect to the others and to the unit of measure previously established.

iii) To assign to the dummy attribute f_0 the real number zero, to the least attribute f_1 the real number one, and to each element f_i a real number $\pi(f_i)$, that we call *rate of differentiation*, which gives the ratio of f_i riskiness respect to f_1 riskiness.

In such a way the DM assigns each attribute a real number going from one (being $\pi(f_1)=1$) to infinite. The higher the riskiness of f_i is, with respect to f_1, the bigger will be the differentiation between the attributes within the acquired information, and the bigger will be the value assigned to the attribute and the *order* in the information system. In such a way an interval scale is obtained on R and the following condition is satisfied:

$\forall\, f_i, f_j \in F,\ \pi(f_i) > \pi(f_j) \Leftrightarrow f_i$ is ranked before f_j.

A greater acquisition of information brings the decisional process towards an increase in the anti-entropy (decreasing the entropy). In such a way, calling $V(f_i)$ the rate of evaluation referred to the factor, we put

$$(5) \qquad \frac{1}{\pi(f_i)} = V(f_i), \quad \textit{in short} \quad V_i.$$

The nearer V_i is to 0, the higher the riskiness of the attribute is. We have $1/\pi(f_1)=V_1$ and for $\pi(f_i) \Rightarrow \propto$, $V_i \Rightarrow 0$; $V_1=1$ is the evaluation of the attribute which is judged less risky in the decision making process. The value $V_i=0$ is an anti-ideal one (*Omega point* [12]), towards which the process proceeds in high model of riskiness. In such a case we have an attribute whose riskiness will be so high that the effects of all the other attributes for the decision will be null. We call $E_i=\log V_i$, entropy of information provided by the attribute f_i and $R_i=1/(1-E_i)$ its rate of resolution.

Given a first level attribute f_i, the DM, on an axis $\mathbf{X^2=S}$ by a dummy attribute s_{p0}, the worst attribute s_{p1}, the unit of measure $(s_{p1}-s_{p0})$, spaces the conditional attributes s_p/f_i taking into account the riskiness of the ones with respect to the others, and assignes the attribute s_p a real number $\pi(s_p)$ by the ratio from the preference of s_p with respect to s_{p0} and the preference of s_{p1} with respect to s_{p0}, using the latter as measure unity. We call $\pi(s_p)$ the rate of differentiation for that attribute. In such a way an interval scale is obtained on $\Re$ with the following condition:

$\forall\, s_p, s_q \in S_i,\ \pi(s_p) > \pi(s_q) \Leftrightarrow s_p$ is ranked before s_q.

The division of each attribute in subattributes is an increase of information inside the decisional process. This step enables us to reach more complexity but also more order. We are nearer the Omega point compared with the former step and we measure this approach by the new anti-entropy of the system.

Given f_i, for each attribute $\in S$ the partial evaluation is:

$$(6) \qquad \frac{1}{\pi(s_p)} = V(s_p / f_i)$$

and relating to the attribute f_i to which it belongs the global evaluation is

$$(7) \qquad V(s_p, f_i) = \frac{1}{\pi(f_i)} \frac{1}{\pi(s_p)} = V(f_i)\ V(s_p / f_i), \text{ in short } V_{ip}.$$

The nearer the evaluation V_{ip} is to zero, the more risky the attribute s_p/f_i is. We multiply $V(f_i)$ by $V(s_p/f_i)$ taking into account the increase of the approach at the anti-ideal point that the new level adds to the process. For decisional purposes we suppose that a subattribute belonging to a more risky attribute, increases its undesirability: the distance from the anti-ideal point is smaller. We call $E_{ip}= \log V(s_p, f_i)$, entropy of information provided by the joint attribute $(f_i s_p)$ and $R_{ip}=1/(1-E_{ip})$ its rate of resolution.

MEASURE OF INFORMATION FROM POSTULATES

The measure of information defined in section 3, by formula (1a) can be derived directly from the following postulates which are necessary by a useful measure of information.

Postulate 1). Given a product set F×S, the measure $I(s_p;f_i)$ of the information provided by f_i about s_p is once-differentiable function $\Phi(x,y)$ of two variables $x=V(f_i)$ and $y=V(s_p,f_i)$: $\Phi(x,y) = \Phi[V(f_i),V(s_p,f_i)]$.

Postulate 2). Given a product set F×S×T, the joint measure $I(t_h,s_p,f_i)$ of the information provided by s_p and f_i about t_h is the same function $\Phi(x,y)$ in which, however, $x=V(s_p,f_i)$ and $y=V(t_h,s_p,f_i)$.

Postulate 3). The measure of the joint information provided about t_h by the pair $s_p f_i$ satisfies the relation

$$(8) \qquad I(t_h;s_p,f_i) = I(s_p;f_i) + I(t_h,s_p;f_i).$$

Postulate 4). Given two independent sets F×S and F'×S', that is, sets for which $V(f,s,f',s')=V(f,s)\ V(f',s')$, the measure $I(f_i f_j,s_p s_q)$ of the information about the pair $s_q s_p$ provided by the pair $f_i f_j$ satisfies the relation:

$$(9) \qquad I(f_i f_j,s_p s_q)=I(f_i s_p)+I(f_j s_q).$$

The postulates are also sufficient to specify the functional form of the desired measure, apart from a constant multiplier which detemines the size of the unit of information. The derivation of the functional form of the measure involves two main steps.

The first step consists in showing that the function $\Phi(x,y)$ in order to satisfy the postulates [(2), (3)] must be of the form:

(10) $\Phi(x,y) = Z[x]-Z[y]$.

The second step consists in showing that Z(x) must be proportional to log V(x) in order to satisfy postulate (4). Selection of a negative constant will yield finally the equation (1): $I(s_p;f_i) = \log\left[\frac{V(s_p,f_i)}{V(f_i)}\right]$.

Let, for the sake of brevity, $v_0=V(f_i)$, $v_1=V(s_p,f_i)$, $v_2=V(t_h,s_p,f_i)$, and $w_0=V(f_j)$, $w_1=V(s_q,f_j)$. Postulates (1), (2) and (3) imply that the function $\Phi(x,y)$ must satisfy the equation

(11) $\Phi(v_0,v_2) = \Phi(v_0,v_1) + \Phi(v_1,v_2)$.

Differentiation of this equation with respect to v_1, yields:

$$\left[\frac{\partial\Phi(x,y)}{\partial x}\right]_{\substack{x=v_1\\y=v_2}} + \left[\frac{\partial\Phi(x,y)}{\partial y}\right]_{\substack{x=v_0\\y=v_2}} = 0\cdot$$

Since the last equation must be satisfied for all possible values of v_0, v_1 and v_2, the first term must be independent of v_1 and the second must be independent of v_0. In addition, the two partial derivatives must be equal in magnitude and opposite in sign for x=y. It follows that:

$$\frac{\partial\Phi(x,y)}{\partial x} = \left[\frac{\partial Z(u)}{\partial u}\right]_{u=x} \quad \text{and} \quad \frac{\partial\Phi(x,y)}{\partial y} = -\left[\frac{\partial Z(u)}{\partial u}\right]_{u=y}$$

where Z(u) is some once-differentiable function of the single variable u. Then, integration of these equations with respect to x and y, yields

(12) $\Phi(x,y)=Z(x)-Z(y)+k$

where k is an arbitrary constant. Finally, the substitution of eq. (12) into eq. (11) shows that k=0, thereby completing the proof of the eq. (10).

The second step of the derivation of eq. (1) proceeds as follows. Postulate (4), expressed by formula (9), requires the function Z(u) to satisfy the equation:

(13) $Z(v_0,w_0) - Z(v_1,w_1) = Z(v_0) - Z(v_1) + Z(w_0) - Z(w_1)$

where use is made of the fact that, because of postulate (4), $V(f_i,f_j)=V(f_i)V(f_j) = v_0w_0$ and $V(f_i,f_j/s_p,s_q) = V(s_p/f_i)\,V(s_q/f_j) = v_1\,w_1$.

Differentiation of eq. (13) with respect to v_0 and to w_0 yields respectively:

$$w_0\left[\frac{\partial Z(u)}{\partial u}\right]_{u=v_0w_0} = \left[\frac{\partial Z(u)}{\partial u}\right]_{u=v_0} \quad \text{and} \quad v_0\left[\frac{\partial Z(u)}{\partial u}\right]_{u=v_0w_0} = \left[\frac{\partial Z(u)}{\partial u}\right]_{u=w_0},$$

so that $v_0\left[\frac{\partial Z(u)}{\partial u}\right]_{u=v_0} = w_0\left[\frac{\partial Z(u)}{\partial u}\right]_{u=w_0}$

for all possible non-zero values of v_0 and w_0. It follows that $dZ(u)/du = k_1 1/u$ where k_1 is an arbitrary constant, and, by integration, $Z(u) = k_1 \log u + k_2$, where k_2 is a second arbitrary constant. Substitution of this equation into eq. (12), with k=0, finally yelds:

$\Phi(x,y) = k_1 \log x/y$.

The value of k_1 may be selected alone on the basis of convenience.

EVALUATION AND ENTROPY OF THE ACTIONS

Stating that there are only two levels of attributes of the risk, we proceed now for each attribute s_p/f_i on the space F×S, with the evaluation of each action t_{ip}^h, h=1,2,..n, in the space $\mathbf{T}=\mathbf{X}^3$. The preference structure on T, as ordinal function, is modeled quantifying the attractiveness of the elements $t_h \in T$, by the procedure mentioned above.

The value $\pi(t_{ip}^h)$ will be the rate of differentiation for the h-th action. The higher is the riskiness of t_{ip}^h with respect to t_{ip}^1, the bigger will be the differentiation between the actions inside of the acquired information, as well as the value assigned to the action t^h. In such a way an interval scale is obtained on T by the following condition:

$\forall\, t_{ip}^h, t_{ip}^k \in A,\ \pi(t_{ip}^h) > \pi(t_{ip}^k) \Leftrightarrow t_{ip}^h$ is ranked before t_{ip}^k.

The values $\pi(t_{ip}^h)$, $\pi(t_{ip}^k)$ are real numbers expressing the riskiness of one action with respect to the other. Now we define the partial evaluation as:

$$\frac{1}{\pi(t_{ip}^h)} = V(t^h / f_i, s_p) = V(t_{ip}^h).$$

Now, assuming the hypothesis that an investment having a riskful attribute and relative riskful sub-attribute, increases its indesirability, we compute the global value of such evaluation by the formula:

$$(14) \qquad V(t_h; s_p f_i) = \frac{1}{\pi(V_{ip})} \frac{1}{\pi(V_{ip}^h)} = V(s_p / f_i)\, V(t_{ip}^h), \quad \text{in short } V_{ip}^h.$$

We call entropy of the action the value $E_{ip}^h = \log V(t_h; s_p f_i)$.

We multiply those values taking into consideration the belonging of the action to the attribute and sub-attribute. The nearer the value V_{ip}^h is to zero, see fig. 2, the smaller is the value of the entropy, and the more we approach the anti-ideal point by the value $R_{ip}^h = 1/(1-E_{ip}^h)$, that we call rate of resolution relating to the h-th action. The nearer the R_{ip}^h position is to the plane (X^1, X^2), the higher the riskiness of h-th action is.

In the space $(0, X^1, X^2, X^3)$, considering also the evaluation of the p-th and i-th attribute, we can say that the shorter the way to reach the point (0,0,0) is, the greater the differentiation inside the decisional process is, and the smaller the desirability of the action under consideration is.

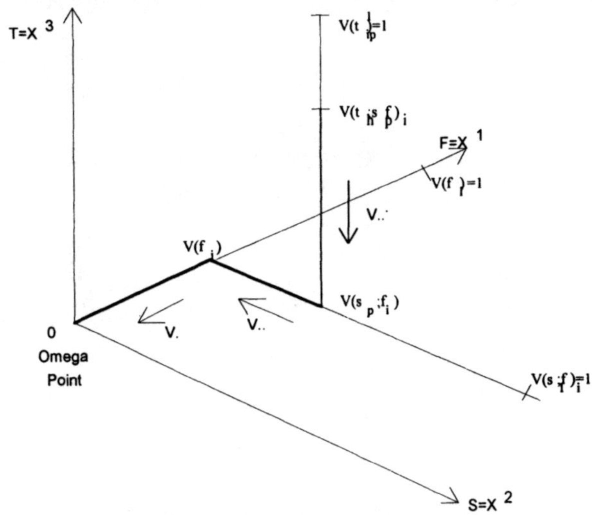

Fig. 2 _ Evaluations fo the different level attributes

THE OUTRANKING RELATION

Now if we consider for the h-th action all the positions acquired under the different attributes of the i-th first level attribute, we have a polygonal representing its profile. The more the polygonal is pressed on the plane (X^1,X^2), the more the action is riskful according to those attributes.

Joining the vertices of the poligonal with the ideal point Ω, we obtain a pyramid in which the smaller is its volume L_i^h, the higher is the riskiness of the h-th action by the i-th attribute under consideration. The total riskiness for the h-th action is obtained multiplying the volumes with respect to all the first level attributes:

$$E^h = \prod_{i=1}^{m} L_i^h . \tag{15}$$

The outranking relation will follow immediately. Given two actions, the h-th and the k-th, we say that the former is less risky than the latter, if and only if, E^h is bigger than E^k:

$$for\ \forall\ a^h, a^k \in A, a^h \succ a^k \Leftrightarrow E^h > E^k . \tag{16}$$

AN EXAMPLE

Considering the portfolio selection, it is convenient to divide the problem in an hierarchical way. The objective is to find the ranking of the common basic shares, and to choose the attributes which we want to take into account for making the evaluations towards the risk. This is a very important step because of the very little quantitative information: most of the information is expressed in a qualitative way. In order to better tackle the problem, we divide the attributes in two levels, see fig. 3.

Goal	I level attributes		II level attributes	
Choosing Shares	Extrinsic Factors	EXF	Economic	EEC
			Political	EPO
			Social	ESO
			Technological	ETE
	Intrinsic Factors	INF	Profitability	IPR
			Size	ISZ
			Technological Control	ITC
			Business Philosophy	IBP
	Investor 's Objectives	IVO	Profit	OPR
			Control	OCO
			Security	OSE
			Excitement	OEX

Fig. 3_ Goal and attributes of different levels

As first level attributes (the space F) we have taken in consideration the following ones: Extrinsic factors (EXF), Intrinsic factors (INF) and Investor's objectives (IVO): to make some comparisons, [we use an example from [8]].

Relating to the space S, in the first set we consider as subattributes the economic (EEE), political (EPO), social (ESO) and technological (ETE) factors; in the second set, we use factors as profitability (IPR), size (ISZ), technological control (ITC) and business philosophy (IBP); finally in the third, profit (OPR), control (OCO), security (OSE) and excitement (OEX), see fig. 4.

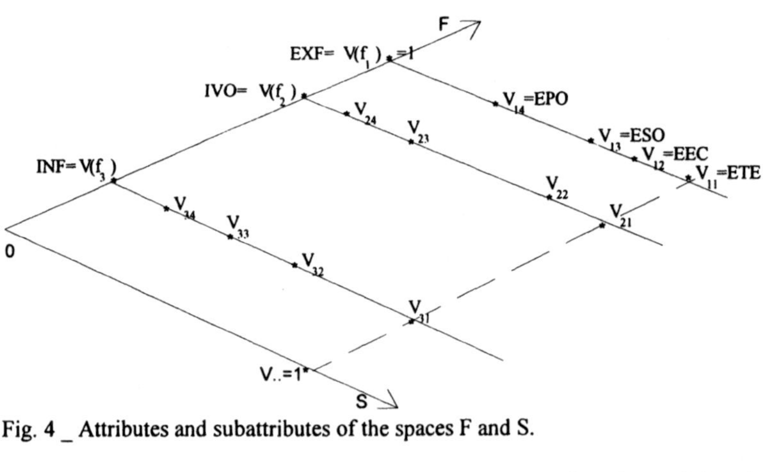

Fig. 4 _ Attributes and subattributes of the spaces F and S.

In each space the DM differentiates the attributes: after determining the least risky attribute, he positions the other attributes distancing them according to their increasing riskiness, on the axis $\mathbf{F=X^1}$, for the first level attributes. From these expressed preferences, the rates of differentiation $\pi(f_i)$, the evaluations $V(f_i)$ and the rates of resolution R_i are derived by formula (1), as shown in Tab. 1.

Tab. 1 _ Rates of the first level attributes.

Rates	EXF f_1	INF f_2	IVO f_3
$\pi(f_i)$	1	4.7	1.38
V_i	1	0.21	0.72
R_i	1	0.39	0.75

Subsequently, following the same procedure, for each set, in the space F×S, after determining the least risky attributes, the DM expresses his undesirabilities graphically according to the previous scheme on the corresponding segments parallel to the axis $S=X^2$. From the preferences expressed are derived the rates of differentiation, the evaluations and rates of resolution, as shown respectively in Tab. 2 and Tab. 3.

With reference to the table 2, we can say, for example, that for the factor INF, if the worst subcriterion IPR has value 1, the best ITC has a preferability which, in comparison, is 5 times higher. The values of the preceding table, from the highest to the lowest, show the increasing order of preferability of the sub-attribute associated with the attribute.

Tab.2 _ Rates of differentiation and partial evaluations for the subattributes

EXF f_1			INF f_2			IVO f_3		
s_i/f_1	$\pi(s_{i1})$	$V(s_i/f_1)$	s_i/f_2	$\pi(s_{i1})$	$V(s_i/f_2)$	s_i/f_3	$\pi(s_{i1})$	$V(s_i/f_3)$
EEC	1.2	0.83	IPR	1	1	OPR	5.5	0.18
EPO	2.63	0.38	ISZ	1.2	0.83	OCO	1	1
ESO	1.43	0.70	ITC	5	0.20	OSE	2.5	0.40
ETE	1	1	IBP	2.7	0.37	OEX	1.47	0.68

Tab.3 _ Joint evaluations and rates of resolution for the subattributes

EXF f_1			INF f_2			IVO f_3		
s_i/f_1	$V(s_i,f_1)$	R_{1p}	s_i/f_2	$V(s_i,f_2)$	R_{2p}	s_i/f_3	$V(s_i,f_3)$	R_{3p}
EEC	0.83	.84	IPR	0.21	.39	OPR	0.13	.33
EPO	0.38	.51	ISZ	0.17	.36	OCO	0.72	.75
ESO	0.70	.74	ITC	0.04	.24	OSE	0.29	.45
ETE	1	.28	IBP	0.08	.28	OEX	0.49	.58

The values in tab. 3 express the riskiness of the second level attributes, also taking into account the riskiness associated with the respective first level attribute to which they belong.

Now the same procedure is used to evaluate the actions for the different subattribute. In the vertical segments drawn for each attribute in consideration, see fig. 5, the DM chooses the worst action, and afterwards all the others taking into account their different riskiness.

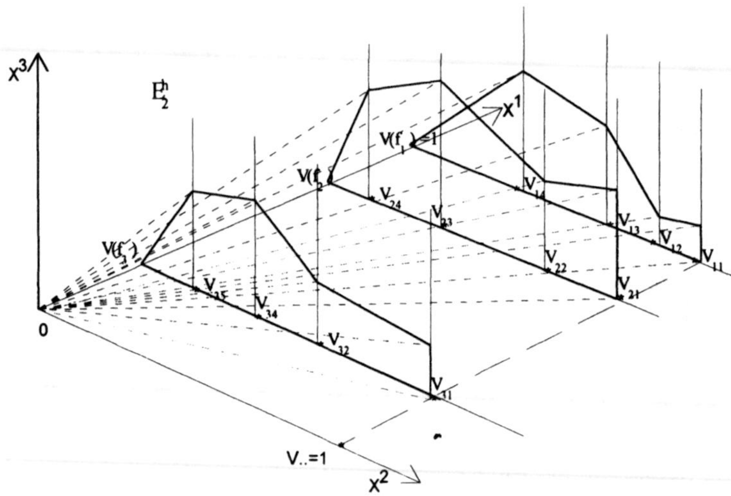

Fig. 5 _ Evaluations of the h-th alternative under different subattributes.

From the values of rates of differentiation and the relating values of the partial evaluation, and using formula (5), the global evaluations and the rates of resolution are derived, as represented in Tab. 4.

Tab.4 _ Rates for the actions relating to each subattributes

		INVESTMENTS											
Attr/Sbattr.		T_1			T_2			T_3			T_4		
f_i	s_{ip}	V^1_p	V^1_{ip}	R^1_{ip}	V^2_p	V^2_{ip}	R^2_{ip}	V^3_p	V^3_{ip}	R^3_{ip}	V^4_p	V^4_{ip}	R^4_{ip}
EXF	EEC	.34	.28	.44	1	.83	.84	.55	.46	.56	.55	.46	.56
	EPO	.84	.32	.47	.46	.17	.36	1	.38	.51	.27	.10	.31
	ESO	1	.70	.74	.25	.18	.37	.68	.48	.58	.21	.15	.35
	ETE	.34	.34	.48	.52	.52	.60	1	1	1	.18	.18	.37
INF	IPR	.78	.78	.80	1	1	1	.23	.23	.40	.65	.65	.70
	ISZ	.54	.45	.56	.54	.45	.56	.19	.16	.35	1	.83	.84
	ITC	.25	.05	.25	.40	.08	.28	1	.20	.38	.44	.09	.29
	IBP	1	.37	.50	.24	.09	.29	.62	.23	.40	.38	.14	.34
IVO	OPR	.62	.11	.31	.73	.13	.33	.30	.05	.25	1	.18	.37
	OCO	.41	.41	.53	.26	.26	.43	.71	.71	.74	1	1	1
	OSE	1	.40	.52	.53	.21	.39	.78	.31	.46	.42	.17	.36
	OEX	.65	.44	.55	1	.68	.72	.38	.26	.43	.36	.24	.41

As the riskiness expressed for the attributes increases, the values computed decrease in coincidence with an ever-increasing order in the decisional system and thus with the decrease in its entropy and rate of resolution.

Through the undesirabilities of each action in the various attributes and thus by the calculation of the volumes of the pyramids, which have as base the profiles defined and as height the distances of the factors from the point Ω with minimum entropy, the values of partial and total rates of resolution are derived, see tab. 5.

Tab. 5 _ Partial and total rates of resolution relating to each attribute and action

	I N V E S T M E N T S			
ATTRIBUTE	T_1	T_2	T_3	T_4
EXF	.132	.132	.192	9.325E-02
INF	1.66E-02	1.342E-02	.0135	.158
IVO	.0736	8.107E-02	.0716	7.91E-02
R^h x 10E-04	1.615	1.439	1.862	1.167

The values of the total resolution have been calculated, see last row in Tab. 5 and the outranking relation, as shown in formula (16), supplies the following order of preferability from the less to the more risky action:

$$T_3 \succ T_1 \succ T_2 \succ T_4 .$$

The first three actions are significantly different, but a clear distinction with the last one T4 is evident. The computed index of resolution is an absolute one and this is a very important characteristic of the model proposed, because it enables identification of the convergence of undesirabilities on the various actions and their comparison with others that can be inserted into the process.

CONCLUSIONS

The final ranking has reassumed the fitness of the acquired information from the DM. Moreover, it has allowed the integration of the qualitative and quantitative data, and the concept of attractiveness or riskiness has permitted the DM to express, in a simple way, all the undesirabilities related to the levels on which the problem has been hierachized.

The graphic formulation of judgements implicitly and completely obsorbs the possible thresholds of preference-indifference with a linear decisional process, comprehensible for the DM and without exhausting pairwise comparisons and use of adjectives which would have the effect of losing objectivity in evaluation.

The concept of the ideal point Omega, having maximum anti-entropy, is suitable for the understanding of the optimality of the acquired information and the expressed undesirabilities towards risk. Some extensions of the model are possible for cases with more hierachical levels, for example, with more involved decision makers or more joined risky situations. Some inserted devices allow to adjust the proposed model for problems not conceived in a hierarchical structure.

For the application of the model, a software, by an interactive dialogue, brings the DM to the formulation of undesirabilities and to the achievment of the related ranking.

REFERENCES

[1] _ BANA e COSTA C., VANSNICK J.C.: Sur la quantification des jugements de valeurs: l'approche MACBETH, Cahier du Lamsade, Juillet, 1993.

[2] _ BRANS J. P., VINCKE Ph.: A Preference Ranking Organisation Method, Management Science, 31/6, pp. 647-656, 1985;

[3] _ FANO R.: *Trasmission of information*, John Wiley & Sons, Inc. New York, London.

[4] _ FANTAPPIE' L.: Sull'interpretazione dei potenziali anticipati della meccanica ondulatoria e su un Principio di finalità che ne discende, Rend. Acc. d'Italia, serie 7.a, vol. 4°, fasc. 1-5, 1942.

[5]_ MARTEL J. M., AZONDEKON H. S., ZARAS K., Preference relatives in multicriterion analysis under risk, Document de travail, n.91-35, Faculté des Sciences de l'AdministrationUniversité Laval, Quebec.

[6]_ ROBERTS F.S.: *Measurement Theory with Applications to Decision Making, Utility and the Social Science*, Addison-Wesley, London 1979.

[7] _ ROY B.: *Methodologie multicritere d'aide a la decision*, Economica, Paris, 1985;

[8] _ SAATY T. L., *The analytic hierarchy process*, McGraw Hill Publishing Company, 1980.

[9] _ SCHÄRLING A. _ *Décider sur plusiers critères*, Presses Polytechn. Romandes, Lausanne, 1985.

[10] _ SHANNON C., WEAWER W.: *The Mathematical Theory of Communication*, Urbana, University of Illinois Press, 1949.

[11] _ SLOWINSKY R., ROY B.: Criterion of distance between technical programming and socio-economic priority, Cahier du Lamsade N° 95, Paris, March 1990.

[12] _ TEILHARD DE CHARDIN P.: *Le phénomène humain*, Edition du Seuil, 1955.

[13] _ VANDERPOOTEN D.: The Construction of Prescriptions in Outranking Methods, in *Reading in Multiple Criteria Decision Making*, Bana e Costa C. A. Ed. Springer-verlag, 1990.

[14] _ VINCKE P.: *L'aide multicritere a la decision*, Editions de l'Universite' de Bruxelles, 1989:

[15] _ WEBER M. , EISENFUHR F. , WINTERFELDT von D.: "The effects of splitting attributes on weights in multiattribute utility misurament", Management Science , 1988, vol. 34 , pp.431-445.

MULTICRITERIA DECISION MAKING AND PORTFOLIO MANAGEMENT WITH ARBITRAGE PRICING THEORY

Christian Hurson [1], Nadine Ricci-Xella [2]
[1] University of Rouen, CREGO
76821 Mont Saint Aignan cedex
France
[2] Faculté d'Economie Appliquée
C.E.T.F.I.
Aix en Provence, France

Abstract: This paper proposes to combine Arbitrage Pricing Theory (APT) and multicriteria decision making to model the portfolio management process. First APT is used to construct some efficient portfolios to estimate its expected return and to identify influence factors and risk origins. Then, two multicriteria decision methods: ELECTRE TRI outranking method and the MINORA interactive system are used to select attractive portfolio, using APT factors as selection criteria. This methodology is illustrated by an application to the French market.

Key-words: Finance, Arbitrage Pricing Theory, Multiriteria Decision Making, Multidimensional Risk Analysis, Empirical Test, French Market.

INTRODUCTION

Portfolio management is a matter of compromise between risk and return; thus it is fundamental to understand the theoretical and empirical relation that exists between these two concepts. It is Markowitz who gave the start of portfolio financial modern theory, proposing in 1952 his famous Mean-Variance model. Following this model, any investor works toward two ends: maximisation of the expected return and minimisation of the risk (measured by the variance). Based on the same principle, numerous problems were developed later. Among these developments, the more significant are expected utility theory, stochastic dominance, and two equilibrium models: the Capital Arbitrage Pricing Model (CAPM, Sharpe 1964) and the Arbitrage Pricing Theory (Ross 1976). Except in APT, in this approach, that can be named "classical approach," the conception of risk is unidimensional. An analysis of risk nature in portfolio management shows that this one comes from various origins and then its nature is multidimensional. APT proposes a different approach for financial market equilibrium. It is a description of equilibrium more general than the CAPM. The single hypothesis of APT is that the return of any financial asset is influenced by a limited number of common factors and a factor specific to the financial asset. Then, the advantage of APT is that it recognizes the multidimensional nature of risk and that it does not impose a restrictive comportment to the investor as in the MEDAF. The APT can be efficiently used to determine the expected return of portfolio and the different origins of risk. However, it does not answer to the question: how to manage portfolio selection?. A possible solution is to use multicriteria decision making which is an advanced topic of operational research conceived to manage this kind of problem. Furthermore, multicriteria decision making, presents the advantage to be able to take into account the preferences of any particular decision maker.

This paper proposes to use an APT model to construct a set of efficient portfolios, determine their expected return and identify the various common factors that influence the financial market. Then, two multicriteria decision methods, the ELECTRE TRI (Yu 1992) and the MINORA interactive system (cf. Siskos et al. 1993 or Spyridacos and Yannacopoulos 1994) are used to perform portfolio selection managing various risk origins; this is done using the risk attached to each factor (measured by the factor sensibility) as a portfolio selection criterion. This paper is divided in three sections and a conclusion. The first section presents Arbitrage Pricing Theory, multicriteria decision making and the link existing between them. The second section develops our methodology: the APT model performed and the multicriteria decision making methods used. The third section presents an application of this methodology to the French market: data base, results and comments. Finally, in the conclusion we summarise the obtained results and give some future direction.

1. ARBITRAGE PRICING THEORY AND MULTICRITERIA DECISION MAKING

1.1 The Equilibrium Model

At the begin of the development of the modern financial theory, the aim was to evaluate the future, hence risky, monetary flows. We have three approaches:

- Fundamental analysis with Discounted Cash Flow model where security prices are the expected actualised cash flows. The problem of this method is to characterise theses cash flows in an infinite horizon,
- Chartist's analysis which is based on the observations of past phenomena.
- The modern approach, based on Markowitz model (1952) about portfolio selection and its principal developments, namely CAPM and APT models.

1.1.1. The Capital Arbitrage Pricing Model (CAPM)

The CAPM developed by Sharpe (1963), Mossin (1966) and Lintner (1965) is an equilibrium model that linearly links the expected return of an asset i, $E(R_i)$, to a single risk origin represented by the market portfolio, R_M:

$$E(R_i) = R_F + [E(R_M) - R_F]\, \beta_i \text{ with } \beta_i = cov(R_i\,,\, R_M) / \sigma^2(R_M) \qquad \text{(Eq. 01)}$$

APT is that it recognizes the multidimensional nature of risk and that it does not impose a restrictive comportment to the investor as in the MEDAF. The APT can be efficiently used to determine the expected return of portfolio and the different origins of risk. However, it does not answer to the question: how to manage portfolio selection?. A possible solution is to use multicriteria decision making which is an advanced topic of operational research conceived to manage this kind of problem. Furthermore, multicriteria decision making, presents the advantage to be able to take into account the preferences of any particular decision maker.

This paper proposes to use an APT model to construct a set of efficient portfolios, determine their expected return and identify the various common factors that influence the financial market. Then, two multicriteria decision methods, the ELECTRE TRI (Yu 1992) and the MINORA interactive system (cf. Siskos et al. 1993 or Spyridacos and Yannacopoulos 1994) are used to perform portfolio selection managing various risk origins; this is done using the risk attached to each factor (measured by the factor sensibility) as a portfolio selection criterion. This paper is divided in three sections and a conclusion. The first section presents Arbitrage Pricing Theory, multicriteria decision making and the link existing between them. The second section develops our methodology: the APT model performed and the multicriteria decision making methods used. The third section presents an application of this methodology to the French market: data base, results and comments. Finally, in the conclusion we summarise the obtained results and give some future direction.

1. ARBITRAGE PRICING THEORY AND MULTICRITERIA DECISION MAKING

1.1 The Equilibrium Model

At the begin of the development of the modern financial theory, the aim was to evaluate the future, hence risky, monetary flows. We have three approaches:

- Fundamental analysis with Discounted Cash Flow model where security prices are the expected actualised cash flows. The problem of this method is to characterise theses cash flows in an infinite horizon,
- Chartist's analysis which is based on the observations of past phenomena.
- The modern approach, based on Markowitz model (1952) about portfolio selection and its principal developments, namely CAPM and APT models.

1.1.1. The Capital Arbitrage Pricing Model (CAPM)

The CAPM developed by Sharpe (1963), Mossin (1966) and Lintner (1965) is an equilibrium model that linearly links the expected return of an asset i, $E(R_i)$, to a single risk origin represented by the market portfolio, R_M:

$$E(R_i) = R_F + [E(R_M) - R_F]\, \beta_i \text{ with } \beta_i = cov(R_i , R_M) / \sigma^2(R_M) \qquad \text{(Eq. 01)}$$

where R_F is the return of the risk free asset, β_i is the well-known beta that mesures the sensitivity of an asset to the market and $\sigma^2(R_M)$ is the variance of the market portfolio.

The principal objection to this model is that it does not resist well to empirical tests. Especially, note:

- Roll's critic (1977) about the impossibility to test the model without the exact composition of the market portfolio;
- Existence of anomalies (size effect, seasonal effect,..);
- Non stationarity of risk parameters.

These critics suggest that an equilibrium relation does not exist under the form described by the CAPM. Thus, we must choose another process of return generation. These difficulties lead to a new formulation issued from linear algebra, the Arbitrage Pricing Theory or Ross's APT (1976), as an alternative model to CAPM.

1.1.2. The Arbitrage Pricing Theory (APT)

The APT proposes a multifactor relation between return and risk under less restrictive hypotheses than CAPM. It supposes that the return of any asset, R_i (i=1,2..n), is equal to the expected return, $E(R_i)$ plus an unexpected return ($\Sigma_k b_{ik} F_k + e_i$), where F_k is a common factor, b_{ik} is the sensibility coefficient of asset i to factor k and, e_i the specific risk component of the asset i. This gives the following equation, known under the name of equation I of APT.

$$R_i = E(R_i) + \Sigma_k b_{ik} F_k + e_i. \qquad \text{(Eq. 02)}$$

The common risk factors represent the surprises that meet the investors and which cannot be avoided. One estimates that on a long term, on average, these surprises compensate themselves.

To determine the APT fundamental relation, Ross developed an economic argument that is: at equilibrium we cannot have any arbitrage portfolio. This arbitrage portfolio must verify three conditions:

(a) no change in wealth: $\Sigma_i w_i = 0$, (Eq. 03)
(b) no additional systematic risk: $\Sigma_i w_i b_{ik} = 0$, (Eq. 04)
(c) no complementary return: $\Sigma_i w_i E(R_i) = 0$. (Eq. 05)

where w_i is the proportion of the asset i in the arbitrage portfolio.

Combined with linear algebra, it follows from this reasoning that the proportion's vector $\underline{w}$ of any asset in the portfolio is orthogonal to: (a) a one vector $\underline{1}$, (b) the vector of sensibilities $\underline{B}$, and (c) the vector of expected return $\underline{E}$. Then, the expected return can be expressed as a linear combination of a constant multiplied by one, plus a constant multiplied by b_{i1}, plus a constant multiplied by b_{i2} and so on, until b_{iK}. Hence, we obtain the APT fundamental relation called equation II :

$$E(R_i) = \lambda_0 + \Sigma_k b_{ik} \lambda_k \qquad \text{(Eq. 06)}$$

where λ_0 is the intercept of the pricing relationship (zero-beta rate) and λ_k the risk premium on the k-th factor.

The above equality works because Ross assumes that the arbitrage portfolio error term variances become negligible when the number of assets, n, increases. Here, Ross applies the great number law to justify the convergence of a serie of random variables to a constant; but this law is not sufficient and the residual term is not necessarily equal to zero (because this law refers to the degeneration of the residual joint distribution). Taking into account these difficulties, several authors tried to use other derivations in order to replace the approximation with an equality; this is done using two types of argument: an arbitrage argument or an equilibrium argument.

■ *The arbitrage argument*

Concerning arbitrage argument, Huberman (1982) uses a Strict Factor Structure that considers the arbitrage as the capacity to create sequences of arbitrage portfolios, then in this case the pricing error is upper bounded. After Ingersoll (1982) took the direction of asymptotic arbitrage portfolio to obtain a smaller upper bound without the necessity of a diagonal covariance matrix for returns. Chamberlain-Rothchild (1983) derived an upper bound for all assets in the frame of Approximate Factor Structure. This upper bound is the product of the $(K+1)^{th}$ eigenvalue of the covariance matrix of idiosyncratic returns with an arbitrage coefficient along the efficient frontier. In these structures, the pricing error is finite, but its range has an unknown nature. Hence, following arbitrage argument direction, investors insure themselves against their specific risk holding many securities in small quantities. Nevertheless, in this purpose we do not know how many securities will be sufficient and in which quantities, because these derivations evaluate the asset together but nobody knows how APT model evaluates each asset separately.

■ *The equilibrium argument*

To answer to the latter problem, the literature develops the equilibrium argument where an investor will be characterised by a certain level of risk aversion, that can be defined in two ways:

- either it is related to the Marginal Rate of Substitution, and in this case Connor (1982), Dybvig (1982) or Grinblatt-Titman (1982) develop a lower bound or an upper bound of the pricing error, which is a function of the risk aversion level, the supply of security per capita and the residual risk.
- or it is characterised by the detention of a well-diversified portfolio following Shanken (1982, 1985) or Gilles-Leroy (1991) critics. Then Connor (1984)

proposed a competitive equilibrium of APT that can be applied either in a finite economy or in an infinite economy. In Connor's study, the diversification mechanism in finite economy is not the same than in infinite economy because investors hold a well-diversified portfolio[1]. Finally, Chen-Ingersoll (1983) add that this diversified portfolio is optimal for at least a risk averse investor who maximises his utility.

The APT based on equilibrium, linking itself to competitive market, offers a conceptual base stronger than the APT based on arbitrage. Nevertheless, as said before, the model needs enough securities without knowing both the number and the nature of risk factors (cf. Ricci-Xella 1994).

1.2. The Link between APT and Multicriteria Decision Making

Thus, APT seems to be a better model than CAPM because in APT there exists several risk sources. Note that the APT is a normative model that imposes restrictive hypotheses, even if they are less restrictive than in the CAPM. Then, to make the model more realistic, it is necessary to weaken some hypotheses as homogeneous anticipation, e.g., risk appreciation. It would be particularly interesting to adapt the model to the preferences of the investor who in practice is a real person or group of persons. Effectively, any investor is confronted with a given risk in a particular situation. Then he has objectives and a risk attitude that are specific to him, which should be taken into account.

Identifying several sources of risk, the APT pleads for a portfolio management policy able to manage and take into account a multicriteria choice. Then, multicriteria decision making furnishes an interesting methodological framework to our study. Linking the multicriteria evaluation of asset portfolio and the research of a satisfactory solution to the investor's preferences, the multicriteria decision making methods allow to take into account the investors' specific objective. Furthermore, these methods do not impose any normative scheme to the comportment of the investors. The use of multicriteria methods allows to synthesise in a single procedure the theoretical and practical aspects of portfolio management, then it allows a non normative use of theory.

Note that the application of the above principles is difficult because of the complexity of multicriteria problems on the one hand and the use of criteria from different origins and of conflictual nature on the other hand. Furthermore, multicriteria decision making will facilitate and favour the analysis of compromise between the criteria. It equally permits to manage the heterogeneity of criteria scale

---

1. - In *finite economy*, investors eliminate the risk exploiting the singularity of the residual covariance matrix, and they hold a particular combination of securities in such way than the specific risk cancels each other out;
   - In *infinite economy*, investors diversify holdings on several securities in infinitesimal proportion: this well-diversified portfolio allows to be insured against specific risk, that avoiding it.
   These diversification mechanisms allow to distinguish those economies but, empirically there is no way to differentiate it if we only observe the return of the portfolio and the evaluation of the securities.

and the fuzzy and imprecise[2] nature of the evaluation that it will contribute to clarify.
The originality of multicriteria decision making equally offers, systematizing the decision process, the possibility to obtain a gain of time and/or to increase the number of assets considered by the practitioner. Moreover, in a market close to efficiency, as are all the big markets, it's the good and fast use of all available information that ensures informational efficiency of capital markets and will permit the investor to compete.

Now, before to present our methodology, it is necessary to briefly present what is Multicriteria Decision Making.

1.3 Multicriteria Decision Making

Multicriteria decision making is not only a set of methods conceived to manage multicriteria problems. It is an activity that begins with the problem formulation (choice of a set of criteria, or construction of a multiobjective problem under constraints) and finishes with the proposition of a solution. This activity contains five steps:

Step 1: decision set elaboration and choice of a problematic (ranking, sorting, choice or description of alternatives)

Step 2: elaboration of a coherent family of criteria, that is a set of criteria that present the properties of exhaustiveness, non redundancy and cohesion.

Step 3: modeling of decision maker preferences, choice of an aggregation method (a family of methods)

Step 4: choice of a method and application

Then we can see, once again, the complementarity of APT and multicriteria decision making. Actually, identifying the set of independent factors that best explain the return, the APT brings to multicriteria decision making the coherent family it needs.
Lets see now the other steps of multicriteria decision making. About the decision set, we decide to use portfolio rather than securities because the evaluation of securities with APT does not give reliable result, which is not the case with portfolio. After that, we had to construct a set of portfolio, this was done generating portfolio following their variance in order to obtain various risk levels.

About the problematic, the ranking and the sorting of these portfolios retain our attention. The ranking because it answers to a common and natural preoccupation of analysts. We decided to use the sorting of portfolio in three categories ("good" portfolio, "bad" portfolio and uncertain portfolio). This way of acting fits well with the purpose of portfolio management: which portfolios are interesting, which portfolios are not interesting and which portfolios must be studied further.

---

2. Here the words fuzzy and imprecise refer to: (a) the delicacy of an investor's judgement (the human nature and the lack of information), that will not always allow to discriminate between two close situations, on one hand, and (b) on the other the use of a representation model, which is a simplification of reality that expresses itself in an error term.

Then we have to choose the methods we will apply. In order to do this let present the different approaches that exist in multicriteria decision making. The specialists of multicriteria decision making distinguish three families of methods and two approaches or schools. These three families are the Multi-Attribute Utility Theory (MAUT), the outranking methods and the interactive methods.

MAUT is an extension of the expected utility to the case of multiple criteria. Then, MAUT aggregates the criteria in a unique function that represents the decision maker preferences and that must be maximised. This approach takes an interest in the forms of the utility function, in the aggregation conditions and in the construction methods.

The outranking methods exploit an outranking relation that represents the decision maker's preferences securely established, following the available information. An important characteristic of the outranking relation is that it admits intransitivity and incomparability in the preferences of the decision maker.

The interactive methods are iterative and consist of a succession of calculation and dialogue steps. A dialogue step allows to collect some information about the decision maker's preferences. A calculation step uses this information to found a compromise solution, then this one is proposed to the decision maker in a new dialogue step. If the decision maker is satisfied, the method stop, otherwise a new information is asked to the decision maker in order to determine an other compromise.

Obviously this classification is not perfect, since some methods cannot be easily attached to one of these families and some others can belong to two families. An other classification, more recent, distinguishes the descriptive methods and the constructive methods.
The descriptive approach comes from physics. It supposes that the decision maker has a stable and "rational" preference system that can be described by a utility function and that it is possible to reason objectively on a clear problem. The utility function must be a good approximation of the decision maker's preference system to found the "best solution". In this approach, there are no possibilities for hesitations and incomparabilities.

The constructive approach is more recent. It is born from the numerous critics made on the descriptive approach. This approach considers that in a decision process the intervention of human judgement sets the descriptive approach not appropriated (cf. Roy 1987, 1992). The decision maker's preferences are not stable, not very structured and conflicting. There is no objective reasoning, the perception of the problem has an influence on its modelling and its resolution, and inversely. Then, it is normal that the problem formulation and the decision maker's preferences can evolve during the resolution, and that the model used accepts hesitations and incomparabilities.

One can see that MAUT belongs in the class of the descriptive approach, while the outranking methods to the constructive one. But the interactive methods do not found clearly their place in one of these approaches. Most of them use a unique synthesis criterion that can be assimilated to a utility function. Nevertheless, its use is different. In an interactive method the synthesis criterion is used locally to determine a new proposition. Then, an interactive method could belong, following the cases, to the descriptive or the constructive approach (cf. Roy 1987). It would belong to the constructive approach if it admits enough trials, errors and backups to allow a free exploration by the decision maker of the whole range of solutions, and his control of the decision process.

The reality is probably a combination of the descriptive and constructive hypothesis, but experience and some studies conduct to think that the constructive approach is more close to reality. Only a constructive method can help the decision maker to solve his problem without any normativity. Concerning portfolio management, we think that this approach will favours the analyse of compromises and the understanding of the complex relation between return and risk. Then, the multicriteria decision methods we choose to use, the MINORA interactive system and the ELECTRE TRI outranking method, belong in the class of the constructive approach.
Let us note that the supposition of an existence of an order is the same as the supposition of the existence of a utility function, thus of the hypothesis of transitivity and comparability of preferences. Then, the outranking methods do not seem to be well adapted to solve ranking problems, so the use of an interactive method as MINORA seems to be a better way. However, for sorting problems an outranking method as ELECTRE TRI seems to be well adapted and presents the advantage to admit incomparabilities and intransitivities.

2. THE METHODOLOGICAL FRAMEWORK

In this section we will present our methodological framework. In a first paragraph we will present the implementation of our APT model, and in a second paragraph the multicriteria methods used in this study.

2.1 Methodology and empirical test of APT

Notice that the existence of factor structure in the equation I of APT (eq. 02) is a hypothesis, while the equation II of APT (eq. 06) is an implication. To test APT on ex-post data, we suppose that reliable information does not vary from one period to another and is sufficient and available to determine the real anticipation.
After that, we will present the development of APT empirical tests that we use; then we will expose our methodology.

2.1.1. *The tests on APT consist of the identification of the number and nature of risk factors.*

■ Concerning the determination of *the number of factors*: it was initially motivated by models derived by Merton (1973) in the intertemporal capital asset pricing context which suggests the existence of at least two factors in economy; by Long (1974) or Cox-Ingersoll-Ross (1985) who indicated that the existence of several factors is an attractive alternative.
In APT, one generally proposes two methods:
- either one determines the sufficient number of factors, using data analysis like Principal Component Analysis or using Maximum Likelihood Factor Analysis as in Roll-Ross (1980), Dhrymes-Friend-Gultekin-Gultekin (1984, 1985), ...
- or to prespecify the number of factors to test APT validity as in Chen (1983), Oldfield-Rogalski (1981), ...

■ Concerning the *identification of the factors*:
- One often uses the Fama-MacBeth's technics (1973) about CAPM relating these factors to exogenous variables[3]. Effectively in the first version of the APT the nature of the factors was unknown, then without any attractive sense for commercial practitioner. Roll and Ross (1980) were among the first to look specifically for APT factors.
- Afterwards, the following version of APT gave an economic interpretation to the factors easily comprehensible, then acceptable, by portfolio managers. Following Roll-Ross (1983), because more than the half of the realised return is the result of non anticipated variations, the systematic forces which influence returns are those that cause variations of interest rate and revenues. The most famous study was made by Chen, Roll and Ross (1983), who derived the common factors from a set of data and then tested them for their relationship to fundamental macroeconomic variables, such unanticipated changes in inflation, changes in expected industrial production, unanticipated changes in risk premia and unanticipated changes in the slope of the term structure of interest rates.
- Finally, the mesoeconomic[4] APT is an interesting solution but it is expensive to collect monthly accounting and financial data.

That is why we decide to explore the macroeconomic version of the APT which determines a prespecified number of factors.

---

3. Or endogeneous with completude tests, in order to study if reflected risk in the covariance matrix of return are valuable and no other anomalies appear (size effect, week-end effect, moments of distribution,...). We also suggested taking mesoeconomic variables with financial and accounting components (cf. Ricci-Xella's thesis).
4. We use the term mesoeconomic to signifie that the risk factors are both micro and macroeconomic variables (cf. Ricci-Xella 1994).

2.1.2. *The methodology to obtain persuasive (reliable) exogenous risk factors is the following:*

■ Our *sample* consists of return on Paris Stock Exchange firms (SDIB-SBF) (from November 1983 through September 1991), avoiding simultaneous quotation in France (spot market and forward market until October 10, 1983). For each asset, we take alteration of capital into account (contribution of capital, payment of dividend, ...) and we correct it[5]. To determine the return, we consider the actual quotation and the number of exchange shares during the month.

Macroeconomic variables came from O.E.C.D. and I.N.S.E.E. They are presented in Table 1. We determine unexpected changes in those factors, and we add the market portfolio to obtain, theoreticaly, an equality in the pricing relationship. From the 28 variables of origin, some will be eliminated because they are too correlated. The others were used to create eleven non anticipated or non expected macroeconomic variables which we employed.

Table 1: Definition of macro-economic variables

N°	Symbol	Name
MACRO1	MP_t	Monthly growth of industrial production
MACRO2	UP_t	Annual growth of industrial production
MACRO3	UEI_t	Non expected inflation
MACRO4	DEI_t	Variations of expected inflation
MACRO5	UPR_t	Risk premium variation (by bound)
MACRO6	$CACEXC_t$	Risk premium variation (by cac240)
MACRO7	TS_t	Variation of the interest rate term structure
MACRO8	XM_t	Commercial balance variation
MACRO9	HT_t	Growth rate of money
MACRO10	ATL_t	Market index variation (logarithmic)
MACRO11	VCI_t	Variation of consumption prices index

■ In the first *step,* the test consists to determine the number of significant factors. We calculate the matrix of sensibilities using time-series to regress odd asset returns on unexpected macro-economic variables plus a market index portfolio. In the second step, a cross-sectional regression on even portfolio returns using the sensitivities as independent variables (we assume the beta constant over the period of time) in which the parameters being estimated are λ_0 (zero-beta or the return on the riskless asset) and λ_k (the risk premiums). Finally, the third step suggests to use a battery test (F-test, t-student, Root Mean Square Error, R^2,...) to determine the number of pertinent factors.

5. We avoid having two assets with two distinct Sicovam' codes, 2 codes for 1 name, double quotation on spot and forward market, doubloon on particular days as holidays, strike days, bomb alerts, computer problems and no more than 4 sucessive missing quotations.

Over the 16 possibilities[6] of grouping assets into portfolios, we adopt 28 portfolios of 6 assets each as advised by Gibbon-Ross-Shanken (1986). After multiple possible combinations, we retain the best model that corresponds to an APT of 11 macroeconomic factors (cf. Table 4). These portfolio are generated using monthly logarithmic returns (cf. Ricci-Xella 1994, file axicml5). The eleven macroeconomic variables are normalized and we add the French market index that we called CAC240. This index is calculated using the general index of the SDIB-SBF (Société de Diffusion des Informations Boursières-Société Française de Bourse) that contains at maximum the 250 most important french securities. Concerning the riskless asset we retain the short term interest rate at the end of the period (or PIBOR-3months) that we give monthly.

2.2. Multicriteria methodology

The two Multicriteria decision methods used here are, the MINORA interactive system and the ELECTRE TRI outranking method.

2.2.1 *The Interactive Multicriteria Decision Making System MINORA*

MINORA is an interactive multicriteria decision making system that ranks, from the best to the worst a set of alternatives, following several criteria. For this purpose the MINORA system uses the UTA ranking algorithm from Jacquet-Lagrèze and Siskos.(1982). From the ranking of a subset of well-known alternatives made by the decision maker, UTA uses ordinal regression to estimate a set of separable additive utility functions of the following form :

$$u(g)=u_1(g_1) + u_2(g_2) +...+ u_k(g_k) \qquad \text{(Eq. 7)}$$

where $g=(g_1,...,g_k)$ is the performance vector of an alternative and $u_i(g_i)$ is the marginal utility function of criteria i, normalised between 0 and 1. The ordinal regression is performed using linear programming (for more details see, for example, Despotis et al., 1990). In MINORA the interaction takes the form of an analyses of inconsistencies between the ranking established by the decision maker and the ranking issued from the utility function estimated by UTA. Two measures of these inconsistencies are used in MINORA: (1) The F indicator which is the sum of the deviations of the ordinal regression curve (global utility versus decision maker's ranking), e.g. the sum of estimation errors. (2) The Kendal's t that gives a measure, from -1 to 1, of the correlation between the decision maker's ranking and the ranking resulting from the utility function. At optimality, when the two rankings are similar, the F indicator is equal to 0 and the Kendal's t is equal to 1.

6. The number of created portfolios is given by the Greater Common Denominator of asset in the portfolio, thus $168 = 2^3 * 3^1 * 7^1$. We take the product of the entire number of exposants plus one thus $(3+1)*(1+1)*(1+1) = 16$ collectings.

The interaction is organised around three questions presented to the decision maker : (1) Is he ready to modify his ranking ? (2) Does he wish to modify the relative importance of a criterion, its scale or the marginal utilities (trade off analysis) ? (3) Does he wish to modify the family of criteria used : to add, cancel, modify, divide or join some criteria? (4) Does he wish to modify the whole formulation of the problem ? These questions send back to the corresponding stages of MINORA and the method stops when an acceptable compromise is determined. Then the result (a utility function) is extrapolated to the whole set of alternatives to give a ranking of them (cf. Siskos et al. 1993 or Spyridacos and Yannacopoulos 1994).

The MINORA system presents two main advantages :

- It furnishes a ranking of portfolio, that is a natural preoccupation frequently used by portfolio managers.
- The form of the interactivity. All the originality of the MINORA system can be found in the inconsistencies analysis in an interactive way. It allows to help the decision maker to construct his own model in a non normative way and organises, in a unique procedure, all the activity of decision making, from the model formulation to the final result (a ranking of the alternatives from the best to the worst in the case of MINORA). In the same time the decision maker is constantly integrated to the resolution processes and can control its evolution at any moment.

Finaly, notice that MINORA method had been used successfully to solve numerous management problems, particularly in portfolio management (cf. Zopounidis 1993, Zopounidis et al. 1995, 1997, Hurson and Zopounidis 1993, 1995, 1996).

2.2.2. *The Outranking Multicriteria Decision Making Method ELECTRE TRI*

ELECTRE TRI is an outranking method specially conceived for sorting problems, it is used to sort the portfolio in three categories : attractive portfolio, uncertain portfolio (to be studied further) and non attractive portfolio. ELECTRE TRI deals only with ordered categories (complete order). The categories are defined by some reference alternatives or reference profiles (one down profile and one up profile) which are themselves defined by their values on the criteria. Next we can define the categories C_i, i=1,...,c, where C_1 is the worst category and C_c the best one. We can also define the profiles r_i, i=1,...,c-1, where r_1 and r_{c-1} are the lower and the upper profile respectively. Then, the profile r_i is the theoretical limit between two categories C_i and C_{i+1} and r_i represents a fictitious portfolio which is strictly best than r_{i-1} on each criterion. In ELECTRE TRI, the information asked from the decision maker about his preferences takes the form, for each criterion and each profile, of a relative weight and an indifference, preference and veto threshold.

To sort the portfolios, ELECTRE TRI compares each of them to the profiles using the concepts of indifference, preference and veto threshold in order to construct a concordance index, a discordance index and finely a valued outranking relation as in ELECTRE III method (cf. Roy and Bouyssou 1993). This valued outranking relation, $s_s(a,b)$ measures from 0 to 1 the strength of the relation "a outranks b " (a is at least as good as b). This valued outranking relation is transformed in a "net" outranking relation in the following way: $s_s(a,b) \geq l \Leftrightarrow aSb$, where S represents the net outranking relation, a and b two portfolio and λ a "cut level" ($0.5 \leq l \leq 1$) above which the relation a outranks b is considered as valid. Then the preference P, the indifference I and the incomparability R are defined as follows:

aIb ⇔ aSb and bSa,
aPb ⇔ aSb and no bSa,
aRb ⇔ no aSb and no bSa.

In ELECTRE TRI there are two non total compensation sorting procedures (the pessimistic one and the optimistic one), to assign each alternative into one of the categories defined in advance. In the sorting procedure, portfolio a is compared at first to the worst profile r_1 and in the case of aPr_1, a is compared to the second profile r_2, etc., until one of the following situations appears:

(i) aPr_i or aIr_{i+1} and r_{i+1} Pa,

(ii) aPr_i and r_{i+1} Ra, r_{i+2} Ra,.....r_{i+m} Ra, r_{i+m+1} Pa.

In situation (i) both the procedures assign portfolio to the category i+1. In situation (ii), the pessimistic procedure classifies portfolio a to category i+1, while the optimistic procedure classifies portfolio a to procedure i+m+1. When the value of l gradually decreases, the pessimistic procedure becomes less compulsive and the optimistic procedure less permissive. Evidently the optimistic procedure tends to classify portfolio a to a higher possible category, in contrast to the pessimistic procedure that tends to classify portfolio a to the lower possible category. In general, the pessimistic procedure is applied when a policy of prudence is necessary or when the available means are very constraining. The optimist procedure is applied to problems where the decision maker desires to favour the alternatives that present some particular interest or some exceptional qualities. In portfolio management the optimistic procedure will be well adapted to an optimistic investor with a speculative investment policy, for example, while a prudent investor, following a passive investment policy, will prefer the pessimistic procedure.

ELECTRE TRI manages incomparability in such a way that it will point out the alternatives which have particularities in their evaluations. In cases where some alternatives belong to different categories in both procedures, the conclusion is that they are incomparable with one or more reference profiles (more the number of categories between the two affectations will be important more the "particularities" of the alternative will be important). This is because these alternatives have good values for some criteria and, simultaneously, bad values for other criteria; moreover, these particular alternatives must be examined with attention. In this way the notion of incomparability included in ELECTRE TRI brings an important information to the decision maker and for this reason the best way to employ ELECTRE TRI is to use the two affectation procedures and to compare the results.

The advantages of ELECTRE TRI are the following:

- ELECTRE TRI by sorting the portfolio is well adapted to the purpose of portfolio management (acceptable portfolio, portfolio to be studied further and unacceptable portfolio).
- ELECTRE TRI, as all the methods of the ELECTRE family, accepts intransitivity and incomparability. In ELECTRE TRI this is done in such a way that the method will point out the alternatives that have particularities in their evaluation.
- The ELECTRE family uses techniques easy to understand for the decision maker.

The methods from the ELECTRE family are very popular, they have been used with success in a great number of studies and in portfolio management by Martel et al. (1988,), Szala 1990, Khoury et al. (1993) Hurson and Zopounidis (1995, 1996), Zopounidis et al. 1997).

3. AN APPLICATION TO THE FRENCH MARKET

In this section we present the application of our methodology to the French Market. First we present the formulation of the multicriteria problem and then the applications of ELECTRE TRI and MINORA.

3.1 The multicriteria formulation

Let us first explain the choice of the APT version used in this study. This version regresses 11 normalised macro-economic variables plus the market index. These variables were regressed on the monthly logarithmic return of a set of portfolios that was generated following their capitalisation. Comparing to the theoretic APT model presented in the precedent paragraphs, the F_k are macro-economic variables not expected and/or non anticipated. In this version we take only the test of APT on 28 portfolio of 6 stocks (from 16 possibilities). This comes from the recommendations of Gibbons, Ross and Shanken (19865) that are confirmed by the lower value of the Root Mean Square Error, as one can see on the following table.

Table 2: RMSE test

grouping type	Value of RMSE
RR012: 14 portfolios of 12 stocks	0,0001827
RR008: 21 portfolios of 8 stocks	0,0005599
RR007: 24 portfolios of 7 stocks	0,0005946
RR006: 28 portfolios of 6 stocks	**0,0004834**
RR004: 42 portfolios of 4 stocks	0,0008367
RR003: 56 portfolios of 3 stocks	0,0009989
RR002: 84 portfolios of 2 stocks	0,0013316

In addition to the RMSE, we test the various versions of APT using the R^2 adjusted. This test confirms that the chosen version was the best one. All these tests were performed on portfolios generated in function of their capitalisation and their return, using 6 different regressions (with or without a constant term at the first or the second step).

Let us note that, in the majority of our APT tests, the variables that best identify the risk factors are, in decreasing order: the constant component, the consumption price index (Macro6 or Macro11) and the market portfolio (CAC240). Then come some variables that are valuable with a significant level between 0 and 5% (in decreasing order): The risk premium (Macro6), the monthly growth of industrial production (Macro1), the growth rate of money (Macro9). We made an application of the APT with eleven variables. Nevertheless, the criteria corresponding to the variables 2, 3, 4, 5, 7 and 8 had no influence on the result of the multicriteria methods. Furthermore, by cancelling these criteria, the others became more significant. This is why we decide to use in our study the following macro-economic variables: macro 1, 6, 9, 11 and the CAC240.

Now let us see how it is possible to interpret these results. When we test the APT on our sample, we perceive that the variations of the index of consumption prices and the market portfolio are pertinent variables for stocks' evaluation. In order to know if the market portfolio is sufficient we test the CAPM on the one hand and we regress the CAC240 on the macro-economics variables on the other hand, as it was suggested by Chaffin (1987). The results show that the CAC240 is not well explained by these variables. Then they are not necessarily included in the market portfolio and the later complete the macro-economic variables and do not serve as a substitute to them. Let us now give an interpretation to the fact that the variable macro11 takes a preponderant place in the stocks' evaluation. This interpretation is that, on the studied period (1983-1991), France knew a situation of disinflation.

Then we must decide which variables we take as criteria for the multicriteria methods. The sensitivity factors b_{ik}, as measures of risk, will be used as criteria for portfolio selection (as the beta in the CAPM). These sensitivity factors are called beta (thus beta 1 is the sensitivity factor corresponding to variable macro 1) are presented in Table 3 (in order to make it presentable these beta are multiplied by 1000). All these criteria must be maximised, except those concerning inflation.

Table 3: The chosen criteria (sensitivity factors)

	Beta 1	Beta 6	Beta 99	Beta 11	Beta cac240
P1	2.65	9.34	0.86	-0.69	-154.05
P2	4.15	-2.66	4.54	-1.85	1210.42
P3	4.6	3.74	3.92	-1.68	593.63
P4	-2.89	-5.98	-1.99	1.62	1413.65
P5	3.13	3.86	2.33	-2.91	449.65
P6	0.91	4.86	0.53	0.18	292.95
P7	1.44	-9.07	2.07	2.91	1941.5
P8	1.18	3.99	2.69	-1.63	361.5
P9	1.56	11.64	2.15	2.34	-314.95
P10	0.14	0.55	-0.25	-0.28	402.25
P11	-1.12	-1.78	-1.47	-0.42	1356.25
P12	-1.97	1.76	-1.89	0.9	682.01
P13	-2.5	4.65	-0.45	3.49	526.68
P14	0.27	3.21	-0.05	-0.98	269.82
P15	-0.16	6.48	-1.55	0.87	54.17
P16	0.31	17.22	2.52	0.04	-252.06
P17	0.01	1.47	0.36	0.65	829
P18	0.59	2.35	3.09	2.69	747.38
P19	-0.01	-4.5	2.16	-0.66	1376.23
P20	0.35	7.08	0.59	1.92	77.84
P21	0.99	-0.04	0.97	-1.54	931.67
P22	-2.69	1.98	-2.39	-2.38	795.61
P23	0.14	6.44	-0.91	-1.59	342.34
P24	0.55	4.06	0.22	0.78	430.81
P25	-0.64	0.22	-0.97	-0.8	791.08
P26	-0.87	1.36	-0.09	0.7	565.7
P27	-2.62	-1.07	-1.91	0.92	917.73
P28	0.15	2.12	0.12	-0.78	620.64

3.2 Application of ELECTRE TRI

Unfortunately, it was not possible, during this study to manage the multicriteria methods with a "real" portfolio manager. Then, in order to present an application of our methodology, we decide to play his part. As was indicated, the objective is to sort the portfolio in three categories: attractive portfolios (C_3), uncertain portfolios which need further study (C_2), non attractive portfolios (C_1). As the weight of the criteria in ELECTRE TRI must be chosen by the decision maker, considering the absence of a "real" decision maker for this study, we decide to give the same weight to all the criteria. Thus, each criterion is supposed to have the same importance for the decision maker. In addition to these weights, in order to compute the discordance and concordance indices in ELECTRE TRI we have used the reference profiles and the thresholds presented in the following table 4.

Table 4: The reference profiles and their preferential parameters

Criteria / Parameters	Beta 1	Beta 6	Beta 9	Beta 11	Beta cac240
High reference profile	0.56	3.80	0.9	0.7	760
Low reference profile	-0.1	1	-0.3	-0.8	380
Indifference threshold	0.1	0.1	0.1	0.1	10
Preference threshold	0.5	0.5	0.5	0.5	50
Veto threshold	4.6	17.22	4.54	3.49	1941.5

The indifference and the preference profiles are perception thresholds. The indifference threshold gives the value below which the decision maker considers that the difference between two portfolios is not significant. Then, in our application, a difference lower than 0.1 on the criterion beta 1 is considered as not significant. The preference threshold gives the value above which, a difference between two portfolios imply a certain (strong) preference for one of them, considering the criterion examined. For example, in our study, a difference greater than 0.5 between two portfolios on the criterion beta 1, implies considering this criterion alone, a strong preference for one of these two portfolios. Here, the values of these thresholds are the same for the four first criteria because their scales have similar ranges of values. The veto threshold has a different nature. It gives the value above which a difference on the criterion between two portfolios a and b, in favour of a, imply the rejection of the outranking of b by a ("b is at least as good as a"), even if b has better values on the other criteria. In ELECTRE TRI, the portfolios are compared to the preference profiles. Then, the preference threshold has for effect to forbid, in the pessimistic procedure, the sorting of a portfolio in a category if, at least one criterion is in favour of the low profile of this category, with a difference superior to the veto threshold. If this situation appears, the criterion responsible of the veto become decisive. Then, considering that all our criteria have the same type of significance (sensitivity to a macro-economic variable), and that none of them must take a particular importance, we decide not to use veto threshold. This is done fixing the veto threshold to their default value that correspond to the maximum of the criterion (then the vetoes are not active).

The value of l is fixed at its default value 0.67. Table 2 presents the sorting results of the ELECTRE TRI method in optimistic and pessimistic cases. Looking at these results one can remark that the portfolios that belong to the best category (C_3) in both optimistic and pessimistic sortings, are proposed without hesitation to the portfolio manager for selection. The portfolios that belong to the worst category (C_1) in both optimistic and pessimistic sortings are not proposed to the portfolio manager. When the portfolios belong to the uncertain category (C_2) for both optimistic and pessimistic sortings, this means that these have moderate values on all criteria and, consequently, they must be studied further. In the cases where some portfolios belong to different categories in both optimistic and pessimistic sortings, this means that they are incomparable with one or two reference profiles. The portfolio belonging to the categories 2 and 3 can be considered as relatively attractive. Inversely the portfolio belonging to the categories 1 and 2 can be considered as relatively not attractive. The portfolios belonging to the categories 1 and 3 are incomparable with the two profiles. This mean that these portfolios have good values for some criteria and, simultaneously, bad values for other criteria. Also, these portfolios must be examined further as the portfolios of the category C_2. In this way, the notion of incomparability included in the ELECTRE TRI method brings an important information to the portfolio: it points out the portfolios that have particularities in their evaluation and can represent an interesting opportunity or a particular risk.

Table 5: Results of ELECTRE TRI

	Pessimistic procedure	Optimistic procedure
Category 1	P4, P7, P9, P11, P12, P13, P15, P22, P23, P25, P27	P4, P15, P27
Category 2	P6, P10, P14, P16, P17, P18, P19, P20, P24, P26, P28	P6, P10, P11, P12, P13, P17, P18, P20, P24, P25, P26, P28
Category 3	P1, P2, P3, P5, P8, p21	P1, P2, P3, P5, P7, P8, P9, P14, P16, P19, P21, P22, P23

3.3 Application of MINORA

In order to apply the MINORA system, it is necessary to have a reference set of portfolios and a ranking expressed by the portfolio manager on this reference set. The choice of the reference must obey to two principles: 1) include well-known stocks by the portfolio manager; 2) the portfolios of this set must cover all the range of possibilities. Considering the absence of a real portfolio manager in out study, we use only the second principle. To help the research of a reference set, MINORA calculates the links between the different portfolios. Then, to respect the second principle we choose a set of 18 not linked portfolios. To rank these portfolios we used the results of ELECTRE TRI in the following way:

- the portfolios affected to the category C3 in the pessimistic and optimistic procedures take the rank 1;
- the portfolios affected to the category 3 in optimistic procedure and to the category 2 in pessimistic procedure take the rank 2;
- the portfolios affected to the category 3 in optimistic procedure and to the category 1 in pessimistic procedure, and the portfolio affected to the category 2 in the two procedures take the rank 3;
- the portfolios affected to the category 2 in optimistic procedure and to the category 1 in pessimistic procedure take the rank 4;
- the portfolios affected to the category 1 in the two procedures take the rank 5.

The use of this ranking causes homogenization of the results and will facilitate the comparison of the results. Then, the portfolios 1, 2, 3 and 8 take the rank 1, the portfolios 16 and 18 the rank 2, the portfolios 22, 7 and 9 the rank 3, the portfolios 13 and 11 the and 4 and finally the portfolios 15 and 4 the rank 5. Then MINORA system, through the UTA method, provides the following model of additive utility:

$$u(g) = 0.043u_1(\text{ beta 1}) + 0.285u_2(\text{ beta 6 }) + 0.103u_3(\text{ beta 9 }) + 0.285u_4(\text{ beta 11 }) + 0.284u_5(\text{ betacac }) \qquad \text{(Eq. 8)}$$

This utility function is the most appropriate, since it proceeds correctly in the ranking of all the portfolios of the reference set. With this utility function, the two consistency measures have optimum values (that is F=0 and t=1), indicating complete agreement between the portfolio manager and the model of additive utility. The marginal utility functions corresponding to this model are presented in the above figure 1 to 5. In these figures, there are three utility curves (low, middle and high). The middle one corresponds to the above presented model of additive utility and, also, gives the relative weight for the criterion. The two others (the low one is mistaken with the abscissa axis) show the entire range of the possible marginal utility functions, with respect to the portfolio manager's ranking in the reference set.

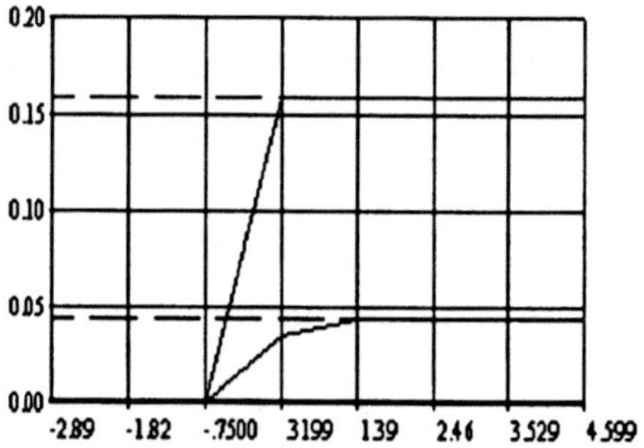

Figure 1: utility curve, Beta 1

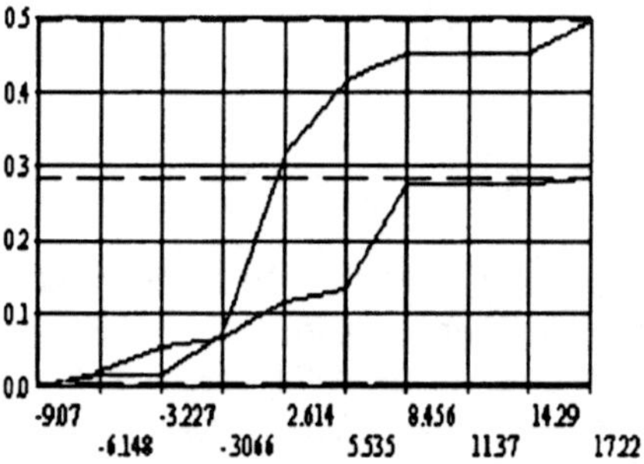

Figure 2: Utility curve, Beta 6

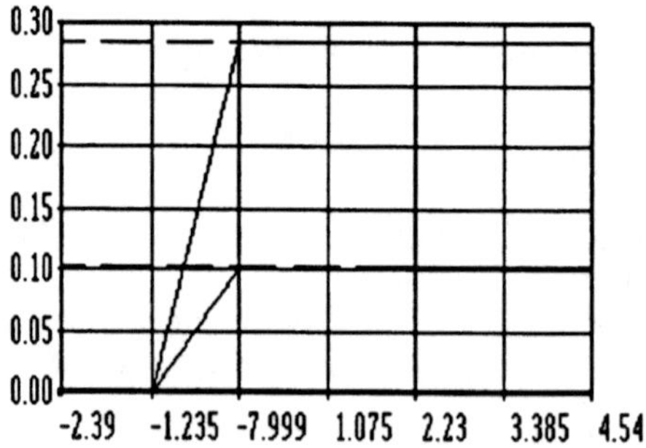

Figure 3: utility curve, Beta 9

Figure 4: Utility curve, Beta 11

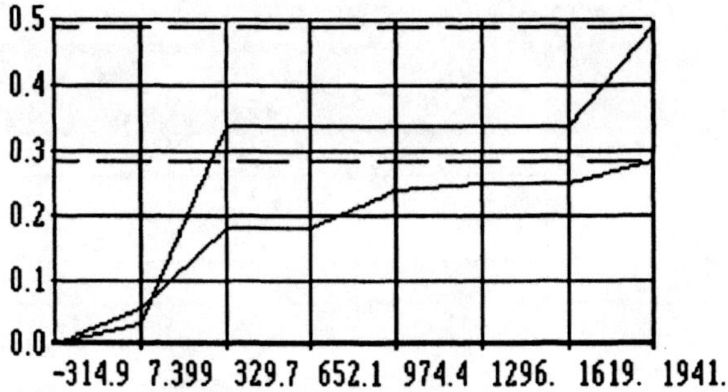

Figure 5: Beta cac240

The observation of the results shows that no criteria, but the criterion Beta 1, have a negligible weight. Nevertheless, this is not sufficient to appreciate the relative importance of a criterion, the later depends also on the discriminatory power of the criteria. The discriminatory power of a criterion depends on the shape of its marginal utility function, this is all the more important since the slope of the marginal utility curve is high (if the curve is flat this mean that all the portfolios have the same value on the criterion and then this criterion has no effect). Then, observing the figure 1 to 5, one can see on the one hand that the criteria beta 6, beta 11 and beta 4 have a strong discriminatory power on all their scale, and on the over hand, that the criteria beta 1 and beta 9 have a discriminatory power only on a part of their scale. For example, the criterion beta 1 discriminates only between the portfolio with a negative value and the portfolio with a positive value, but it does not discriminate between the portfolio with positive values.

Finally, after extrapolation of the utility function to the whole set of portfolio, we obtain the result presented in the table 6 (next page). In this table one can also found the results of ELECTRE TRI for comparison.

Comparing the results of MINORA with those of ELECTRE TRI, one can remark that there is an agreement: the portfolios well ranked by MINORA are in the best categories C_3 by ELECTRE TRI, and vice versa. This agreement asserts the interest of the study and allows the portfolio manager to be confident with the results.

Table 6: Results of MINORA and ELECTRE TRI

Portfolio	Utility	Ranking	Electre pess	Electre opt
P5	0.7330	1	C3	C3
P2	0.5217	2	C3	C3
P3	0.5217	2	C3	C3
P8	0.5217	2	C3	C3
P1	0.5213	2	C3	C3
P21	0.4900	6	C3	C3
P23	0.4899	7	C1	C3
P28	0.4890	7	C2	C2
P19	0.4835	9	C2	C3
P14	0.4784	10	C2	C3
P16	0.4784	11	C2	C3
P18	0.4745	11	C2	C2
P24	0.4744	13	C2	C2
P6	0.4653	14	C2	C2
P17	0.4608	15	C2	C2
P20	0.4537	16	C2	C2
P22	0.4537	17	C1	C3
P7	0.4737	17	C1	C3
P9	0.4365	17	C1	C3
P10	0.4061	20	C2	C2
P26	0.3795	21	C2	C2
P25	0.3772	22	C1	C2
P13	0.3772	23	C1	C2
P11	0.3220	23	C1	C2
P27	0.3175	25	C1	C1
P12	0.3059	26	C1	C2
P15	0.3059	27	C1	C1
P4	0.3059	27	C1	C1

4.Conclusion

In this paper, we present a methodology for portfolio management that exploits the complementarity and the respective advantages of APT and multicriteria decision making. The APT enables us to efficiently evaluate the return of portfolios and, by identifying the relevant common factors of influence, gives the way to perform a multicriteria management of risk. Then, multicriteria decision making provides an original and efficient framework to conduct a portfolio selection using the criteria identified by the APT.

The use of multicriteria decision making methods allows to take into consideration all the relevant criteria, whatever their origins are, for portfolio selection. The similarities in the results that we notice using both MINORA and ELECTRE TRI relays confidence in portfolio selection. Moreover, ELECTRE TRI with the notion of incomparability brings an important information to the portfolio manager, especially when the evaluation of portfolios appears difficult. Furthermore, MINORA systems helps the decision maker to understand his preferences allowing him to construct his portfolio selection model in an interactive manner. This way, the portfolio selection model is undertaken without any normative constraints.

Finally, this methodological framework brings a new knowledge to portfolio selection and helps with the improvement of a scientific and active portfolio management.

Nevertheless, this study is not perfect, and some improvement and development can be proposed. First, it will be interesting to extrapolate this study by using financial and accounting criteria as suggested by Ricci-Xella (1994), in order to evaluate the stocks return better. Furthermore, this will allow to appreciate better the comportment of the portfolio manager confronted with accounting documents or financial announcements. Secondly, it will also be interesting to improve the study by using a better data base (more portfolios on a present period), using three stage least squares or using non-linear risk factors (interest rates are exponential). Third, this methodology must be tested, which means compared with other models (APT macro, classical APT, CAPM,...). Note that multicriteria decision making methods are conceived to exploit the decision maker's experience, then, to be conclusive, a test of our methodology must be done with the participation of a "real" portfolio manager.
Finally, portfolio management includes other decisions problems such as international diversification, the use of other financial instruments (bonds, futures, forwards,...) or management control. Then, it will be interesting to extend this decision making methodology to this kind of problems, in order to determine optimal investment choices.

REFERENCES

Chamberlain G., Rothchild M. "Arbitrage, factor structure and mean-variance analysis on large asset markets », *Econometrica*, 51, September 1983, 1281-1301.

Chen N.F., Ingersoll J. "Exact pricing in linear factor models with finitely many assets : A note », *Journal of Finance*, 38, June 1983, 985-988.

Chen N.F. "Some empirical tests of the theory of arbitrage pricing », *The Journal of Finance*, December 1983, 1393-1414.

Connor G. "A unified pricing beta », *Journal of Economic Theory*, 34, 1984, 13-31.

Connor G.,"Arbitrage pricing theory in factor economies », PHD Yale University USA, 1982.

Cox J., Ingersoll Jr, Ross S. "An intertemporal general equilibrium model of asset prices », *Econometrica*, March 1985, 363-384.

Despotis D.K., Yannacopoulos D., Zopounidis C. "A review of the UTA multicriteria method and some improvement", *Foundations of Computing and Decision Sciences*, 15, 2, 1990, 63-76.

Dhrymes P. J., Friend I., Gultekin M. N., Gultekin N. B. "A critical re-examination of the empirical evidence on the arbitrage pricing theory », *The Journal of Finance*, 39, June 1984, 323-346.

Dhrymes P. J., Friend I., Gultekin M. N., Gultekin N. B. "New tests of the APT and their implications », *The Journal of Finance*, July 1985, 659-674.

Dybvig P. "An explicit bound on individual asset's deviations from APT pricing in finite economie », Working Paper Yale University USA, 1982.

Fama E., F. MacBeth J. D. "Risk, return, and equilibrium : Empirical tests », *Journal of Political Economy*, 81, May-June 1973, 607-636.

Gibbons M. Ross S. Shanken J. "A test of the efficiency of a given portfolio », *Econometrica*, 57, September 1989, 1121-1152.

Gilles C., LeRoy S, "On the arbitrage pricing theory », Working paper, seminars at Carleton University USA, May 1990 see also *Economic theory*, 1, 1991, 213-229.

Grinblatt M., Titman S. "Factor pricing in a finite economy », Working Paper UCLA USA, N° 11-82, July 1982.

Huberman G. "A simple approach to the arbitrage pricing theory », *Journal of Economic Theory*, 28, October 1982, 183-191.

Hurson Ch., Zopounidis C. "Return, risk measures and multicriteria decision support for portfolio selection », proceedings of the second Balkan conference on operational research B. Papathanassiou et K. Paparizos (Eds), Thessaloniki, 1993, 343-357.

Hurson Ch., Zopounidis C. "On the use of Multicriteria Decision Aid methods for portfolio selection", *Journal of Euro-Asian Management*, 1, 2, 1995, 69-94.

Hurson Ch., Zopounidis C. "Méthodologie multicritère pour l'évaluation et la gestion de portefeuilles d'actions', *Banque et Marché*, 28, Novembre-Décembre 1996, 11-23.

Hurson Ch. *La gestion de portefeuilles boursiers et l'aide multicritère à la décision*, Thèse de Doctorat, GREQAM., Université d'Aix-Marseille II, 1995.

Ingersoll J. "Some results in the theory of arbitrage pricing », Working Paper University of Chicago USA, N° 67, 1982.

Jacquet-Lagrèze E., Siskos J. "Assessing a set of additive utility functions for multicriteria decision making: The UTA method", *European Journal of Operational Research*, 10, 1982, 151-164.

Khoury N., Martel JM., Veilleux M. "Méthode multicritère de gestion de portefeuilles indiciels internationaux", *L'Actualité Economique*, 69, 1, Mars 1993.

Lintner J. "The valuation of risky assets : The selection of risky investments in stock portfolios and capital budgets », *Review of Economics and Statistics*, 47, February 1965, 13-37.

Long J. "Stock prices, inflation and the term structure of interest rates », *Journal of Financial Economics*, July 1974, 131-170.

Markowitz H. M. "Portfolio selection », *Journal of Finance*, March 1952, 77-91

Martel JM, Khoury N., Bergeron M. "An application of a multicriteria approach to portfolio comparisons", *Journal of Operational Research Society*, 39, 7, 1988, 617-628.

Merton R.C. "An intertemporal capital asset pricing model », *Econometrica*, 41, September 1973, 867-888.

Mossin J. "Equilibrium in a capital asset market », *Econometrica*, 34, October 1966, 768-783.

Oldfield G. S. Rogalski R. J. "Treasury bill factors and common stock returns », *The Journal of Finance*, 36, May 1981, 327-353.

Ricci-Xella N. C. "L'APT est-elle une alternative au MEDAF ? un test empirique sur le marché français », Thèse de doctorat ès Sciences de Gestion, University of Aix-Marseille III, France, December 1994.

Ricci-Xella N. C. "Les marchés financiers, mémoire de DEA", CETFI, Université d'Aix-Marseille III.

Roll R., Ross S. "An empirical investigation of the arbitrage pricing theory », *The Journal of Finance*, 35, December 1980, 1073-1103.

Roll R., Ross S. "Regulation, the capital asset pricing model and the arbitrage pricing theory », *Public Utilities Fortnightly*, May 1983, 22-28.

Roll R. "A critique of the arbitrage pricing theory's tests, Part I : On past and potential testability of the theory », *Journal of Financial Economics*, 4, March 1977, 129-176.

Ross S. A. "The arbitrage pricing theory of capital asset pricing », *Journal of Economic Theory*, 13, December 1976, 342-360.

Roy B., Bouyssou D. *Aide Multicritère à la Décision: Méthodes et cas*, Paris, Economica, 1993.

Shanken J. "Multi-beta CAPM or equilibrium-APT ? : A reply », *Journal of Finance*, 40, September 1985, 1189-1196.

Shanken J. "The arbitrage pricing theory: Is it testable? », *Journal of Finance*, 37, December 1982, 1129-1140.

Sharpe W. F. "A simplified model of portfolio analysis », *Management Science*, 9, January 1963, 277-293.

Siskos J., Despotis D.K. "A DSS oriented method for multiobjective linear programming problems", *Decision Support Systems*, 5, 1989, 47-55.

Siskos J., Spiridakos A., Yannacopoulos D. "MINORA: A multicriteria decision aiding system for discrete alternatives", in J. Siskos et C. Zopounidis (ed.), *Special Issue on Multicriteria Decision Support Systems, Journal of Information Science and Technology*, 2, 2, 1993, 136-149.

Spiridakos A., Yannacopoulos D. "A visual approach to the procedures of multicriteria intelligence decison aiding systems", in J ansen, C.H. Skiadas et C. Zopounidis (ed.), *Advances in Stochastic Modelling and Data Analysis*, Dordrecht, Kluwer, 1995, 339-354.

Szala A. "L'aide à la décision en gestion de portefeuille", Universuté de Paris-Dauphine, U.E.R. sciences des organisations, 1990.

Yu W. "ELECTRE TRI: Aspects méthodologiques et manuel d'utilisation", Document du LAMSADE No. 74, (1992), Université de Paris Dauphine, 1992.

Zopounidis, C. "On the use of the MINORA multicriteria decision aiding system to portfolio selection and management", *Journal of Information Science and Technology*, 2, 2, 1993, 150-156.

Zopounidis, C., Despotis, D.K., Kamaratou, I. "Portfolio selection using the ADELAIS multiobjective linear programming system", *Computational Economics*, 1997 (in press).

Zopounidis C., Godefroid M., Hurson Ch. "Designing a Multicriteria Decision Support System for Portfolio Selection and Management", in: Janssen J, Skiadas CH and Zopounidis C. (Eds), *Advances in stochastic Modelling and Data Analysis*, Dordrecht, Kluwer Academic Publishers, 1995, 261-292.

II. MULTIVARIATE DATA ANALYSIS AND MULTICRITERIA ANALYSIS IN BUSINESS FAILURE, CORPORATE PERFORMANCE AND BANK BANKRUPTCY

THE APPLICATION OF THE MULTI-FACTOR MODEL IN THE ANALYSIS OF CORPORATE FAILURE

Erik M. Vermeulen [1] , Jaap Spronk [1] , Nico van der Wijst [2]
[1] Department of Finance, Erasmus University Rotterdam
[2] EIM Small Business Research and Consultancy, Zoetermeer, The Netherlands

Abstract: This paper describes an application of the multi-factor model to the analysis and prediction of corporate failure. The multi-factor model differs from the more usual methods of failure prediction because failure is conditioned on the values of a series of exogenous risk factors rather than on a series of "internal" financial ratio's. Moreover, the multi-factor model is not primarily aimed at classifying firms in categories, but at modelling the influences of exogenous risk factors, through sensitivities, on the firm's cash flow generating process. The paper presents the general multi-factor model and the conditional failure prediction model as bell as the possibilities to apply the model.

Keywords: Multi-factor model, Corporate failure, Application

1. Introduction

Corporate failure has not failed to attract considerable attention in both the academic world and the business community. Hence, the prediction of corporate failure gave rise to a wide variety of studies that originated in the 1930's and that continues until today. Well known examples from sixties are Beaver (1966) and Altman (1968, 1984), followed by, among many others, Edmister (1972) and Taffler (1982). An excellent survey of studies in the area of failure prediction is given by Dimitras et. al. (1996).

The general approach to predict corporate failure is to apply a classification technique, statistical or not, on a sample containing both failed and non-failed firms, in order to find a relation between the firm characteristics on the one hand and future corporate failure on the other. This relation is subsequently used to obtain insight into the predictability of corporate failure.

As is often mentioned in the literature, the use of these approaches is not without problems and pitfalls. Lev (1974, p. 149) states that the theoretical foundation of these models needs improvement. In his essential paper, Eisenbeis (1977) gives a comprehensive summary of the methodological and statistical problems that limit the practical usefulness of most applications of discriminant analysis in the areas of business, finance and economics. Taffler (1982) mentions as a drawback that "dramatic changes in the UK economy and major changes in the system of company taxation *inter alia* call its subsequent intertemporal validity into question."

To overcome these drawbacks, efforts have been made to introduce new approaches or to incorporate theoretically determined financial characteristics to explain corporate failure. For example, concerning the firm characteristics Casey and Bartczak (1985) as well as Gentry, Newbold and Whitford (1985) incorporated cash flow ratios in their model as theoretically determined characteristics predicting corporate failure [1]. Other classification techniques were introduced, such as recursive partitioning, investigated by Frydman et al. (1985). Lawrence et al. (1992) explicitly take account of the fact that the future economic situation may influence corporate bankruptcy [2]. They adopt economic state variables. such as unemployment rates and retail sales in their model and find a significant influence. Zopounides et. al. (1995) apply a multi-criteria approach to the analysis and prediction of business failure in Greece, with promising results.

In this paper, a new framework for failure prediction is presented. Its most important feature is that the prediction of corporate failure is conditioned on the values of exogenous risk factors. For instance, a firm may be predicted to go bankrupt in case the wage rate increases with more than ten percent. In the framework presented here failure is equated to cash insolvency. Consequently, a model for cash balances is developed in which future cash balances are related to future values of risk factors. The reaction of a firm to a change in the value of a risk factor (its sensitivity to that factor) generally differs from the reaction of other firms. Thus, each firm has its own specific risk profile.

The framework may be applied in several fields. For banks it can be used to support the monitoring of clients, e.g. to see which clients will suffer mostly and, hence, will be a risk for the bank. Similarly, it may help the government in detecting those firms or industries that will be harmed severely under various scenarios of risk factor changes. Furthermore, it may help the firm's management to find ways to prevent the problems caused by risk factor changes, since the model indicates how risks are averted by, for example, a change in the sensitivities. Finally, the framework does not only apply in the case of failure but also in case the analysis is focussed on the risks of reaching a specified target level of cash flow.

The remainder of this paper is organized as follows. In Section 2 we discuss the multi-factor approach and in Section 3 the framework for conditional failure prediction is presented. Section 4 illustrates the framework numerically and Section 5 summarizes.

---

[1] Casey and Bartczak found that operating cash flow data did not improve classification accuracy significantly. Gentry, Newbold and Whitford (1985) found that cash-flow-based funds flows offer a viable alternative for classifying failed and nonfailed firms. A study of Gombala (1987) et al. confirmed the findings of Casey and Bartczak that the ratio cash flow from operations divided by total assets is insignificant in predicting corporate failure.

[2] In this paper the terms "bankruptcy" and "failure" are used interchangeably.

2. The multi-factor model

The view of the firm that underlies the multi-factor model, is that a firm's performance is not only determined by the firm's decisions, but also by its uncertain environment. This uncertain part of performance can be attributed to the changes of risk factors and the firm's sensitivities for these changes. This approach to risk has been applied by many researchers and to a wide variety of problems (see e.g. Berry, Burmeister and McElroy (1988), Hallerbach (1994), and Goedhart and Spronk (1991)).

The model presented here differs from the model for stock returns since performance measures of firms (in contrast with security portfolios) are analyzed. More details about this model as well as other applications can be found in Vermeulen et al. (1993, 1994) and Vermeulen (1994). In the following, first a verbal description of the model will be provided, after which the relation between performance measures and risk factors will be formalized.

A familiar view represents the firm as an organized chain of input, throughput, and output. The inputs have to be paid for, which leads to cash outflows, whereas the outputs generate cash inflows. The cash flows are the difference between cash inflows and cash outflows. The supply of input factors and their prices, as well as the demand for output products and their prices, are uncertain and this causes the cash inflows and cash outflows to be stochastic. The risk factors influence the performance measures of the firm. The magnitude of a risk factor's influence on a performance measure of the firm is called the *sensitivity* of the firm's performance measure to the risk factor [3].

Changes in performance are the combined result of changes of the risk factors and the sensitivities to these changes. The higher the sensitivities, the greater the impact of an unexpected risk factor change on performance will be, and the greater the risks the firm runs [4]. The vector of sensitivities is called the risk profile and it describes the risks the firm faces. Thus, it can be seen as a multi-dimensional risk measure.

The sensitivities are related to the firm's characteristics [5]. For instance, the interest rate sensitivity is highly dependent on the firm's level of debt, and the wage rate sensitivity depends, among other things, on the number of employees. Similarly, the business cycle sensitivity is influenced by the firm's product range, and so on.

---

[3] The risk concept used here is largely in line with that of Cooper and Chapman (1987, p. 2) who define risk as : "exposure to the possibility of economic or financial loss or gain, physical damage or injury, or delay, as a consequence of the uncertainty associated with pursuing a particular course of action."

[4] Of course, not only the sensitivity o f the firm, but also the possible change in value of the risk factor (i.e. the variance o f the risk factors) should also be takers into account when evaluating the risk profile.

[5] Actually, they may also depend on industry characteristics, such as the degree of competition in an industry.

To formalize this view on the firm, a one period model is constructed that can be used to analyze the risk and expected level of a firm's performance [6]. The following relation between performance and the risk factor is assumed (later on, the model will be extended to more risk factors.): [7]

$$\tilde{R}_t = a_t + b_t \tilde{f}_t + \tilde{\eta}_t \tag{1}$$

where

$\tilde{R}_t$ = performance,
a_t = a fixed term,
b_t = the sensitivity,
$\tilde{f}_t$ = the value of the risk factor,
$\tilde{\eta}_t$ = the error term,
t = index for time
$\sim$ = random variable.

The fixed term a_t and the sensitivity b_t differ per firm, because of differences in firm characteristics, such as product range and management style.

From expression (1) the following expression is derived for performance at time t+1; defining Δ as the difference operator, i.e., $\Delta x_t = x_t - x_{t-1}$.

$$\begin{aligned}\tilde{R}_{t+1} &= a_{t+1} + b_{t+1}\tilde{f}_{t+1} + \tilde{\eta}_{t+1} \\ &= a_t + \Delta a_{t+1} + (b_t + \Delta b_{t+1})(\tilde{f}_t + \Delta\tilde{f}_{t+1}) + \tilde{\eta}_{t+1}\end{aligned}$$

which, by assuming that the factor values at time t have been realized and thus are not stochastic anymore, can be rewritten as:

$$= a_t + b_t f_t + \Delta a_{t+1} + \Delta b_{t+1} f_t + b_t \Delta\tilde{f}_{t+1} + \Delta b_{t+1}\tilde{f}_{t+1} + \tilde{\eta}_{t+1} \tag{2}$$

Apart from changes in risk factors, expression (2) leaves room for changes in the fixed term a_t and in the sensitivity b_t. These changes are caused by instruments, which may vary from a change in firm characteristics to the purchase of a financial contract. The fixed term may change because of the costs associated with the application of the instruments. Here we assume that no instruments are used and, thus, that $\Delta a_{t+1} = \Delta b_{t+1} = 0$. Expression (2) then boiles down to:

---

[6] An advantage of the one-period model is that it facilitates easy presentation. Extension to more than one period is possible. Since the main principles of risk management shown in the one period model are equal to those of the multi-period model. we shall concentrate on one period models.

[7] This model differs from the arbitrage pricing theory (APT) as developed by Ross (1976) in several aspects. The APT describes the pricing of assets in a general equilibrium frame work. The approach presented in this article limits itself to postulating a functional relation between a performance measure and a multitude of stochastic risk factors. The multi factor approach does not assume a general equilibrium framework nor intends to derive market prices for the risk factors involved, see Hallerbach (1994, p. 33 ff).

$$\tilde{R}_{t+1} = a_t + b_t f_t + b_t \Delta\tilde{f}_{t+1} + \tilde{\eta}_{t+1} \tag{3}$$

Subtracting (1) from (3), leads to an expression for the change in performance:

$$\Delta\tilde{R}_{t+1} = b_t \Delta\tilde{f}_{t+1} + \tilde{\varepsilon}_{t+1} \tag{4}$$

where
$\tilde{\varepsilon}_{t+1} = \tilde{\eta}_{t+1} - \tilde{\eta}_t$

This expression can easily be extended to more than one risk factor. In that case we get:

$$\Delta\tilde{R}_{t+1} = \sum_{i=1}^{k} b_{it} \Delta\tilde{f}_{it+1} + \tilde{\varepsilon}_{t+1} \tag{5}$$

This model is further extended by Vermeulen et al. (1993) who explain the sensitivities in terms of firm characteristics:

$$b_{it} = \sum_{m=1}^{M} fc_{tnm} \gamma_{im} \tag{6}$$

where
fc_{tnm} = firm characteristic *m* of firm *n* at time *t*,
γ_{im} = the influence of firm characteristic *m* on the sensitivity to risk factor *i*.

Substituting expression (6) into expression (5) leads to:

$$\Delta\tilde{R}_{t+1} = \sum_{i=1}^{k} \left(\sum_{m=1}^{M} fc_{tnm} \gamma_{im} \right) \Delta\tilde{f}_{it+1} + \varepsilon_{t+1} \tag{7}$$

The parameters γ_{im} can be estimated using panel techniques. In order to estimate expression (7) the data of various firms can be pooled, i.e., in to estimate γ_{im} only one regression is needed in which data of a sample of firms over a number of a time periods are used. [8]

[8] Moreover, estimating expression (5) directly is more difficult because of the possibly low number of observations, since in practice most financial firm characteristics are only observed at a yearly basis for smaller firms, and quarterly for larger corporations.

3. The conditional failure prediction model

To predict bankruptcy, future cash balances are modelled on the basis of the multi-factor model that was discussed in Section 2, which gives a detailed view on future cash flows.

For reasons of simplicity we equate bankruptcy and cash insolvency. Other definitions of bankruptcy are also possible, e.g. the inability to repay a loan, to exhaust normal overdraft facilities or to reach any other specified (target) level of cash flows. Cash insolvency occurs when the cash flows become negative and the deficit flow is large enough and lasts long enough to exhaust the cash balances. This relationship can be expressed in general terms as follows:

$$\widetilde{CB}_t = CB_{t\text{-}1} + N\widetilde{O}CF_t + \widetilde{CF}_t \tag{8}$$

where

$\widetilde{CB}_t$ = cash balances at time t,

$N\widetilde{O}CF_t$ = non operational cash flow at time t,

$\widetilde{CF}_t$ = the cash flow at time t.

As shown in the preceding section, the uncertain cash flows depend on the uncertain development of the risk factors. In order to analyze future (operational) cash flows we apply the framework developed earlier and arrive at a model in which the cash flow of year t can be explained by last year's cash flow and by the changes of some risk factors. Thus, total risk is broken down into the separate influences of the underlying risk factors. In that way, not only the magnitude of risk is obtained. but also the source of risk and the firm's exposure to this specific kind of risk. This detailed analysis may help in preventing bankruptcy, since it shows which risks to concentrate on. Moreover. if the decision maker provides a set of future factor values. he obtains an indication of failure conditional on this specific set.

The empirical multi-factor model used to estimate the sensitivities, which was formulated in expression (7), can be rewritten in cash flow terms as:

$$\widetilde{CF}_t = CF_{t\text{-}1} + \sum_{i=1}^{k} b_i \Delta\widetilde{f}_{it} + \widetilde{\varepsilon}_t \tag{9}$$

where

$\widetilde{CF}_t$ = the stochastic cash flow in year t,

$\widetilde{CF}_{t\text{-}1}$ = the realized cash flow last year,

$\Delta\widetilde{f}_{it}$ = the change in risk factor i,

b_i = the sensitivity to risk factor i,

$\tilde{\varepsilon}_t$ = an error term for which the usual assumptions hold.[9]

After substitution of equation (9) in (8) we obtain:

$$\widetilde{CB}_t = CB_{t-1} + N\widetilde{OCF}_t + CF_{t-1} + \sum_{i=1}^{k} b_i \Delta\tilde{f}_{it} + \tilde{\varepsilon}_t \tag{10}$$

If the stochastic specification of the non operating cash flows and the risk factors was precisely known, then the probability density function of $\widetilde{CB}_t$ could be obtained and thus its expected value and variance. Then, clearly the probability of failure could be calculated, i.e., the probability that $\widetilde{CB}_t < 0$. In practice, information of such a precise nature on the uncertain risk factors is, of course, not available.

However, in order to obtain insight into the odds of bankruptcy, a specific set of factor values can be chosen and it can be investigated whether or not such a scenario would lead to cash insolvency. In other words, when we neglect the error term. net cash balances are involuntary reduced to zero and cash insolvency occurs if the realized values CB_t, $NOCF_t$ and Δf_{it} are such that

$$CB_t = CB_{t-1} + NOCF_t + CF_{t-1} + \sum_{i=1}^{k} b_i \Delta f_{it} < 0 \tag{11}$$

In that case the plane distinguishing between bankrupt and non-bankrupt firms is:

$$CB_t = CB_{t-1} + NOCF_t + CF_{t-1} + \sum_{i=1}^{k} b_i \Delta f_{it} = 0 \Rightarrow \tag{12}$$
$$CB_{t-1} + NOCF_t + CF_{t-1} = -\sum_{i=1}^{k} b_i \Delta f_{it}$$

Hence, the set of firms which are expected to become cash insolvent is:

$$S^- = \left\{ CB_{t-1} + NOCF_t + CF_{t-1} + \sum_{i=1}^{k} b_i \Delta f_{it} < 0 \right\} \tag{13}$$

And the set of firms which are not expected to become cash insolvent is:

$$S^+ = \left\{ CB_{t-1} + NOCF_t + CF_{t-1} + \sum_{i=1}^{k} b_i \Delta f_{it} \geq 0 \right\} \tag{14}$$

---

[9] i.e. zero expected value, constant variance, no autocorrelation in the residuals and independence between the residual term and the independent variables.

Thus, whether or not a firm belongs to a particular set depends on (a) its last year's cash balances, (b) its cash flow (c) its expected non-operational cash flows, (d) its sensitivities and (e) the factor values that were provided by the decision maker.

In order to predict corporate failure, these characteristics and the future course of the risk factors must be known. The sensitivities can be estimated using expression (7). After estimation of the sensitivities. for each possible combination of future risk factors it can be seen whether or not the firm fails. In this way, a failure prediction is obtained that is conditional on the scenario of future factor values. Good candidates for possible scenarios should be provided by the decision maker. who may use a macro-economic model for this purpose.

4. Illustration

In this section, the use of the conditional failure prediction model is demonstrated with a constructed example. Of course. demonstration on the basis of a set of relevant data of failed and non-failed firms would be preferable. This would also give insight into the efficiency of the method. relative to other approaches. For lack of suitable data we have to limit ourselves to a . constructed application. This may give, however, a clear understanding of the advantages of the method, by demonstrating how corporate failure is related to certain risk factors. This may prove to be important in the early detection of financial difficulties: a timely early warning almost inevitably is also a detailed rather than a general warning.

To begin with, the model presented in Section 3 is restricted to one risk factor, which simplifies equation (12) to

$$CB_{t-1} + \widetilde{NOCF}_t + CF_{t-1} = -b\Delta\tilde{f}_t \qquad (15)$$

For four fictitious firms the values of CF_{t-1} and b are given in Table 1. Furthermore, we assume (CB_{t-1} + $NOCF_t$) to be equal to zero and the decision maker wants to investigate the consequences of a drop in the factor value by four. Hence, $\Delta f_t = -4$.

Table 1: The cash flows and sensitivities of some firms

	Cash flow	Sensitivity, b	Cash flow/b
Firm A	30	3	10
Firm B	5	1	5
Firm C	15	5	3
Firm D	2	1	2

Note: amounts are in millions of guilders.

The so-called *critical line* in the (CF_{t-1},b)-plane is:

$$CF_{t-1} = 4.b \qquad (16)$$

Figure 1 presents this critical-line as well as the position of the four firms. The vertical axis presents last year's cash flow (CF_{t-1}), and the horizontal axis the sensitivity to the risk factor (b). All firms are plotted in accordance with their (CF_{t-1},b)-values. The firms below the critical line, such as Firm C, are expected to become cash insolvent.

In spite of the availability of forecasting models, the risk factors may still be subjected to much uncertainty. Then it is interesting to investigate not only one particular change of the risk factor, but also other possible changes. Thus. one obtains different critical lines for different changes in the risk factor. Assuming the following possible percentage changes of the risk factor: Δf = 0, -4, -8, -12, the lines in Figure 2 were drawn.

Figure 2 indicates which firms have negative cash flows given particular changes of the risk factor. For instance, it is easily seen that Firm a will become cash insolvent if the change of the risk factor is between -8 and -12, whereas Firm D already performs badly if the risk factor changes by -4.

Using this method, it is easily computed what the percentage of firms suffering cash insolvency will be, given a particular change in the risk factor. For example, if the risk factor decreases 12, we see that all firms become cash insolvent, whereas this percentage reduces to 50% (2 of the 4 firms) if the risk factor drops by 4. These figures (the change of the risk factor and the percentage of firms that have negative cash flows given that particular change of the risk factor) are printed in the graph, separated by a comma.

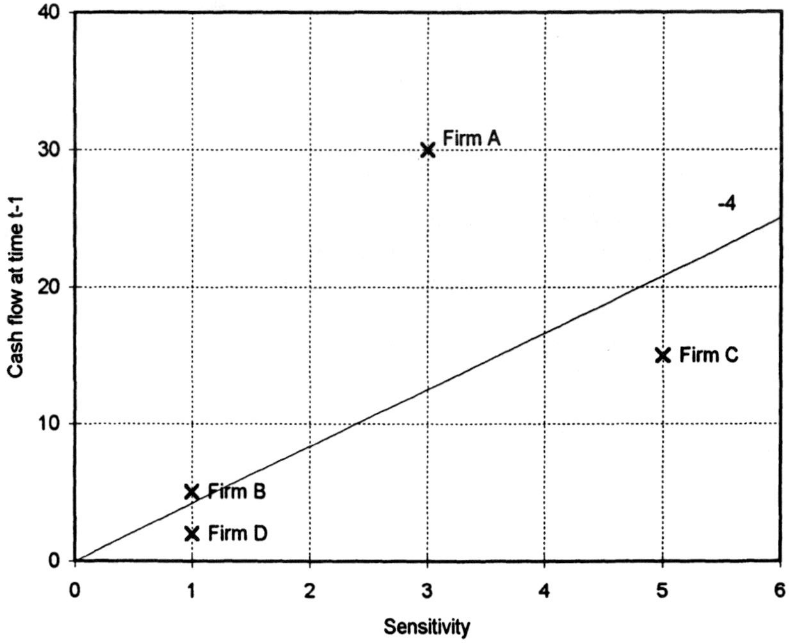

Figure 1: The critical line

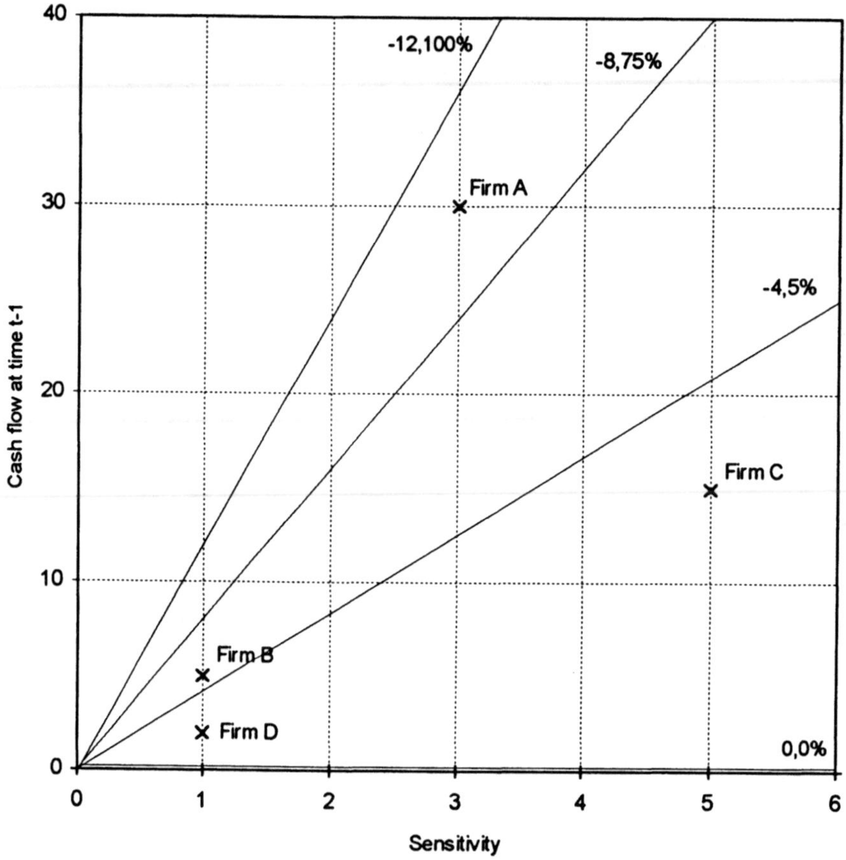

Figure 2: The failure dial

In case the conditional failure prediction model includes a multitude of risk factors (as was seen in expressions (13) and (14)), visualization remains possible. We will demonstrate this for two risk factors. assume that the sensitivities to these risk factors have been estimated as given in Table 2. Now, in order to calculate the odds of corporate failure, a guess of the values of both risk factors must be known. Given the firms' cash flows and sensitivities, for each firm a critical line is drawn in Figure 3.

Table 2: The cash flow and sensitivities of some firms

	C.fl.	Sens. b_1	Sens. b_2	C.fl./b_1	C.fl./b_2
Firm A	30	3	2	10	15
Firm B	6	1	2	6	3
Firm C	15	3	1	3	15
Firm D	2	1	1	2	2

Note : amounts are in millions of guilders.

The horizontal axis of Figure 3 represents possible values of the first risk factor, the vertical axis those of the second risk factor. The critical line simply denotes all factor values for which the expected cash flow is zero, thus the general expression for the critical line is:

$$CF_t = CF_{t-1} + b_1 \Delta f_1 + b_2 \Delta f_2 = 0 \tag{17}$$

For example, the critical line of Firm B is:

$$6 + 1 * \Delta f_1 + 2 * \Delta f_2 = 0 \tag{18}$$

In Figure 3 the values on the axes are negative. If the realized factor values appear to be to the left of the critical line. the firm is expected to go bankrupt. For instance, if both risk factors take the value -5, then Firm B fails, since $CF_{t-1} + b_1 \Delta f_1 + b_2 \Delta f_2 = 6 + (1 * -5) + (2 * -5) = -9 < 0$.

Given the critical lines, failure polylines can be drawn, which are the two-dimensional equivalents of the failure dial. The failure polylines divide the plane of the risk factors in various sets. The first set refers to those combinations of risk factor values for which no firm goes bankrupt. The second set indicates those combination of risk factor values for which only one firm goes bankrupt, and so on. The last set indicates combinations of risk factors for which all firms go bankrupt.

Figure 4 presents the failure polylines. Should the combination of the values of the risk factors lie to the left of the failure polyline, then this line indicates what percentage of firms goes bankrupt. For instance, are the changes of the first and second risk factor -5 and -8 respectively, then all firms will go bankrupt. Similar to the failure dial, the failure polylines indicate what percentage of firms will go bankrupt for the specified combination of risk factor values.

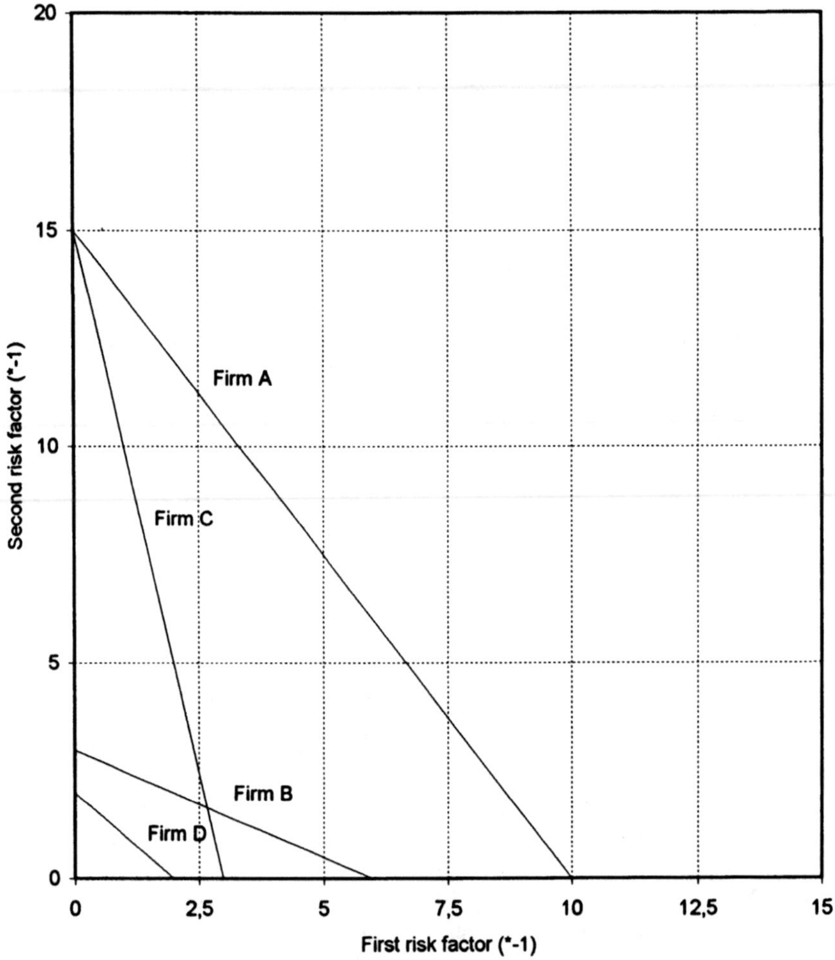

Figure 3: The critical lines in a two-factor space

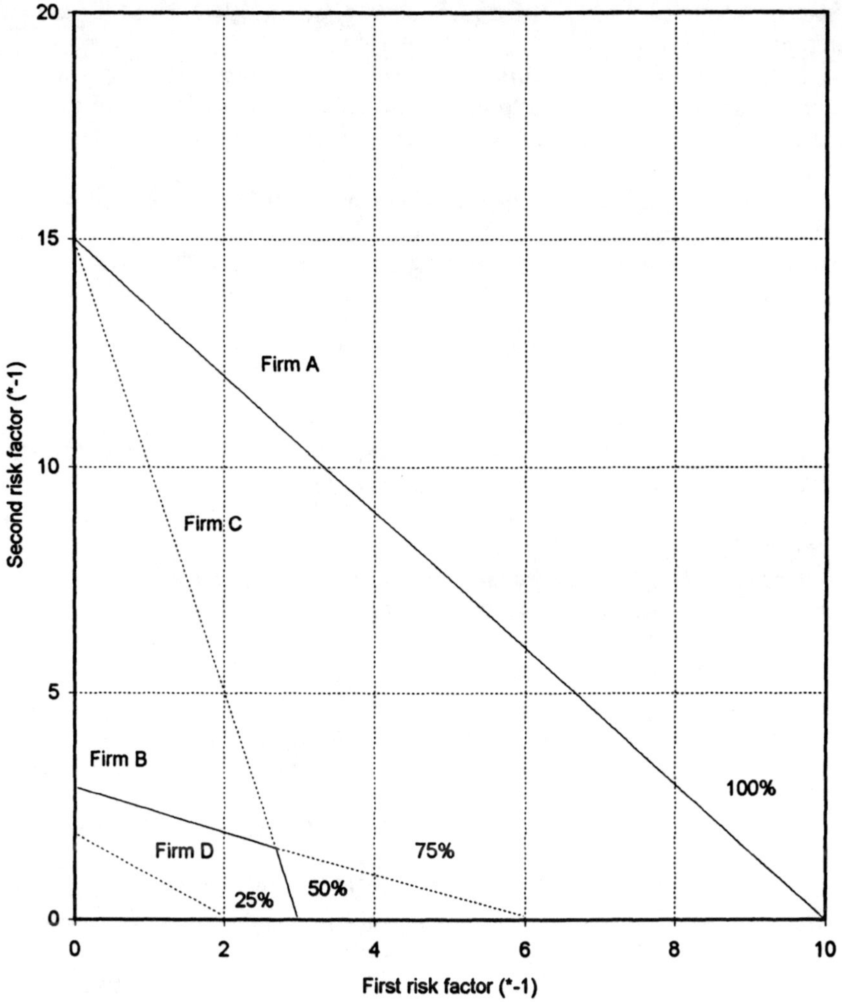

Figure 4: The failure polylines

5. Summary

In this paper, a framework for conditional failure prediction is presented. The future operational cash flows are analyzed by a multi-factor model, in which corporate failure is explicitly related to the future course of risk factors. Since this future course of risk factors is not certain, the prediction of corporate failure is conditional on the (predicted or assumed) future values of the risk factors. This framework not only helps in indicating *that,* but also in explaining *why* a firm is predicted to go bankrupt. Consequently, it provides also clues to prevent bankruptcy.

In the more fortunate situation that the firm is not predicted to go bankrupt, the framework still may be of help in managing the firm's cash flow and risks. In that case, the firm's management is searching for ways to reach a specified target cash flow level.

In this paper, we also presented several ways to visualize this relation between the future cash flow and the possible risk factor values. In a nutshell, these visualizations provide a clear picture of the odds a firm runs. Such a picture may, among other applications, be useful for banks when they monitor their clients.

References

Altman, E.I. (1968), "Financial ratios, discriminant analysis and the prediction of corporate bankruptcy", *The Journal of Finance* 23, 589-609.

Altman, E.I. (1981), "The success of business failure prediction models: .-In international survey", *Journal of Banking and Finance* 8(2), 171-198.

Beaver, W.H. (1966), "Financial ratios as predictors of failure", in: *Empirical Research in Accounting: Selected Studies,* supplement to *Journal of Accounting Research 5,* 179-199.

Berry, M.A., E. Burmeister and M.B. MIcElroy (1988), "Sorting out risks using known APT-factors", *Financial Analyst Journal,* March-April 1988, 29-42.

Casey, C. and N. Bartczak (1985), "Using operating cash flow data to predict financial distress: some extensions", *Journal of Accounting Research* 23(1), Spring (1985).

Cooper, D.F. and C.B. Chapman (1987), *Risk analysis for large projects, models methods and cases,* John lViley & Sons.

Dimitras, A.I., S.H. Zanakis and C. Zopounidis (1996), "Survey of business failures with an emphasis on prediction methods and industrial applications", *European Journal of 0perational Research* 90, 487-513.

Edmister, R.O. (1972), "An empirical test of financial ratio analysis for small business failure prediction", *Journal of Financial and Quantitative Analysis* 7, March, 1477-93.

Eisenbeis, R.A. (1977), "Pitfalls in the application of discriminant analysis in business, finance and economics", *The Journal of Finance* 32(3), 875-900.

Frydman, H., E.I. Altman and D. Kao (1985), "Introducing recursive partitioning for financial classification: the case of financial distress", *Journal of Finance* XL/1, 269-291.

Gentry, J.A., P. Newbold and D.T. Whitford (1985), "Classifying bankrupt firms with funds flow components", *Journal of Accounting Research* 23(1), Spring 1985.

Goedhart, M.H. and J. Spronk (1991), "Multi-factor financial planning: an outline and illustration", in A. Lewandowski and V. Volkovich (eds.), *Multiobjective problems of Mathematical Programming,* Springer Verlag, Berlin, 176-199.

Gombola, M.J., M.E. Haskins, J.E. Ketz and D.D. Williams (1987), "Cash flow in bankruptcy prediction", *Financial Management*, Winter 1987, 55-65.

Hallerbach, W.G. (1994), *Multi-attribute portfolio selection: a conceptual framework*, Ph.D. thesis, Erasmus University Rotterdam, The Vetherlands.

Lawrence, C.L., L.D. Smith and M. Rhoades (1992), "An analysis of default risk in mobile home credit", *Journal of Banking and Finance* 16, 299-312.

Lev. B. (1974), *Financial statement analysis: a new approach*, Prentice-Hall, Inc., Englewood Cliffs, New Jersey.

Ross, S.A. (1976), "The arbitrage theory of capital asset pricing", *Journal of Economic Theory* 13, 341-360.

Taffler. R.J. (1982), "Forecasting company failure in the UK using discriminant analysis and financial ratio data", *Journal of the Royal Statistical Society A* 145, 342-58.

Vermeulen, E.M. (1994), *Corporate risk management: a multi-factor approach*, Ph.D. thesis, Erasmus University Rotterdam. The Netherlands.

Vermeulen, E.M., J. Spronk and D. van der Wijst (1993), "A new approach to firm evaluation", *Annals of Operations Research* 45, 387-403.

Vermeulen, E.M., J. Spronk and D. van der Wijst (1994), "Visualizing interfirm comparison", *Omega, International Journal of Management Science* 22(4).

Zopounidis, C., A.I. Dimitras and L. le Rudulier (1995), "A multicriteria approach for the analysis and prediction of business failure in Greece", *Université de Paris-Dauphine, Document du Lamsade* no.132.

MULTIVARIATE ANALYSIS FOR THE ASSESSMENT OF CORPORATE PERFORMANCE: THE CASE OF GREECE

Yannis Caloghirou, Alexandros Mourelatos, Lefteris Papagiannakis
Laboratory of Industrial & Energy Economics
National Technical University of Athens, Dpt. of Chem. Eng., Div. II
Zografou Campus, GR-157 80, Greece.

Abstract: Principal component analysis is integrated with cluster analysis to assess the performance of two different branches of the greek manufacturing sector, i.e. the pharmaceutical and the olive and seed oil industries. The factors of size, profitability and financial status are extracted by applying the principal component analysis. Cluster analysis is then applied to form homogenous groups of firms corresponding to a set of similar characteristics and patterns of behaviour. Common patterns that influence corporate performance in the period 1981-91 are established and compared between these two sectors. Results show that a good financial status is a sufficient but not a necessary condition for a good profitability record in the case of both sectors. It also appears that size affects the profitability of a greek pharmaceutical firm whereas there is no evidence of a clear relationship between these two factors in the case of the olive and seed oil industry. The results are interpreted in the framework of the relevant Greek state and European Union policies and are in agreement to other empirical research findings.

Keywords: Multivariate analysis; Corporate performance; industry.

INTRODUCTION

Economic performance of a firm is easily assessed by analyzing various financial statements such as the balance sheet, the income statement, the annual report to the shareholders, etc. Although the fundamental principles of accounting lead quite often to arbitrary estimation of various items of costs (such as depreciation costs and inventory costs) financial statements do give valuable information to many parties interested in the performance and soudness of a firm provided that the same principles are constantly applied.

Interfirm comparison is a more difficult task for two main reasons. First of all, two firms in the same industrial sector may use quite different approaches to evaluate various items of the balance sheet and the income statement. This significantly changes the values of various performance criteria especially in periods of high inflation. Additionally, the huge dataset that includes hundreds of firms to be compared and tenths of variables to describe the different aspects of economic performance can not easily be handled by simple statistical methods.

The problem of different accounting principles is encountered in all financial analysis techniques and it is hard to give a unique and generally accepted solution. Computerization of modern industrial systems and the establishment of a national accountancy system partially solves the problem. The problem of large dataset is easier to be anticipated and many researchers have applied quite different approaches. A first approach is the computation of various ratios based on data taken from the financial statements, [1-2]. Again the computerization of industrial systems has facilitated the collection and processing of data, making the performance assessment based on financial ratios a popular approach. However, the simultaneous examination of many ratios is a difficult task and necessitates the participation of experts to extract a useful piece of information. The basic disadvantage of such a procedure is that it does not provide a method of firms' classification.

Another approach is the application of multicriteria approaches as the multi-dimensional character of all modern operating systems is better reflected on the conceptual basis of multicriteria analysis (MCA), [3-5]. In MCA, the rankings obtained by the simultaneous consideration of multiple ratios contain more information with regard to the totality of the financial indices. Furthermore, the comparative examination of the multicriteria and unicriteria rankings can reveal those ratios, which, in a specified set of firms, constitute the most relevant indices of corporate success, [6]. However, the results of these studies are hardly utilized by practitioners due to the complexity of MCA and the arbitrary selection of weights.

An alternative approach is the use of multivariate analysis as it is now widely recognized that business phenomena can not be expressed and predicted by taking into account only a small number of financial ratios. Therefore, elaborate statistical models have been developed to forecast the business future particularly with respect to bankruptcy risk, [1,7-8]. Moreover, multivariate models have also been applied to determine industry and corporate performance, [9-12]. The main difference between multicriteria and multivariate methods is related to the way the various weights are assigned to the relative performance criteria. Thus, in a multicriteria method the decision maker subjectively determines the relevant weights, whereas in a multivariate method weights are estimated rather objectively out of the dataset.

This paper integrates principal component analysis with cluster analysis to assess the overall performance of an industrial sector. The analysis extends in two different greek sectors, the pharmaceutical industry and the olive and seed oil industry. Reasons of variations and similarities that influence the corporate performance in the period 1981-91 are highlighted and compared between the two sectors. The results are further interpreted in the framework of the relevant greek state and EU policies. Section 2 presents the methodological approach and Section 3 gives a brief description of the two greek industrial sectors. Section 4 presents the results of the integrated multivariate analysis that is proposed. Finally, Section 5 draws the main conclusions.

METHODOLOGICAL APPROACH

The method uses data extracted from the financial statements that are yearly published by ICAP SA in the years 1981, 1987 and 1991, [13]. Figure 1 depicts the methodological approach. The selection of various financial ratios to express the most important aspects of a firm's performance is a crucial step.

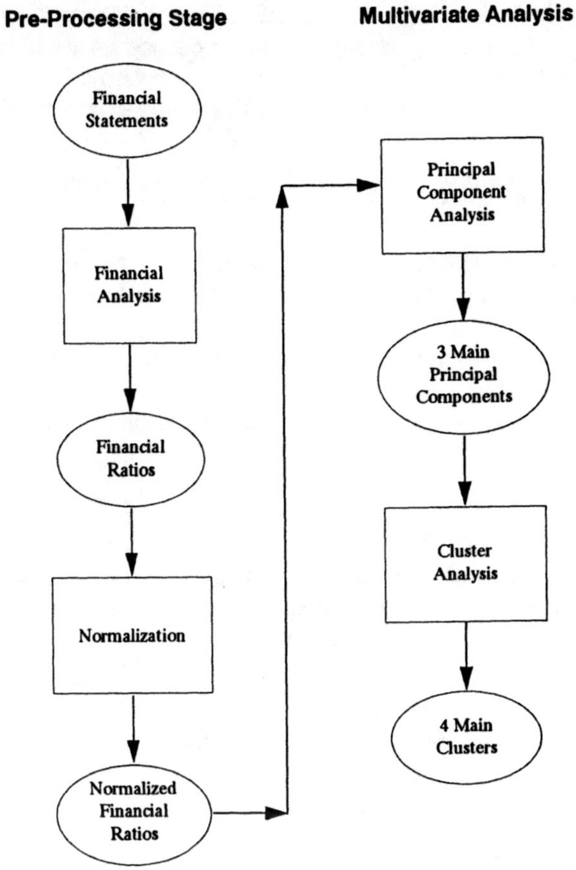

Fig. 1. Flow chart of the methodological approach.

The variables that are used to define the type of a performance criterion are drawn from experts opinion and sectoral analyses. The most important criteria of corporate performance are:

1. *Size Factor*. As there is no unique variable to express the size of a firm, it is calculated using simultaneously the variables of total assets, net worth, shareholders' equity and turnover. In the greek case, the figures for turnover refer to the year 1991.
2. *Financial Factor*. The choice of the proper ratio between debt and equity is a crucial problem in corporate financial planning. It has widely been recognized that the ideal firm is one that neither shows absolute aversion to external finance nor receiving excessive loans, [14]. The ratio:

$$FF1 = \frac{Net\ Worth}{Total\ Assets} \tag{1.}$$

express a tendency for the contribution of own funds to finance the total assets of the firm. The dynamic changes in the economic environment impose a balance between external and internal funds so there is no ideal value for this ratio. Furthermore, indices referring to short-term solvency reflect another interesting aspect of capital structure. The ratio:

$$FF2 = \frac{(Net\ Worth) + (Long\text{-}Term\ Liabilities)}{(Net\ Fixed\ Assets) + (Stocks)} \qquad \textbf{(2.)}$$

enables the analyst to trace the formation of a sort of "permanent" working capital. High values of this ratio reflect a rather safe financial position of the company.

3. *Profitability Factor*. In most cases this is the ultimate criterion of success as profit maximization is always the main objective of managerial efforts. In this paper, the profitability of a firm is measured by the combined use of the ratios:

$$PF1 = \frac{Net\ Profits}{Net\ Worth} \qquad PF2 = \frac{Gross\ Profits}{Total\ Assets} \qquad \textbf{(3.)}$$

   The selection of gross profits instead of net profits as the numerator of the second profitability ratio is imposed by the application of principal component method. As the variables should be independent of each other, the use of net profits in both ratios could cause statistical problems.

Having calculated the financial ratios of all the firms in a discrete time period, a normalization procedure is then undertaken. For each firm, the values of every variable are divided by the average value of the sector. The reason is that the values of all the variables, after the normalization procedure, are of the same order eliminating a substantial variability imposed by the different units that each variable is measured (eg. the size of a firm, that is better expressed on the basis of financial data, can be used together with financial ratios that measure the profitability and financial status of a firm).

The normalized financial ratios are used in the principal component analysis to extract the three factors that are used to explain the information given by the eight variables (see Fig. 1). An excellent description of the principal component method can be found in Zopounidis and Skiadas, [15]. The principal component analysis calculates the scores of every firm for the three factors examined (size, financial status and profitability). Plotting the scores of all the firms on two-dimesional diagrams, one can draw useful conclusions regarding the relationships that exist among the three factors. The main contribution of the principal component analysis is that it eliminates the number of graphs that ones has to analyse in order to detect relationships among the three factors. Principal component analysis is applied by using PONTOS.[1]

---

[1] PONTOS is a statistical package that has been developed by the National Technical University of Athens, Department of Chemical Engineering, in collaboration with University of Utah, laboratory of CMARC and runs on a PC-platform. It can handle a very large number of cases and variables. It also incorporates advanced data pre-processing techniques such as normalization and vardia (a graphical rotation technique to easily locate the direction of new factors).

After having extracted the three main factors, a qualitative procedure is followed in order to test the null hypothesis, i.e. that there exists a strong relationship (either a positive or a negative one) between two factors. The criterion for the acceptance of the null hypothesis is the presence of a strong pattern; the majority of the firms fall on the south-western to the north-eastern direction of the two-dimensional diagram (for a positive relationship or to the opposite direction for a negative relationship) that is formed by the two factors. Therefore, the selection of those firms that belong to each of the four quarters of the diagram is crucial.

The factors that have been calculated by the principal component analysis do not divide the firms into homogenous clusters. They rather divide the firms into quarters that show the relative distance of these firms from the typical firm of the industrial sector (the firm that has average values for all the performance criteria). Therefore, cluster analysis is further applied to the results of the principal component analysis to form homogenous clusters of firms (for a description of cluster analysis see Calithrakas-Kontos, Dimitras and Zopounidis, [16]). The scores of the two factors used to form a diagram are fed to the cluster analysis. The results are the number of the firms that belong to each quarter of the diagram. By counting the number of firms in any direction one can safely identify the existence of a pattern between the two factors (alternatively the application of the chi-square test provides the statistically accepted procedure for the acceptance or rejection of the null hypothesis).

Cluster analysis is applied by using STATISTICA V4.3 by StatSoft SA, [17]. The factors calculated by the principal component analysis are by definition statistically uncorrrelated. Therefore, the euclidean distance and the simple linkage technique can be safely used for the formation of the clusters. The examination of three discrete time periods can reveal the change of the type of the relationship between the two factors. Finally, by examining two different sectors one can analyse the different behaviour of the firms active in each market and the different market conditions.

THE GREEK CASE

The Pharmaceutical Industry

Pharmaceutical industry is an important branch of the chemical industry in Greece. In the time period 1981-91, 10% of the firms active in the sector had more than 200 employees. Eighty-eight (88) pharmaceutical firms operate in Greece, a quite large number by international standards especially for a small country with a population of ten million people. There are some international companies which import and produce pharmaceuticals in Greece and many local producers.

Pharmaceutical consumption, as percentage of gross domestic product in Greece, is one of the highest among developed countries and it is in the range of 1.51% in the year 1989, [18]. Pharmaceutical consumption has increased in the period 1981-86 with an annual growth rate of 7.0% in constant prices and in the period 1987-90 with an annual growth rate of 23.8% in current prices which is well above the consumer price index, [19]. Pharmaceutical production in Greece has also shown a considerable rate of growth during the period 1981-91. Especially in the period

1981-87, the annual growth rate in constant prices, expressed in greek drachmas, was 6.5% whereas during the period 1986-91 the corresponding annual growth rate was 9% in current prices, expressed in ECUs. Imports increased, too, and constituted the 50% of output in 1990 whereas exports constituted the 12.6% of output in the same year, [18].

This significant increase of pharmaceutical production in the period 1981-91 was not linked with an increase of investment activity in the sector. Total fixed assets increased with an annual rate of 16.5% in current prices whereas net fixed assets (total assets minus depreciation) increased with an annual rate of 10% in current prices, which is well below the industrial wholesale index. It can be argued that in the period 1981-91 the discrepancy between production and investment activity is due to: a) the small returns on shareholders' equity recorded, b) the fact that production increase in the 1980's was achieved through the use of the excess productive capacity formed in the 1970's and the increase of stocks, [19].

The share of shareholders' equity to total asets has increased from 29% in 1981 to 33% in 1991. Financing of current assets and stocks by short-term liabilities has slightly decreased from 65% in 1981 to 60% in 1991. On the contrary, the contribution of long-term liabilities and shareholders' equity in the financing of net fixed assets and stocks has increased from 66% in 1981 to 79% in 1991.

There is a tradition of state intervention in the sector which is expressed in three different dimensions. First, quality of drugs is controlled and licenses for new drugs are awarded by the National Organization of Drugs (NAD), a firm that is 100% owned by the Ministry of Health and Social Insurance. Second, Ministry of Commerce controls the prices of final products mainly for two reasons: a) the government's policy to provide cheap drugs to the vast majority of the population and b) the government's policy to restrict the large deficits of insurance institutions that are fully controlled by the state. Furthermore, the general economic policy, and in particular the income and monetary policies, influence the wage rate, the cost of money and other items of production cost.

This intervention resulted in the decrease of profit margins of greek pharmaceutical firms. Empirical research has shown that large firms in order to keep their profit margins use small firms as sub-contractors. Small firms are now working on a take-it or leave-it scheme which leads to very small margins and low profitability, [19]. Additionally, large firms are now importing more to decrease production costs forcing small firms to further lower they profit margins.

To sum up, in the period 1981-91 the greek pharmaceutical industry has been affected by the general economic crisis. Although has consumption been increased causing a significant production increase, investment activity did not follow this trend. There has been a tradition of state intervention in the greek pharmaceutical industry which has accelerated the establishment of sub-contracting relations. These relations could explain why large firms show large profits whereas small firms, which are higly dependent on the large firms, have low profit margins.

The Olive and Seed Oil Industry

The olive and seed oil industry is an important branch of the food industry in Greece. In 1988, the number of establishments in the sector accounted for the 16% of the total number of food firms. The total number of olive and seed oil firms is equal to eighty-six (86) a reasonable figure for a country with significant but fragmented agricultural production.

Considering the general trends in the industry during the period 1981-91, it has to be mentioned that in physical terms: a) total production of olive and seed oil remained almost constant at 375,000 tonnes, b) final consumption of olive oil remained also constant at 300,000 tonnes, c) exports have increased from 25,000 tonnes in 1981 to 100,000 tonnes in 1991 whereas imports reached a maximum level of 50,000 tonnes, almost due to imports of seed oil.

It is also worth-mentioning that Greece has the highest per capita consumption of olive oil in the world reaching the level of 20 kg. Furthermore, the consumption of seed oil has significantly increased in the last years reaching the level of 40% of the total olive and seed oil consumption in 1991.

Financial statement analysis of the firms active in the sector shows: a) a decrease of investment activity, and b) an increase of the percentage of current assets to total assets which were mainly financed by current liabilities. This is justified by the following evidence: a) although the share of shareholders' equity to fixed assets has been increased, the amount of total assets financed by shareholders' equity has decreased from 26% in 1981 to 10% in 1991, b) short-term liabilities have increased their contribution to current assets from 90% in 1981 to 125% in 1991, c) long-term liabilities contributed 22% of the foreign funds in 1981 and only 13% in 1991, and d) the ratio of depreciation to gross fixed assets has increased from 27% in 1985 to 40% in 1991.

Government regulation of the sector is achieved through the mechanism of income subsidies to greek farmers in the framework of the Common Agricultural Policy of the EU. It is believed that these income subsidies are partially earned by small firms which manage to compete with the large ones on a rather equal basis. Furthermore, in the 1980's a policy supporting the establishment of cooperative organizations in the production of olive oil was promoted by the greek state. Among the cooperative organizations active in the sector, the five largest are now heavily debted deteriorating the overall performance of the sector.

Overall, during the period 1981-91, the greek oilve and seed oil industry experienced the gradual integration of the greek market to the european market. Although did local consumption remain almost constant, the share of local production directed to the european market has significantly increased. Additionally, the share of seed oil consumption increased in respect to the total consumption. As the olive and seed oil market is basically a price-competion market, small firms show high profitability and compete with large ones by exploring the mechanism of income subsidies of the greek farmers.

RESULTS OF THE MULTIVARIATE ANALYSIS

The Pharmaceutical Industry

Principal component analysis is applied in the case of the greek pharmaceutical industry to assess the corporate performance in the years 1981, 1987 and 1991. The sample size is the total number of firms active during the whole period 1981-91 (that is 56 firms). Factor loadings for the greek pharmaceutical industry are shown in Table 1. The first factor corresponds to size as the loadings of the first four variables, that is total assets, net worth, shareholders' equity and turnover, are significantly higher than the loadings of the other four variables. Only seven firms in the years 1981 and 1991 and eight firms in the year 1987 can be considered as large ones (that is the scores of the size factor are greater than one).

The financial factor has loadings significantly higher for the fifth and the sixth variables whereas the last identified factor is the profitability factor with higher loadings for the variables of net profits to net worth and gross profits to total assets. There are only few large firms with high values of the financial factor (see figure 2). The financial factor for the majority of large firms get low values. Small firms have rather high values of the financial factor. Low values of the financial factor correspond to both large and small firms in the sector. The same evidence is drawn from Fig. 3 that depicts the number of firms in the four clusters that are formed for every possible pair of the three factors in the years 1981, 1987 and 1991. There are twenty-four large firms with low values of financial factor whereas there are twenty-seven small firms with high values of the same factor. Financial factor is negatively related to size in the case of the greek pharmaceutical industry in 1981.

Table 1. Factor loadings for the case of the greek pharmaceutical industry.

VARIABLE	1981			1987			1991		
	SIZE FACTOR	FINANCIAL FACTOR	PROFITA-BILITY FACTOR	SIZE FACTOR	FINANCIAL FACTOR	PROFITA-BILITY FACTOR	SIZE FACTOR	FINANCIAL FACTOR	PROFITA-BILITY FACTOR
Total Assets	0.93	-0.09	0.03	0.90	-0.21	-0.11	0.94	-0.20	0.07
Net Worth	0.95	0.19	0.03	0.87	0.21	0.26	0.83	0.22	0.24
Shareholders' Equity	0.94	0.05	-0.05	0.74	-0.04	-0.37	0.83	0.43	-0.14
Turnover	-	-	-	-	-	-	0.94	-0.21	0.14
Net Worth to Total Assets	0.03	0.88	0.16	-0.01	0.93	0.16	-0.03	0.92	-0.10
Net Worth plus Long-Term Liabilities to Net Fixed Assets plus Stocks	0.04	0.90	0.00	-0.04	0.93	-0.09	0.04	0.89	-0.06
Net Profits to Net Worth	0.16	0.33	0.68	-0.03	0.30	0.81	0.29	-0.06	0.72
Gross Profits to Total Assets	-0.14	-0.08	0.87	-0.08	-0.20	0.83	-0.06	-0.08	0.79

Another significant relationship is established between profitability and size. Thirty small firms have low values of the profitability factor whereas only four large ones. On the other hand, there are twenty-one large firms and twenty small ones with high values of the same factor. Therefore, evidence show that profitability is positively related to size in the case of the greek pharmaceutical industry in 1981.

Finally, profitability factor is negatively related to the financial factor. The majority of firms with low values of the profitability factor have high values of the financial factor (i.e. twenty-six firms), whereas the most of firms with low values of the financial factor have high values of the profitability factor (i.e. twenty-nine firms). A negative relationship between size and financial factor probably exists in the year 1987 (see Fig. 3). Twenty-three large firms have low values of the financial factor whereas small firms have both low and high values of the financial factor. Furthermore, a positive relationship between size and profitability still exists. Thirty-two small firms show low values of profitability whereas large firms have either low or high values of profitability. However, the relationship between financial and profitability factors is not valid any more.

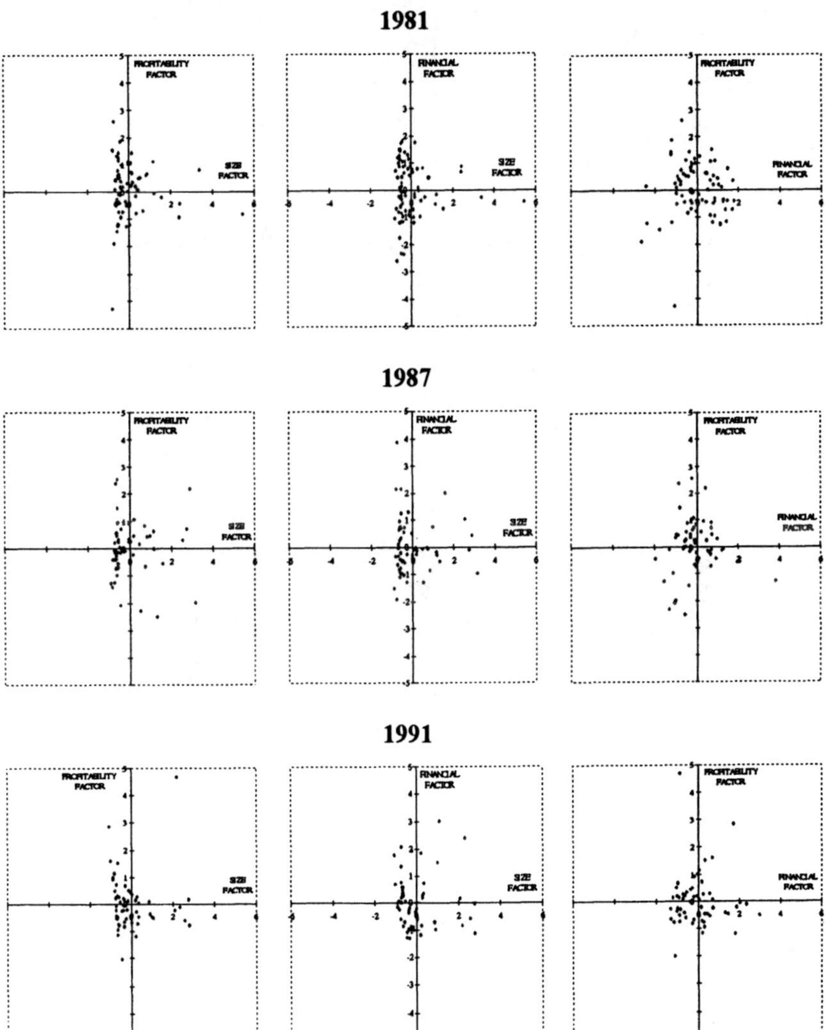

Fig. 2. Results of the principal component analysis for the greek pharmaceutical firms.

The positive relationship between size and profitability still exists in the year 1991 (see Fig. 3). Thirty-two small firms show low values of profitability. On the other hand, nine large firms have large values of profitability and only one large firm has low values. However, the relationship between financial and profitability factors is not valid any more. Furthermore, the negative relationship between size and financial factors is not valid any more, too.

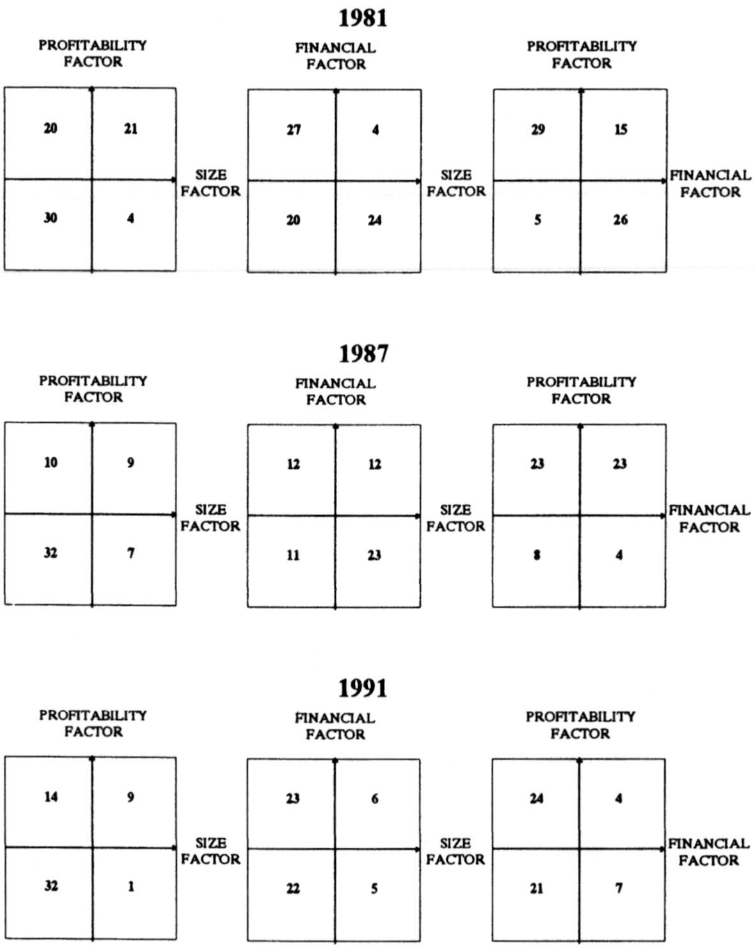

Fig. 3. Results of the cluster analysis for the greek pharmaceutical firms.

Briefly, there is a strong evidence that size is a crucial parameter that determines the profitability record of a pharmaceutical firm. In the period 1981-87, size also determined the financial status of a pharmaceutical firm. It seems that small firms tried to self-finance total assets whereas large firms used long-term loans and current liabilities. After 1987, however, size did not influence financial status. Only in 1981 there was a relationship between profitability and financial status. It seems that a good financial status is a sufficient but not a necessary condition for a good profitability. The multivariate analysis showed that there were pharmaceutical firms with low values of the financial factor and high values of the profitability factor but there were also firms with low and high values of the profitability factor which both had low values of the financial factor.

The Olive and Seed Oil Industry

Principal component analysis is also applied in the case of the greek olive and seed oil industry in the years 1981, 1987 and 1991. The sample size is the total number of firms active during the whole period 1981-91 (that is 58 firms). Factor loadings for olive and seed oil industry are shown in Table 2. As in the case of the pharmaceutical industry, the three identified factors are the size factor, the financial factor, and the profitability factor. Only four firms in the years 1981 and 1987 and six firms in the year 1991 can be considered as large ones (that is the scores of the size factor are greater than one).

Table 2. Factor loadings for the case of the greek olive and seed oil industry.

VARIABLE	1981			1987			1991		
	SIZE FACTOR	FINANCIAL FACTOR	PROFITA-BILITY FACTOR	SIZE FACTOR	FINANCIAL FACTOR	PROFITA-BILITY FACTOR	SIZE FACTOR	FINANCIAL FACTOR	PROFITA-BILITY FACTOR
Total Assets	-0.73	0.09	0.00	-0.79	-0.18	0.10	1.13	-0.03	-0.01
Net Worth	-0.99	0.03	0.00	-0.98	-0.16	0.04	0.84	-0.02	-0.01
Shareholders' Equity	-0.66	-0.09	0.00	-0.60	0.05	-0.11	0.33	-0.02	-0.01
Turnover	-	-	-	-	-	-	0.90	0.02	0.01
Net Worth to Total Assets	-0.03	-0.77	0.01	0.18	-0.82	0.35	-0.04	-0.83	-0.28
Net Worth plus Long-Term Liabilities to Net Fixed Assets plus Stocks	-0.01	-0.76	0.03	0.12	-0.89	0.13	-0.01	-0.86	-0.20
Net Profits to Net Worth	0.00	0.01	0.84	0.04	-0.40	-0.73	0.00	0.29	-0.79
Gross Profits to Total Assets	0.00	0.02	0.84	0.01	-0.18	-0.78	0.00	0.21	-0.81

Fig. 4 depicts the distribution of olive and seed oil firms on the three planes that are formed by the size, the financial, and the profitability factors in the years 1981, 1987 and 1991. There is no clear relationship among the three factors and all the firms are evenly distributed on the three planes. The same conclusion is drawn out of Fig. 5 that depicts the results of cluster analysis for olive and seed oil industry in the years 1981,1987,1991.

There is no obvious relationship between size and financial factors in the year 1987. However, only two large firms have low values of the financial factor whereas twenty-five small firms have high values (see Fig. 5). A negative relationship exists between size and profitability. Twenty-three small firms show high values of the profitability factor whereas nineteen large firms have low values. Additionally, there are only two large firms with high values. Finally, a negative relationship exists between financial and profitability factors. There are twenty-one firms with high values of the profitability factor and low values of the financial factor whereas firms with high values of the financial factor have both low and high values of the profitability factor.

There is no obvious relationship between size and financial factors in the year 1991. Although three large firms get high values for the financial factor, there were twenty-five small firms with low values of the financial factor (the opposite from the distribution in 1987). Now, there is a positive relationship between size and profitability (in 1987 there was a negative one). Thirty-three small firms show low values of profitability whereas twenty large firms have high values. On the contrary, there are only three large firms with low values and three small firms with high values.

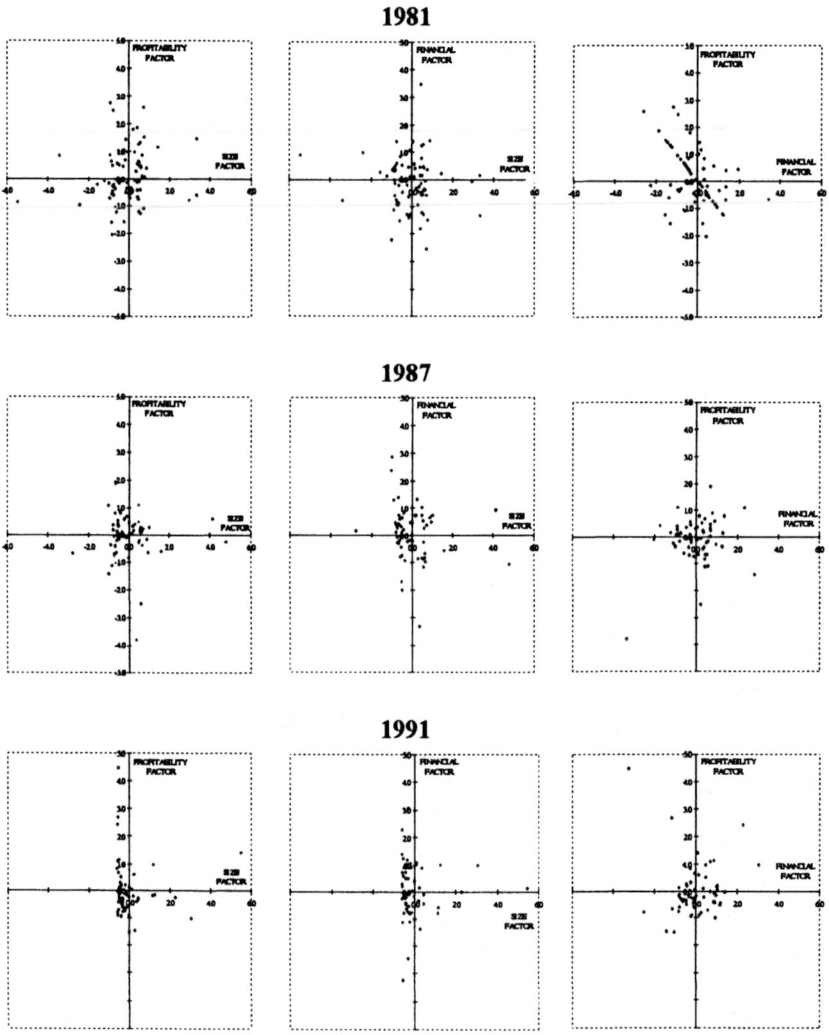

Fig. 4. Results of the principal component analysis for the greek olive oil firms.

Finally, another positive relationship exists between financial and profitability factors (in 1987 there was a negative one). There are thirty-one firms with low values of the profitability and the financial factors and ten firms with high values of the same factors. There are only two firms with low values of the financial factor and high values of the profitability factor.

In short, there is no evidence that size influences financial factor in the case of olive and seed oil industry. In 1987 and 1991, size could explain the profitability record of a greek olive and seed oil firm with no clear direction of influence. In 1987 large firms usually had low profitability and small firms usually showed high profitability but in 1991 the opposite was true. The same was true for the relationship between financial and profitability factors. In 1987 firms with high values of the financial factor showed low values of the profitability factor but the opposite was demonstrated in 1991.

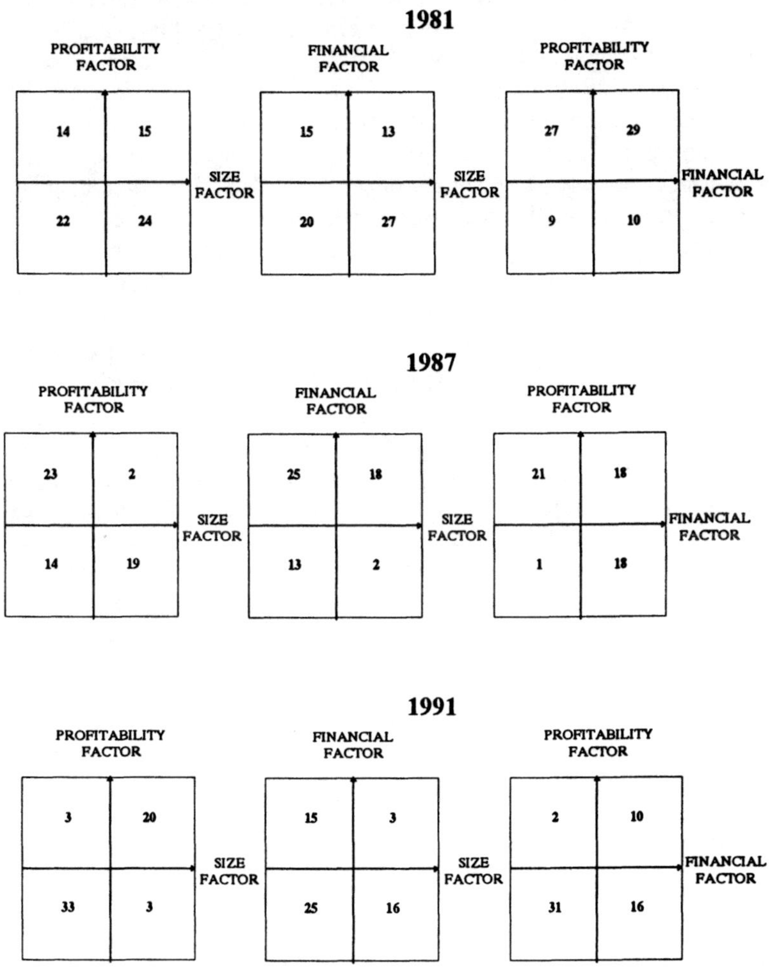

Fig. 5. Results of the cluster analysis for the greek olive oil firms.

CONCLUSIONS

Multivariate analysis has been proved a useful tool to assess corporate performance of various industrial sectors. The integration of principal component analysis with cluster analysis enables the decision maker to investigate the existence of relationships among the factors of size, profitability and financial status of a specific industrial sector. Besides, the implications of the national public policy and EU policies on the firm's activity should be examined and analyzed in order to meaningfully interpret the results of multivariate analysis. In the two cases examined, state intervention influences the corporate performance but a different type of intervention leads to different corporate behaviours.

In the case of the greek pharmaceutical industry a clear positive relationship between size and profitability has been established in all the three years examined. It appears that a critical size of a firm exists in order to achieve high profitability whereas small firms have on the average low profitability. This result seems to comply with the results of other empirical researches showing that price controls, imposed by the greek governments, force large firms to protect their profit margins by using small firms as sub-contractors on a take-it or leave-it scheme. Such a positive relationship, however, was not constantly identified in the case of greek olive and seed oil industry. In 1981 there was no relationship at all whereas in 1987 there was a negative relationship between size and profitability. It seems that income subsidies of farmers, an objective of the EU Common Agricultural Policy, are partially earned by small firms which achieve significant profit margins.

There is no well-established relationship between the financial and profitability factors in the case of the two sectors. There were firms with low values of the financial factor and high values of the profitability factor but there were also firms with low and high values of the profitability factor which both had low values of the financial factor. The application of a complementary method (multi-criteria analysis) by the authors has led to a similar conclusion in the case of the greek pharmaceutical industry, [20-21]. Finally, financial factor is not related to size for the case of greek olive and seed oil industry whereas for the case of greek pharmaceutical industry financial factor is negatively related to size in the years 1981 and 1987, only.

REFERENCES

1. Barnes, P. The analysis of financial ratios: A review article. J. Bus. Fin. Acctng. 1987,14(4):449-461.
2. Lev, B. *Financial statement analysis*. Prentice-Hall, Englewood Cliffs, NJ, 1974.
3. Smith, P. Data envelopment analysis applied to financial statements. Omega 1990,18(2):131-138.
4. Zopounidis C, Dimitras AI. Evaluating bankruptcy risks: A multicriteria decision support approach. Proceedings of the EURO XIV Symposium, Israel,1995.
5. Zopounidis C, Dimitras AI. The forecasting of business failures: Overview of methods and new avenues. In *Applied stochastic models and data analysis*, Janssen J, Skiadas C, eds, World Scientific Publ. Co, 1993.
6. Diakoulaki D, Mavrotas G, Papagiannakis L. Determining objective weights in multiple criteria problems: The CRITIC method. Computers Ops. Res. 1995,22(7):763-770.
7. Dambolena, IG. Prediction of corporate failures. Omega 1983,11(4):355-364.
8. Keasey K, McGuiness P, Short H. Multilogit approach to predicting corporate failure - Further analysis and the issue of signal consistency. Omega 1990, 18(1):85-94.
9. van der Wijst, D. Modelling interfirm comparisons in small business. Omega 1990, 18(2):123-129.

10. Meric I, Eichhorn BH, Welsh C, Meric G. A multivariate comparison of the financial characteristics of French, German and UK manufacturing firms. Proceedings of the EURO XIV Symposium, Israel,1995.
11. Baourakis G, Matsatsinis NF, Siskos Y. The contribution of data analysis models to agricultural marketing. In *Applied stochastic models and data analysis,* Janssen J, Skiadas C, eds, World Scientific Publ. Co, 1993.
12. Janssen D, Janssen J, Troussart J. Principal component analysis of the economic behaviour of the building activities industrial linkages in Belgium during the last important economic crisis. In *Applied stochastic models and data analysis*, Janssen J, Skiadas C, eds, World Scientific Publ. Co, 1993.
13. ICAP. *Financial directories of Greek industrial and commercial firms*, Athens, 1980-93.
14. Sizer, J. *An insight into management accounting*, Pelican, 1987.
15. Zopounidis C, Skiadas C Principal component analysis of the economic behaviour of the Greek industry. Proceedings of the 2nd symposimum on the future of Greek industry, Technical Chamber of Greece, Vol 1 (in greek), 1989.
16. Calithrakas-Kontos N, Dimitras AI, Zopounidis C. Cluster analysis of the economic behaviour of the Greek chemical production industry. Draft Paper, Technical University of Crete, Department of Production Engineering and Management (in greek), 1994.
17. StatSoft SA, *User's Manual of Statistica*, NY, 1993.
18. Mossialos E, Kanavos P, Abel-Smith B. The impact of the single European market on the pharmaceutical sector. In *Cost containement, pricing and financing of pharmaceuticals in the European Community: The policy-makers' view,* Mossialos E, Ranos C, Abel-Smith B, eds, LSE Health & Pharmetrica SA, Athens, 1994.
19. Papagiannakis, L. *The Greek pharmaceutical industry.* Farmetrica (National Organization of Drugs). Athens, 1990.
20. Diakoulaki D, Mavrotas G, Papagiannakis L. A multicriteria approach for evaluating the performance of industrial firms. Omega 1992, 20(4):467-474.
21. Caloghirou Y, Diakoulaki D, Mavrotas G. An objective multicriteria evaluation of industrial firms. Proceedings of the 2nd Balkan conference on OR, Thessaloniki 1993.

STABLE SET INTERNALLY MAXIMAL : A CLASSIFICATION METHOD WITH OVERLAPPING

Alain Couturier [1], Bernard Fioleau [2]
[1] Conservatoire National des Arts et Métiers (CNAM) Nantes
25 boulevard Guy Mollet 44071 NANTES Cedex 03, FRANCE

[2] Faculté des Sciences Économiques et de Gestion de Nantes
110 Boulevard Michelet 44322 NANTES Cedex 03 FRANCE

Abstract: When the classification methods are applied to the companies' positioning they are linked, either assigned to company in a group whose characteristics are predefined (economically should company rather than a company in financial difficulties) or to separate sub-groups within a population of companies. The stable set internally maximal (SSIM) allows, based on financial characteristics of a group of companies from the same sector, to obtain a classification of these in non-separated sub-groups. Such a classification improves the preciseness of the companies' positioning within the economy activity.

Keywords: ssim, classification, overlapping, company typology

INTRODUCTION

This document presents the results obtained by applying the concept of stable set internally maximal (SSIM) to the field of company listing.

The researcher's interest has for a long time been focused on the companies' problem of its positioning or of its classification within its sector of activity. It is commonly known that in the same sector there are different types of companies with dramatically varying financial structures and varying performances. It is therefore obvious that financial analysis can be useful when analysing sub-sets of companies having the same homogeneous profile.

Furthermore for many countries whose economic context is generally qualified as difficult or even critical, the bankruptcy risk is a permanent reality. For example, in France the bankruptcy rate for the past 5 years at least, has been between 2% and 3% of the total number of companies. That is to say, about 40,000 bankruptcies per year.

Following Altman's works (1968) many researchers used the discriminant linear analysis (LDA) to analyse companies, in particular to evaluate the bankruptcy risk. The numerous scoring functions that have been devised since then are all generally in the form of a linear combination with a limited number of financial ratios. They are now part and parcel of the tools which financial analysis use, not only to identify the company's present situation but also when analysing its future (Zavgren (1983), Dimitras, Zanakis, Zopounidis (1996)).

The LDA statistical technique is based on several assumptions that are liable to falsify the obtained results, if they are not checked. The LDA's use in the financial field has raised many reservations not only on the theoretical side (Eisenbeis (1977), Krzanowski (1977)) but also on the practical side (Joy, Toffelson (1975) Zavgren (1983), Malécot (1986)). This reservations concern in particular:
- the difficulty, from a mathematical point of view, to satisfy the two necessary assumptions for the LDA's implementation: multinormality of the criteria and equality in the criteria's spread within samples. As these two assumptions concern financial ratios they are rarely verified.
- the major problems from the practical point of view concern the calculation of the classification rate of errors, the selection of significant ratios and the choice of the company's assignment to one of the two classes.

In order to answer these critics and also to improve the tool other avenues have since been explored, however there is no conclusive evidence that the results obtained are significantly better. Several scoring functions have been built from discriminant non-linear analysis techniques (Altman et al. (1977), Bardos (1986)). Neural analysis has not only been used for bankruptcy risk but also for typological reasons (Bell et al. (1990), Varetto, Marco (1993)).

Discriminant linear analysis creates certain statistical problems : independence, multinormality, equality of the variance-covariance matrixes. The choice of the numerous criteria characterising a company is decisive, as regards the results obtained with a certain singularity risk for the latter ones. This problem is well summarised by Lev (1974) when he says that " there is no well-defined theory of corporate failure to direct design and performance of empirical studies. Lacking such a theory, researchers adopt a trial-and-error process of experimenting with a large number of measures... As expected, results of such unguided research efforts are often inconsistent, and almost impossible to generalize".

The company's position is usually based on the interpretation of comparative values taken from a set of criteria, generally financial ratios. These are considered as representative of its global situation. The non-homogeneity of these criteria can nullify any notion of scalar products, therefore correlation coefficient or distance.

The classification problems not only concern the search to belong to a predetermined class (among a possible two or more) but also they expose groups with homogeneous characteristics (within a population). In both cases it concerns separate sub-sets. In the classification field, apart from the remarkable contribution of Varetto and Marco, the contributions of Coats, Fant (1992), Hansen, McDonald, Stice (1992) and Mutchler (1989) must also be mentioned. Nevertheless, the companies real situation often seems more complex in the sense that certain of their characteristics bring them nearer certain groups while others push them away.

We propose to use in this article the ssim method to reveal non-separate groups of companies with varying profiles. The results obtained by this classification can be compared with those obtained from a discriminant function. As the ssims offer the particularity, for any company, to belong to several groups, this allows the financial analyst to fine-tune its position within the economic activity centre.

The ssim presented in the first part expose, using a set of characterised individuals by n criteria, non-separate sub-groups. The second part will analyse the results obtained using this method on a sample of French companies.

1. STABLE SET INTERNALLY MAXIMAL (SSIM)

Definition :

Let a set E and r = (E, R) a symmetric contrasting relation[1] (of antagonism, repulsion, difference, etc.) on E

X, sub-set of E is a ssim, if :

a) X is internally stable that is to say :

$$X \cap r(X) = \emptyset$$

b) X is maximal that is to say :

$$X \cup r(X) = E$$

where r(X) the set of elements in contrast has at least one element of X

Comments :

1 - $R \subset E \times E$ is the set of pairs in opposition.

$(x, y) \in \Re \quad \leftrightarrow \quad x\,r\,y$: x and y are opposed.

2 - If E represents a set of companies, the relation x r y can then be interpreted as : two companies x and y are opposed, therefore they can not belong to the same group (or class) [2].

3 - The ssims do not make up a partition of E : the same company can belong to two ore more ssims.

---

[1] It is possible to foresee a wider definition of ssims lifting the symmetric constraint.

[2] E can also be a set of products, securities, market characteristics, etc.

Example 1 :
Let E = { a, b, c, d, e } and r = (E,R) the relation defined by the following table (the relation being symmetrical, only the superior triangle is represented) :

	a	b	c	d	e
a	0	0	1	1	0
b		0	0	0	1
c			0	1	0
d				0	1
e					0

0 characterises the absence of opposition between the two companies, inversely 1 implies an opposition. For example, in the above table it can be seen that a and b are not opposed whereas a and c are.

The ssims are the sets : { c, e } { b, c } { b, d } {a, e } { a, b }

The set { c, e } is a ssim as :

$$X \cap r(X) = \emptyset \qquad \text{and} \qquad X \cup r(X) = E$$

with r(X) = r ({ c, e }) = { a, b, d }

1.1. Principle

The algorithm is based on the following comment :

For each element $x \in E$
. either x is not taken
. either x is taken but the elements in opposition to x are not taken

Therefore the following expression is used to calculate :

$$T = \prod_{i} \{ x + r(x) \} \qquad (1)$$

where r(x) is the monomial of the opposite elements to x
This expression is expanded taking into account the simplifications :

$e + e = e$
$e * e = e$
$e_1 + e_1 * e_2 = e_1$

+ and * are the Boolean operators: or, and.

The ssims are complementary to the terms of T. By taking the preceding example :

$$T = (a + cd)(b + e)(c + d)(d + e)$$

For example, the first term, (a+cd), must be used in the following way: either it is kept which means that neither c nor d (noted cd) are kept or a is not kept. The choice can be described as follows:

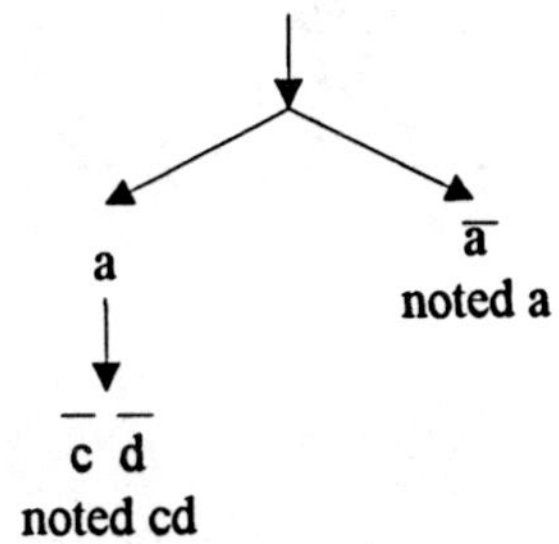

Comment : If the set E contains *n* elements, the T set contains no more than 2^n elements.

1.2. Algorithm and illustration

1.2.1. Notations

Elements of the unit are represented by small letters: x, y, ...
Units are represented by capital letters: A, E, ...
Unit groups are marked in the following way: ε, ℓ, ...

Let A a group of elements and ε a group of ssim of A.

1.2.2. Algorithm

For all new elements x and all E ssim of ε, three possibilities must be foreseen:

a) no element y of E is opposed to x: E is replaced by $E \cup \{x\}$
b) all the elements of E are opposed to x: a new ssim must be added $\{x\}$
c) certain elements of E are opposed to x: the E ssim is partitioned into two subgroups
   - E_1, all the elements of E are not opposed to x
   - E_2, all the elements of E are opposed to x,
   A new ssim is added, $E_1 \cup \{x\}$

The new group ε is simplified: all the ssims included in the other ssims are deleted.

1.2.3. Illustration

The algorithm above is applied to example 1:

Processing of the element	Results: ε	Case
a	{{a}}	b)
b	{{a,b}}	a)
c	{{a,b},{b,c}}	c)
d	{{a,b},{b,d},{b,c}}	c)
e	{{a,b},{a,e},{b,d},{b,c},{c,e}}	c)

2. APPLICATION

The ssim method has been tested on a sample of French companies belonging to the general mechanical engineering economic activity sector[3]. The objective in this case is to test the method's aptitude to extract sub-groups of companies whose financial characteristics are clearly different.

The sample is composed of 165 companies in activity having the following characteristics :

- ≥ 50 staff,
- legal form : limited liability company or private limited company
- fiscal year : 1995

Only 165 companies complied with these three criteria, the reason being is the choice of a too recent fiscal year. It must be underlined that no company was excluded from the sample and that the ratios used were not transformed in any way[4].

14 criteria were calculated for each company (expressed as financial ratios) which synthesised the main aspects of their financial situation, their activity, their efficiency and of their profitability. These ratios are listed below in the table 1.

On the other hand, according to the score obtained by each company the financial risk can be assessed by using a bankruptcy risk scale which is associated

[3] Sectors 25 and 26 of the NAP of the INSEE (French national institute of statistics and economic surveys). This sample was obtained from the SCRL company's "Diane" database.

[4] Especially the ratios were not subject to any boundary markings. This practice as well as the exclusion of companies considered at the outset as "abnormal" generally has as objective, to obtain normally distributed ratios. Normality tests carried out on several ratios show, that in practice, it is very difficult to obtain such a result : Fioleau (1992).

to it. This score was calculated with the help of the Conan Holder function (1979) ; which, with the financial analysis score utilised at the Bank de France, is one of the most frequently used in France. The Conan et Holder score function and the risk scale which is associated to it are presented in appendix 1.

1- $\frac{\text{Net Worth}}{\text{Fixed Assets}}$	8- $\frac{\text{Gross Profit}}{\text{Sales}}$
2- $\frac{\text{Long Term Debt}}{\text{Net Worth}}$	9- $\frac{\text{Interest Expenses}}{\text{Added Value}}$
3- $\frac{\text{Net Worth}}{\text{Total Assets}}$	10- $\frac{\text{Gross Profit}}{\text{Debts}}$
4- $\frac{\text{Inventories}}{\text{Sales}}$	11- $\frac{\text{Net Worth} + \text{Long Term Liabilities}}{\text{Total Assets}}$
5- $\frac{\text{Receivables}}{\text{Sales}} * 360$	12- $\frac{\text{Quick Assets}}{\text{Current Liabilities}}$
6- $\frac{\text{Accounts Payable}}{\text{Purchases}} * 360$	13- $\frac{\text{Interest Expenses}}{\text{Debts}}$
7- $\frac{\text{Added Value}}{\text{Fixed Assets}}$	14- $\frac{\text{Wages \& Changes}}{\text{Added Value}}$

Table 1 : LIST OF RATIOS USED

Comments :

1 - The ratios 10 to 14 are those that appear in the Conan Holder function score. The ssims were searched for using these five ratios, the other criteria allowed to characterise the exposed groups. The scores obtained by each company also provided information on their financial health.

2 -Finally, it can be seen that all the selected companies do not constitute a random sample but a group where it could be assumed that the comparable character of the company's activity's is likely to produce similar performances.

In order to exclude the distribution tail ends and to take into account the possible non linearity, each criteria was broken down into five classes. The choice of the five classes that were used allows to find the same black and white variations as in cartography (Bonnin (1975)). In this respect, Saporta (1990) noted that "it is possible to resort to re-coding. The most frequent process consist in making a variable qualitative by breaking it down into classes which also allows to utilise the possible non linearity".

Graphic 1 shows the distribution of criteria 10. Apart from the two extreme classes, it is possible to, following the required granularity, obtain three classes or more. The breakdown in regular fixed size intervals along the length of the distribution is not pertinent, as there is a risk that it will separate the coherent groups and create non-homogeneous groups.

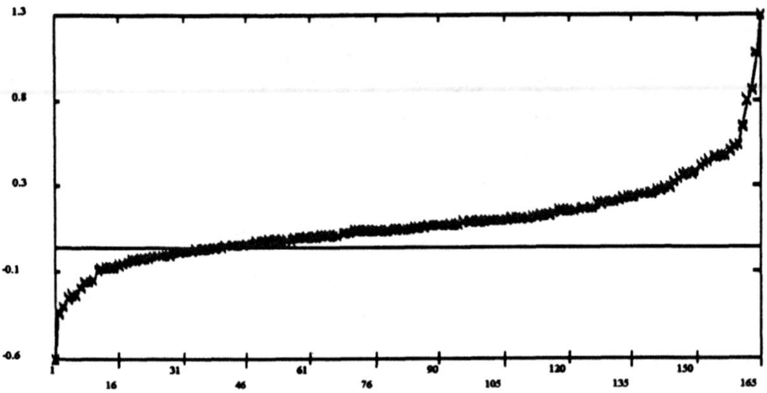

Graphic 1 : Criteria 10

For calculation reasons - a relation of n individuals gives a maximum of 2^n monomial - the companies' number, on which the ssims were calculated, was limited. In order to extract these companies it was necessary to carry out an analysis of the principal components based on re-coded data. For each axis, among the 165 companies that were taken into account, it was the companies with a quality projection superior to 90% that were chosen. 30 companies, which can be considered as representative of the 165, were thus extracted from this analysis.

If it is noted that $x = (x_1, \ldots x_5)$ and $y = (y_1 \ldots y_5)$ represent two companies they would be considered opposite if :

- their distance is superior to a threshold s_1

or

- for at least one criteria i, the difference between x_1 and y_1 is superior to a threshold s_2

The distance used is the sum of the absolute values differences :

$$d(x,y) = \Sigma | x_i - y_i | \qquad (2)$$

Several processes were carried out for different threshold values s_1 and s_2. It can be seen that for weak values of s_1 (s_1<0.5) the groups have only one company and for the values s_1 > 3.5 the 30 companies are broken down into two groups that are not apart.

The threshold used for s_1 is 1.5. Once it is fixed, a threshold for s_2 at 0.25 allows simultaneously to regroup the companies with similar profiles and exclude those that are separated by at least one of the criteria.

	$s_1 = 1$	$s_1 = 1.25$	$s_1 = 1.5$
s_2 unfixed	X	X	
$s_2 = 0.25$			X
$s_2 = 0.5$		X	X

Table 2: Process achieved

The process chain is broken down as follows:
Step 1: Coding of the data matrix T data: to all the elements t_{ij} of T is associated an element of the group {0, 0.25, 0.5, 0.75, 1}. This is how the matrix X is built.

Step 2: Calculation of the distance matrix: at the chosen thresholds s_1 and s_2 the distance (x, y) is calculated between two recorded companies x and y.

Step 3: For the companies with a null distance, only one of them represents the group; at this level there remains 126 companies.

Step 4: Construction of the opposition matrix using the equation (2)
- x and y are opposed and do not belong to the same group $\leftrightarrow$ r(x, y)=1
- x and y are not opposed and can belong to the same group $\leftrightarrow$ r(x, y)=0

Step 5: SSIM calculation. The algorithm presented in 1.2.2 is applied.

To the thresholds s_1 = 1.5 and s_2 = 0.25 the processing produces 399 ssims. The appendix 2 and 3 supply an extract of the obtained results.

In appendix 2 the ssims 1 to 10 and 390 to 399 are presented. Appendix 3 indicates the ssims to which the companies 1 to 5 and 149 to 163 belong.

Finally, appendix 4 gives the number of ssims associated to each company.

The analysis produces a core of companies, that is to say companies belonging to several ssims. For the thresholds used, the analysis distinguishes between five different cores. Three of these cores bring together companies having very varying profiles. The last two cores bring together companies whose scores are near the centre point value (6.8) of the Conan Holder function.

Therefore in group 1 ratios 10 to 12 are average, 11 is a weak ratio, 13 is a strong ratio and 14 an average ratio.

For group 2 all the criteria present common characteristics, either weak or average.

The groups 3 and 4 are more homogeneous and they differ from the groups 1 and 2 on criteria 11 and group 5 by the criteria 13.

Lastly group 5 has the following characteristics: 10 and 11 are high ratios, 12 is an average ratio, and 13 and 14 are weak ratios.

A deeper analysis of the companies situation using the characteristic ratios of their financial situation, their activity, their efficiency, and the profitability shows that we are faced with very successful companies (group 5) and in serious difficulty (group 1). The conclusions of the analysis confirm here, without any ambiguity, the information given by the Conan Holder score.

It is not the same for the other groups. In fact, globally the companies belonging to group 2 can be considered as companies whose financial situation seems to have deteriorated. Groups 3 and 4 are composed of companies who are "touch-and-go". When looking at their score it also includes companies in group 2 and companies that are on the limit of groups 4 and 5.

Finally the most discriminating criteria are 11 and 13. Criteria 11 oppose the groups 1 and 2 to the group 3, 4, and 5. Criteria 13 oppose groups 1, 2 and 3 to groups 4 and 5.

CONCLUSION

The preceding application shows that by applying the ssims method to a set of companies being part of the same activity sector it allows to emphasise several sub-groups non apart but whose nucleus has a specific profile. The companies belonging to the intersections allow to predict a possible move from one nucleus to another. Certain characteristics brings them nearer a given group while others, on the other hand, push them away. They play the role of "tangent" companies.

The ssims method can also be applied in other managerial fields, such as the positioning of the products.

For a technical point of view it must be mentioned that :

- contrary to other classification methods (such as neural networks, RTAC, etc.) the results obtained by the ssim method are independent of the data's order.

- a more specific breakdown of the classes would no doubt allow a better definition of the role played by the tangent companies. It can be seen that the increase in the number of classes creates a "hardening" of each nucleus in terms of specific characteristics.

Lastly, future work can be divided in two:

- compare the obtained results by the ssim method to those of other classical or fuzzy classification methods,

- look for more advanced algorithms capable of taking into account a larger amount of data and improve the processing speeds.

REFERENCES

1- Altman E. I. : " Financial ratios, discriminant analysis and the prediction of corporate bankruptcy ", *The Journal of Finance*, Vol XXIII 1968; 589 - 609.
2- Altman E. I., Hadelman R. G., Narayanan P. : " ZETA analysis. A new model to identify bankruptcy risk of corporations ", *Journal of Banking and Finance*, Vol 1. 1977 ; 29 - 54.
3- Bardos M. : " Ratios significatifs et détection du risque. Trois méthodes d'analyse discriminante. " *Cahiers Economiques et Monétaires de la Banque de France*, N° 33. 1986.
4- Bell T., Ribar G., Verchio J. : " Neural nets vs. logistic regression : a comparison of each model's ability to predict commercial bank failure ", *Cash Flow Accounting Conference*, Nice, décembre 1990.
5- Bertin J. : *La graphique et le traitement graphique de l'information*. Flammarion. Paris. 1977.
6- Bonnin S. : *Initiation à la graphique*. Epi Editeurs. Paris. 1975.
7- Coats P., Fant F. : " Recognising financial distress patterns using a neural networks tool ", *Financial Management*, 1992.
8- Conan D., Holder M. : "Variables explicatives de performance et contrôle de gestion dans les P.M.I.", Thèse d'État, CERG, Université Paris Dauphine ; 1979.
9- Couturier A., Fioleau B. : " Une introduction aux techniques de classification à l'aide de réseaux à apprentissage concurrentiel" Actes de la Première Rencontre ANSEG. Saint Nazaire. juin 1994.
10- Dimitras A., Zanakis S., Zopounidis C. : A survey of business failure with an emphasis on prediction methods and industrial applications. European Journal of Operational Researsh 90 1996 ; 487-513.
11- Eisenbeis R.A. : " Pitfalls in the application of discrimant analysis in business, finance, and economics ". *The Journal of Finance*. Volume XXXII. N° 3. 1977 ; 875 - 900.
12- Fioleau B. : "Efficacité et risque d'exploitation. Application aux entreprises du secteur agricole" Thèse de doctorat en Sciences de Gestion. Université de Nantes ; 1992.
13- Gimeno R. : *Apprendre à l'école par la graphique*. Retz. Paris. 1980
14- Hansen J. McDonald J. & Stice J. : " Artificial intelligence and generalised qualitative response models : an empirical test on two audit decision-making domains ", *Decision Science*, Vol 23. 1992. 708-723.
15- Joy O.M. & Toffelson J.O. : " On the financial application of discriminant analysis ". *Journal of Financial and Quantitative Analysis*. Volume 10. N°15. 1975 ; 723 - 739.
16- Krzanowski W.J. : " The performance of Fisher's linear discriminant function under non-optimal conditions ". *Technometrics*. Volume 19. N° 2. 1977.
17- Lev B. : " *Financial statement analysis : A new approach* ". Prentice Hall, Englewoods Cliffs. 1974.
18- Malécot J.F. : " Sait-on vraiment prévoir les faillites d'entreprises ? ". *Economies et Sociétés*. Série SG9, Tome XX, N° 12. 1986.

19- Mutchler J. & Williams D. : " *The relationship between audit technology, clients risk profiles, and the going concern opinion decision* " Working paper, The Ohio State University, Department of Accounting and Information Systems. 1989.
Sakarovitch M. : *Optimisation combinatoire*. Ed. Hermann Orléans France. 1984.
20- Saporta G. : "Introduction à la discrimination. Problématique et méthodes" in "*Analyse discriminante sur variables continues*" Editeur scientifique Gilles Celeux. INRIA Rocquencourt. France. 1990.
22- Varetto F. & Marco G. : " Bankruptcy diagnosis and neural networks ", *International Seminar. European Financial statement data bases : Methods and perspectives*, Bressanone 16-17 september 1993.
23- Zavgren C. : " The prediction of corporate failure : the state of the art ", *Journal of Accounting Litterature*, Vol 2. 1983. 1 - 37.

APPENDIX 1

Conan-Holder discriminant function

$$Z = 16 * \frac{\text{Quick Assets}}{\text{Current Liabilities}} + 22 * \frac{\text{Net Worth + Long Term Liabilities}}{\text{Total Assets}} - 87 * \frac{\text{Interest Expenses}}{\text{Debt}}$$

$$10 * \frac{\text{Wages \& Charges}}{\text{Added Value}} + 24 * \frac{\text{Gross Profit}}{\text{Debts}}$$

Score values	Probability of bankruptcy
- 21.0	100
- 4.8	90
0.2	80
2.6	70
4.7	60
6.8	50
8.7	40
10.7	30
13.1	20
16.4	10

Annexe 2 List of the ssims (abstracts)

[esim 1
ent={1, 20, 27, 28, 32, 47, 49, 85, 96, 97, 140, 142}],
[esim 2
ent={13, 20, 27, 37, 44, 60, 75, 97, 112, 131, 143}],
[esim 3
ent={20, 27, 37, 47, 53, 60, 73, 78, 82, 97, 112, 143}],
[esim 4
ent={1, 20, 27, 28, 36, 47, 49, 85, 96, 97, 140, 142}],
[esim 5
ent={11, 16, 17, 32, 95, 121, 162, 163}],
[esim 6
ent={6, 14, 16, 17, 20, 32, 85, 113}],
[esim 7
ent={1, 20, 27, 28, 32, 49, 53, 84, 96, 97, 140, 142}],
[esim 8
ent={1, 20, 27, 28, 36, 49, 53, 84, 96, 97, 140, 142}],
[esim 9
ent={12, 26, 32, 33, 45, 49, 56, 61, 84, 96, 140, 154}],
[esim 10
ent={1, 28, 32, 45, 47, 49, 53, 61, 96, 97, 140, 142}],

.

.

.

[esim 390,
ent={8, 13, 20, 27, 37, 47, 53, 73, 82, 97, 112, 143}],
[esim 391
ent={25, 52, 92, 116, 139, 152}],
[zsim 392
ent={7, 14, 20, 37, 44, 60, 75, 115, 131, 143}], [393, {1, 127}],
[esim 394
ent={1, 27, 36, 45, 49, 97, 132, 138, 140, 142, 159}],
[esim 395,
ent={17, 20, 32, 131, 143, 163}], [396, {16, 54, 95}],
[esim 397
ent={28, 36, 45, 49, 61, 116, 139, 142}], [398, {9, 48, 55, 78}],
[esim 399
ent={7, 21, 28, 29, 47, 78}]

Annexe 3 List of companies (abstracts)

[ent 1 :
esim={1, 4,7, 8, 10, 11, 14, 16, 17, 18, 19, 21, 23, 24, 25, 26, 28, 29, 30, 31, 33, 35, 37, 38, 58, 66, 68, 69, 72, 78, 81, 83, 84, 85, 105, 109, 110, 113, 114, 121, 126, 129, 132, 142, 161, 169, 171, 184, 208, 227, 265, 283, 288, 289, 290, 299, 300, 302, 306, 315, 318, 320, 321, 326, 328, 332, 341, 373, 378, 385, 388, 393, 394}],

[ent 2 :
esim={39, 47, 51, 56, 57, 59, 60, 65, 88, 91, 93, 143, 151, 155, 160, 178, 181, 182, 183, 189, 194, 200, 203, 217, 218, 222, 223, 236, 249, 251, 262, 273, 275, 297, 316, 345, 352, 353, 355, 360, 361, 371, 376, 381}],

[ent 3
esim={95, 377, 383}],

[ent 4 :
esim={39, 40, 45, 47, 51, 56, 57, 59, 60, 65, 80, 88, 93, 124, 130, 143, 144, 147, 148, 149, 15 «0, 151, 155, 160, 162, 170, 178, 181, 182, 183, 185, 188, 194, 200, 203, 205, 217, 218, 222, 223, 232, 234, 236, 242, 243, 248, 249, 251, 257, 258, 259, 261, 262, 267, 273, 275, 277, 278, 297, 316, 345, 347, 352, 353, 355, 360, 361, 365, 369, 371, 376, 379}],

[ent 5 :
esim={39, 47, 50, 51, 52, 55, 56, 59, 60, 65, 71, 73, 75, 77, 88, 93, 94, 96, 99, 101, 103, 106, 108, 111, 119, 120, 124, 130, 133, 134, 138, 141, 144, 146, 147, 148, 149, 150, 162, 167, 170, 173, 174, 175, 182, 183, 185, 187, 188, 192, 198, 202, 203, 205, 206, 209, 210, 214, 217, 218, 222, 223, 229, 232, 234, 235, 242, 243, 248, 251, 252, 254, 257, 259, 261, 267, 270, 272, 277, 278, 286, 310, 342, 347, 369, 375}],

.

.

.

.

[ent 149 :
esim={325, 344, 358, 365, 379}], [150, {295}], [151, {79, 330, 366}],

[ent 152 :
esim={67, 79, 129, 171, 208, 307, 333, 364, V 367, 383, 391}],

[ent 154 :
esim={9, 15, 16, 24, 30, 53, 58, 70, 83, 89, 98, 110, 114, 117, 136, 142, 263, 282, 285, 288, 292, 299, 326, 330, 341, 358}],

[ent 156 :
esim={275, 352, 357, 376, 381}], [157, {41, 168, 356, 374}],

[ent 159 :
esim={29, 95, 126, 136, 142, 321, 330, 341, 380, 394}],

[ent 160 :
esim={67, 79, 129, 307, 366}], [161, {127, 212}],

[ent 162 :
esim={5, 57, 143, 145, 151, 155, 160, 178, 181, 189, 194, 200, 236, 237, 249, 262, 273, 316, 345, 352, 353, 355, 357, 360, 361, 371, 376, 381}],

[ent 163 :
esim={5, 153, 238, 281, 324, 363, 370, 395}]

Annexe 4 Number of ssims related to each company

[1, 73]	[26, 67]	[55, 13]	[87, 8]	[127, 2]
[2, 44]	[27, 71]	[56, 52]	[89, 17]	[130, 7]
[3, 3]	[28, 110]	[58, 3]	[90, 11]	[131, 16]
[4, 72]	[29, 40]	[59, 2]	[91, 1]	[132, 16]
[5, 86]	[30, 25]	[60, 33]	[92, 1]	[135, 11]
[6, 94]	[32, 123]	[61, 117]	[95, 10]	[137, 17]
[7, 89]	[33, 46]	[62, 12]	[96, 215]	[138, 16]
[8, 22]	[34, 1]	[64, 4]	[97, 127]	[139, 8]
[9, 6]	[36, 54]	[65, 68]	[98, 4]	[140, 110]
[10, 5]	[37, 29]	[66, 121]	[102, 20]	[142, 89]
[11, 15]	[38, 4]	[67, 5]	[104, 32]	[143, 34]
[12, 14]	[39, 8]	[69, 2]	[108, 5]	[144, 3]
[13, 22]	[40, 31]	[70, 56]	[109, 4]	[146, 3]
[14, 55]	[41, 1]	[71, 3]	[110, 25]	[147, 78]
[15, 19]	[42, 11]	[73, 24]	[111, 1]	[149, 5]
[16, 23]	[43, 1]	[75, 21]	[112, 38]	[150, 1]
[17, 96]	[44, 16]	[76, 15]	[113, 17]	[151, 3]
[18, 8]	[45, 80]	[78, 29]	[114, 2]	[152, 11]
[19, 127]	[47, 120]	[80, 2]	[115, 4]	[154, 26]
[20, 105]	[48, 13]	[81, 4]	[116, 23]	[156, 5]
[21, 24]	[49, 110]	[82, 27]	[118, 1]	[157, 4]
[22, 11]	[51, 4]	[83, 5]	[119, 12]	[159, 10]
[23, 30]	[52, 8]	[84, 55]	[121, 13]	[160, 5]
[24, 3]	[53, 78]	[85, 170]	[125, 4]	[161, 2]
[25, 4]	[54, 2]	[86, 13]	[126, 5]	[162, 28]
				[163, 8]

A MULTICRITERIA APPROACH FOR THE ANALYSIS AND PREDICTION OF BUSINESS FAILURE IN GREECE

Constantin Zopounidis [1], Augoustinos I. Dimitras [1] and Loic Le Rudulier [2]

[1] Technical University of Crete
Decision Support Systems Laboratory
University Campus
73100 Chania, Greece

[2] Ecole des Mines de Nancy
Parc de Saurput
54042 Nancy Cedex
France

Abstract: In this paper the multicriteria method ELECTRE TRI is employed to make the discrimination between failed and healthy firms in Greece. An appropriate model was built according to the financial knowledge and past experience. A sample of 30 bankrupt firms matched to a sample of 30 healthy firms is used to evaluate the capability of the method for the prediction of business failure. The results are compared to those derived by a discriminant analysis model. The results using ELECTRE TRI promise satisfactory applications in the domain of financial distress.

Keywords: Business failure, Multicriteria decision aid method, Application

1. INTRODUCTION

The prediction of business failure is a field in which many researchers have been working for the last two decades. As a matter of fact, banks, financial institutions, clients, etc., need such predictions for firms in which they have an interest.

One of the first methods used for the prediction of business failure was multivariate discriminant analysis (DA) proposed by Altman (1968). He proposed a discriminant function with 5 variables for evaluating the risk of business failure. Subsequently, the use of this method has continued to spread to the point where today researchers and practitioners speak of discriminant models of evaluating business failure risk. But, at the same time, the generalization of this method has given rise to numerous studies which criticize it. Eisenbeis (1977) mentioned 7 possible pitfalls in the utilisation of DA: the violation of the distribution assumptions of the variables; inequality in group dispersions; the interpretation of the significance of individual variables; the reduction of dimensionality; the definitions of the groups; the choice of the appropriate a priori probabilities and/or costs of misclassification; the estimation of classification error rates.

Since the study of Altman (1968), several studies proposing other multivariate methods have been used to overcome the disadvantages of the method and to provide higher prediction accuracy. Starting from different views and requirements researchers proposed more sophisticated methods, sometimes already applied to other scientific fields. Among these studies, there are the study of Ohlson (1980) using logit analysis and the study of Zmijewski (1984) using probit analysis. Frydman et al. (1985) first employed the recursive partitioning algorithm to the business failure problem. Mathematical programming methods were used by Gupta et al. (1990). Other methods used were survival analysis by Luoma and Laitinen (1991), expert systems by Messier and Hansen (1988) and neural networks by Altman et al. (1994).

Most of the methods proposed have already been overviewed in the past years, for examination and comparison purposes in some review articles presented. Such reviews were these of Vernimmen's (1978) examining failure models and criticizing their contribution and limits and Scott's (1981) investigating empirical models developed and bankruptcy theories presented mainly on USA studies. Zavgren (1983) surveyed different methods and empirical models proposed for the prediction of corporate failure in USA. Altman in (1984) presented a review of models developed in several countries, Jones (1987) examined the techniques used for bankruptcy prediction in USA and Keasey and Watson (1991) explored the limitations and usefulness of methods used for the prediction of firm financial distress. Dimitras et al. (1996) and Zopounidis (1995) gave a complete review of methods used for the prediction of business failure and of new trends in this area.

However, not only new methods but also new problems affecting the variables involved have surfaced. Up to now, most of the proposed models contain only quantitative variables (financial ratios). But prediction of business failure is also affected by variables of a qualitative character such as quality of management, market trend, market share, social importance, etc. The importance of qualitative variables has been mentioned in several studies like those of Alves (1978), Zopounidis (1987), etc.

To incorporate qualitative variables in the evaluation of business failure risk, multicriteria decision aid methods have been proposed by Andenmatten (1995), Dimitras et. al. (1995), Mareschal and Brans (1991), Zollinger (1982), Zopounidis (1987), and Zopounidis and Doumpos (1997). In addition, these methods allow the decision maker to interact expressing his preferences and past experiences in the building of the failure risk model. The aim of this study is to test the ability of the multicriteria decision aid method ELECTRE TRI, presented by Yu (1992) in predicting business failure, and to compare it with discriminant analysis. Section 2 presents the basic concepts of ELECTRE TRI method. The application of ELECTRE TRI on a sample of Greek firms and the comparison of its results with those obtained with discriminant analysis are presented in section 3. In the concluding remarks, the merits of the proposed multicriteria method are discussed and possible new trends in the field of business failure analysis are given.

2. THE ELECTRE TRI METHOD

The ELECTRE TRI method belongs to the ELECTRE family of multicriteria methods Roy (1991). The particularity of the ELECTRE family (and of the French school) is to refuse the possibility of total compensation between the alternatives performances on the criteria, and then to accept incomparability and intransitivity.

ELECTRE TRI is a multicriteria method specially conceived for sorting problems. Other multicriteria methods conceived for sorting problems have been presented by Massaglia and Ostanello (1991), Roy (1981), and Roy and Moscarola (1977). From a finite set of alternatives evaluated by quantitative and/or qualitative criteria and from a set of categories corresponding to predefined recommendations or norms, ELECTRE TRI proposes two different classification procedures that allow the grouping of alternatives in the prescribed categories. The categories are conceived independently of the set of alternatives and ELECTRE TRI deals with ordered categories (complete order). These categories are defined by some reference alternatives or reference profiles which are themselves defined by their values on the criteria.

Following this, we can define the categories C^i , i = 1,..., k, where C^1 is the worst category and C^k the best one. We can also define the profiles r^i, i = 1, ..., k-1, where r^1 is the lower profile and r^{k-1} the upper. Then the profile r^i is the theoretical limit between the categories C^i and C^{i+1} and r^i is strictly better than r^{i-1} for each criterion.

In ELECTRE TRI, the information asked from the decision maker about his preferences takes the form, for each criterion and each profile, of a relative weight and an indifference, preference, and veto thresholds. Concerning classification, ELECTRE TRI compares the alternatives with the profiles using the classical concepts of concordance index, discordance index and valued outranking relation as ELECTRE III method (cf. Roy and Bouyssou, 1977). Between an alternative *a* and a profile r^i, the concordance index $c_j(a,r^i)$ expresses the strength of the affirmation "alternative *a* is at least as good as profile r^i on criterion j", and for an increasing criterion j is calculated in the following way :

if $g_j(a) \leq g_j(r^i)-p_j(r^i)$, then $c_j(a,r^i) = 0$

if $g_j(r^i) - p_j(r^i) < g_j(a) \leq g_j(r^i)-q_j(r^i)$, then $0 < c_j(a,r^i) \leq 1$ [1]

if $g_j(a) > g_j(r^i)-q_j(r^i)$, then $c_j(a,r^i) = 1$

where $p(r^i)$ and $q(r^i)$ are the preference and the indifference thresholds for criterion and profile respectively. These discrimination thresholds are used in order to take into account the imprecision and/or the uncertainty of the data (criteria evaluations and decision maker's preferences).

A global concordance index $C(a,r^i)$ for the affirmation "a is at least as good as r^i for all the criteria" is then constructed in the following way :

$$C(a,r^i) = \frac{\sum_{j=1}^{n} k_j \cdot c_j(a,r^i)}{\sum_{j=1}^{n} k_j}$$

where k_j is the weight of the criterion j.

The discordance index $D_j(a,r^i)$ expresses the opposition to "a is at least as good as r^i on criterion j" and is calculated in the following way :

if $g_j(a) > g_j(r^i)-p_j(r^i)$, then $D_j(a,r^i) = 0$

if $g_j(r^i)-v_j(r^i) < g_j(a) \leq g_j(r^i)-p_j(r^i)$, then $0 < D_j(a,r^i) \leq 1$ [2]

if $g_j(a) \leq g_j(r^i)-v_j(r^i)$, then $D_j(a,r^i) = 1$

where $v_j(r^i)$ is the veto threshold for the criterion j and the profile r^i.

A credibility degree $\sigma_s(a,r^i)$ for the affirmation "a outranks r^i" is calculated in the following way:

if $F(a,r^i) = \{ j \in F \,/\, D_j(a,r^i) > C(a,r^i) \} = \varnothing$, $\sigma_s(a,r^i) = C(a,r^i)$

if $F(a,r^i) \neq \varnothing$, then $\sigma_s(a,r^i) = C(a,r^i) \cdot \prod_{j \in F} \frac{1 - D_j(a,r^i)}{1 - C(a,r^i)}$

where F is the set of the criteria.

This valued outranking relation $\sigma_s(a,r^i)$ is transformed into a "net" outranking relation as follows :

if $\sigma_s(a,r^i) \geq \lambda$, then a S r^i,

where S represents the outranking relation and λ ($1/2 \leq \lambda \leq 1$) is a "cut level" above which the proposition : "a outranks r^i" is valid.

Then, preference (**P**), indifference (**I**) and incomparability (**R**) are defined in the following way:

---

[1] $c_j(a, r^i)$ is obtained by linear interpolation : $c_j(a,r^i) = \frac{p_j(r^i) - [g_j(r^i) - g_j(a)]}{p_j(r^i) - q_j(r^i)}$

[2] $D_j(a, r^i)$ is obtained by linear interpolation : $D_j(a,r^i) = \frac{g_j(r^i) - g_j(a) - g_j(r^i)}{v_j(r^i) - p_j(r^i)}$

a **I** r^i	means	a S r^i	and	r^i S a
a **P** r^i	means	a S r^i	and	no r^i S a
r^i **P** a	means	no a S r^i	and	r^i S a
a **R** r^i	means	no a S r^i	and	no r^i S a

Note that, if for a criterion j the difference $g_j(a)$-$g_j(r^i)$ [or $g_j(r^i)$-$g_j(a)$] is superior or equal to the value of the veto threshold, then this criterion puts its veto making impossible to state a S r^i (as well as r^i S a).

In ELECTRE TRI, there are two non total compensation procedures (the pessimistic and the optimistic one), so as to assign each alternative into one category among a set of categories defined in advance. In general, the pessimistic procedure is applied when a policy of prudence is necessary or when the available means are very constraining. While the optimistic procedure is applied for problems where the decision maker desires to favour the alternatives that present some particular interests or some exceptional qualities.

In the sorting procedure, firm a is compared at first to the worst profile r^1 and in the case where a **P** r^1 , a is compared to the second profile r^2, etc., until one of thc following situations appears :

- if a **P** r^i and r^{i+1} **P** a or a **I** r^{i+1}, then a is assigned to category i+1 for both pessimistic and optimistic procedures,
- if a **P** r^i and a **R** r^{i+1}, a **R** r^{i+2},..., a **R** r^{i+k}, r^{i+k+1} **P** a, then a is assigned to category i+1 with pessimistic procedure and to category i+k+1 with optimistic procedure.

When the value of λ gradually decreases the pessimistic procedure becomes less constrained than the conjunctive procedure. In this case, it is not necessary that all criteria outrank the profile r^i, but one is satisfied when the majority of criteria outrank this profile. In a similar way, the optimistic procedure becomes more relaxed than the disjunctive procedure. In this case, for an assignment, it is necessary to have not only one criterion which outranks the profile r^i, but a majority rule combined with a mechanism of veto which justify the denial of $r^i > a$ (cf. Roy and Bouyssou, 1993). When the value of λ is equal to 1 the pessimistic and optimistic procedures are identical with conjunctive and disjunctive procedures respectively.

ELECTRE TRI manages incomparability in such a way that it will point out the alternatives that have particularities in their evaluations. In cases where some alternatives are incomparable with one or more reference profiles then they are assigned to different categories by optimistic and pessimistic procedures. This is due to the fact that these alternatives have good values for some criteria and, simultaneously, bad values for other criteria; moreover these particular alternatives must be examined with attention. In this way the notion of incomparability included in the ELECTRE TRI method brings an important information to the decision maker.

3. APPLICATION

In this section, we describe at first the sample and data of the study and, then, the obtained results.

3.1. Sample and data

The sample of firms consisted of 60 industrial firms, named by a1, a2, ..., a60. Firms from a1 to a30 were bankrupted according to the Greek law during years 1985 to 1990. Although the year of bankruptcy is not common for all firms in the sample, they are all considered to be failed in the "zero" year, taken as year of reference. The healthy firms (firms from a31 to a60) were matched to the failed ones according to industry and size (measured by total assets and number of employees). Therefore, two categories were defined to receive these firms :

C^1 : High risk group (failed firms) and

C^2 : Low risk group (non failed or healthy firms).

For each firm data from the balance sheet and income statement were collected for the three years prior to actual failure of the bankrupted firms. No qualitative characteristics were employed because of problems in the collection of them for bankrupted firms in Greece. This sample is considered as an estimation sample. A second sample of 24 firms named was collected in the same way. This second sample was used as a holdout sample to verify the predictive ability of the models provided.

From an initial set of 18 financial ratios calculated, seven of them have been selected, to be employed in the models, using techniques such as principal components analysis, F-test, graphical representation and available financial knowledge (cf. Le Rudulier, 1994). Maybe the proper way to select the criteria would be the use of the preferences of a decision maker (financial analyst) on the available criteria. The selected financial ratios were:

g_1 = Gross profit / Total assets

g_2 = Net income / Total debts

g_3 = Current assets / Short term debts

g_4 = (Current assets - Inventories) / Short term debts

g_5 = Working capital / Total assets

g_6 = Working capital / Currents Assets

g_7 = Total debts / Total assets

The first two criteria are profitability ratios, while the next ones are solvency ratios (liquidity, debt ratios, ...). All the above financial ratios are to be maximized with the exception of g_7 which is to be minimized. This means that the lower the value of g_7 the greater the performance of the firm on this ratio and subsequently the greater chance has the firm to be ranked in the low risk group (category C^2).

3.2. Results

For the application of ELECTRE TRI[3] the profile and the relative thresholds of preference (p), indifference (q) and veto (v) on each criterion were defined by the graphical representation and the previous experience and financial knowledge. The weights (k) for the criteria were taken all equal to 1 for two principal reasons: (1) The seven criteria (g_1, ... g_7) were derived by the principal components analysis and are regarded as more important than the initial set of 18 ratios; (2) in the absence of a real decision maker (financial analyst or credit analyst), it is very difficult to express a preference for a given ratio; moreover, these ratios are considered the most important in their category (i.e. g_1 and g_2 are profitability ratios; g_3, g_4, g_5, g_6 are liquidity ratios; g_7 is debt capacity ratio). For criteria g_1, g_3, g_4, g_5, g_6 the veto threshold was set at the maximum value on the criterion, because of difficulties in definition. Whatever, the conclusions about the ability of this method have to be related to the application to a particular sample for a particular period. The profile r^1 and the relative thresholds are presented in Table 1. This profile has been defined based on widely accepted limits and/or the limits that came out of experience and knowledge of the financial literature. For example, for the criterion g_7 (debt capacity) the value of 80% was determined. For the Greek case, firms with a capacity of debt less than this value are considered to be rather "good". In other case, firms with a capacity of debt superior to this limit are rather "bad". The thresholds are used in order to take into account the imprecision and/or the uncertainty of the data (criteria's evaluations and decision maker's preferences). At this level of analysis, it is necessary to remark that the values of the profile r^1 and the values of the thresholds were also determined by "interactive" use of the software ELECTRE TRI, in order to minimize the "false" assignments. Thus, one observes the dynamic character of the method in the assessment of the sorting model.

Table 1: Profile r^1 and relative thresholds

Criteria	g_1	g_2	g_3	g_4	g_5	g_6	g_7
Profile	20	1	100	60	5	30	80
k	1	1	1	1	1	1	1
q	1	0.05	5	3	0.25	1	1
p	2	0.1	10	6	0.5	3	2
v	max	1	max	max	max	max	15

Setting λ to the value 0.67, the resulted grouping of firms for the optimistic and the pessimistic procedures are presented in the Tables 2 and 3 respectively, where the misclassified firms are in bold. There exist two types of errors: Type I

[3] The authors are indebted to Professor B. Roy for providing the ELECTRE TRI software.

and Type II. The Type I error occurs when a failed firm is classified as healthy while Type II error occurs when a healthy firm is classified to the bankrupt group. For a decision maker the Type I error is the most severe and it should be eliminated as possible. Type II errors results to an opportunity cost for the decision maker. The error rates were calculated and they are presented in Tables 4 and 5 for the optimistic and the pessimistic procedures respectively.

Table 2: Grouping of firms by pessimistic procedure

Group	Firms
C^1	a1 a2 a3 a4 a5 a6 a7 a8 a9 a10 a12 a13 a14 a15 a16 a17 a18 a19 a20 a21 a22 a23 a24 a26 a27 a28 a29 a30 ***a43 a48 a50 a59***
C^2	***a11 a25*** a31 a32 a33 a34 a35 a36 a37 a38 a39 a40 a41 a42 a44 a45 a46 a47 a49 a51 a52 a53 a54 a55 a56 a57 a58 a60

Table 3: Grouping of firms by optimistic procedure

Group	Firms
C^1	a2 a3 a4 a5 a6 a7 a9 a10 a13 a18 a19 a20 a21 a22 a23 a24 a26 a27 a28 a29 a30 ***a43***
C^2	***a1 a8 a11 a12 a14 a15 a16 a17 a25*** a31 a32 a33 a34 a35 a36 a37 a38 a39 a40 a41 a42 a44 a45 a46 a47 a48 a49 a50 a51 a52 a53 a54 a55 a56 a57 a58 a59 a60

Table 4: Misclassification analysis of pessimistic procedure

Type of error	Number of firms	Percentage
Type I	2	6.67 %
Type II	4	13.33 %
Total	6	10.00 %

Table 5: Misclassification analysis of optimistic procedure

Type of error	Number of firms	Percentage
Type I	9	30.00 %
Type II	1	3.33 %
Total	10	16.66 %

In general, misclassifications provided by optimistic procedure ELECTRE TRI resulted from an overestimation of firms' performances. A reduction in misclassification by ELECTRE TRI pessimistic procedure can be remarked. The stability analysis of the model by testing slightly different values for r^1 and the thresholds showed that these results are rather stable.

To reduce the error rates, a third category, named C^3, has been considered. In this group are classified firms for which ranking results between pessimistic and optimistic are different (those firms that, in fact, are incomparable with the profile). This group is considered as "uncertain group" and firms classified in it are considered as firms to be studied further (cf. also Zopounidis, 1987). The three classification groups of the firms presented in Tables 6 and 7 provide the relative analysis of success in classification.

Table 6: Three groups classification of firms by ELECTRE TRI

Group	Firms
C^1	a2 a3 a4 a5 a6 a7 a9 a10 a13 a18 a19 a20 a21 a22 a23 a24 a26 a27 a28 a29 a30 ***a43***
C^2	***a11 a25*** a31 a32 a33 a34 a35 a36 a37 a38 a39 a40 a41 a42 a44 a45 a46 a47 a49 a51 a52 a53 a54 a55 a56 a57 a58 a60
C^3	a1 a8 a12 a14 a15 a16 a17 a48 a50 a59

Table 7: Analysis of the three classification groups provided by ELECTRE TRI

Type of classification	Number of firms	Percentage
Correct classification	47	78.33 %
Type I error	2	6.67 %
Type II error	1	3.33 %
Firms to be studied further	10	16.67 %

Although ELECTRE TRI is not a classical data analysis method, in this application we attempted to verify its discriminant power on firms data of two and three years before failure. The obtained total error rates are summarized in Table 8. There is a clear reduction to the total error rates making the three groups classification more attractive and accurate for the prediction of business failure.

Table 8: Total error of ELECTRE TRI method

	Classification error		
Classification procedure	**year-1**	**year-2**	**year-3**
ELECTRE TRI pessimistic	10.00 %	21.67 %	23.33 %
ELECTRE TRI optimistic	16.67 %	21.67 %	21.67 %
ELECTRE TRI (3 categories)	5.00 %	6.67 %	6.67 %

To test the predictive ability of the model the ELECTRE TRI method was also applied to the holdout sample. The classification accuracy provided is presented in Table 9.

Table 9: Misclassification of ELECTRE TRI grouping on the holdout sample

Type of classification	Number of firms	Percentage
Correct classification	17	70.83 %
Type I error	0	0.00 %
Type II error	1	8.33 %
Firms to be studied further	6	25.00 %

It is important to note that the percentage of misclassifications is approximately the same as the one obtained with the first sample. On the other hand the percentage of firms to be studied further increased slightly. This fact is natural and somehow expected because the method is applied on a new "unknown" sample of firms. The results show that the preferential model is a quite general model for the assessment of failure risk for firms under the same properties as those defined previously and the multicriteria methodology seems to be able to be used for bankruptcy prediction in Greece.

3.3. Comparison between ELECTRE TRI and Discriminant Analysis

The philosophy of the multicriteria method ELECTRE TRI is much different than the one of DA which is a statistical method. ELECTRE TRI works in real time, interacts with the decision maker incorporating his judgements in the model and helps the decision maker to learn about his preferences (see Roy and Bouyssou, 1993). Although DA is much different than ELECTRE TRI, just for comparison reasons a discriminant analysis model was constructed on the data of the basic sample one year prior to bankruptcy, using the 7 ratios selected previously. This model was applied on the data of the two and three years prior to actual failure. Table 10 shows the misclassification analysis of the DA model.

Table 10: Grouping of firms by Discriminant Analysis

Type of error	year-1	year-2	year-3
Type I	33.33 %	46.66 %	43.33 %
Type II	3.33 %	3.33 %	6.66 %
Total	18.33 %	25.00 %	25.00 %

By considering the ELECTRE TRI model results (Table 8) and by comparing it with DA, we can remark that the ELECTRE TRI method gives much better results, particularly for year-2 and year-3. Moreover, most of the firms misclassified by DA are proposed to be studied further by ELECTRE TRI.

As a matter of fact, discriminant analysis does not have the possibility to propose a further study for uncertain firms, and is obliged to classify those firms in one of the two categories, increasing the misclassifications.

The ELECTRE TRI model is able to predict the bankruptcy of a firm with a low percentage of error, even three years before it will happen. Of course, the percentage of uncertain firms is important when we are far from the reference year (year of actual failure).

4. CONCLUDING REMARKS

In this study, the multicriteria decision aid method ELECTRE TRI, is proposed for the prediction of business failure in Greece. This method, especially conceived for sorting out problems, adapts well to the problem of failure prediction.

The results of the application on a sample of industrial Greek firms confirm the ability of the method to classify the firms in three classes of risk (failure / non failure / uncertain), providing a satisfactory degree of accuracy.

Compared to other previous methods, ELECTRE TRI has several advantages:

1. It accepts incomparability, providing an important information to the decision maker for the uncertainty in the classification of some firms;
2. It accepts qualitative criteria (cf. Dimitras et al., 1995);
3. It can contribute in the minimization of the time and costs of the decision making process (ELECTRE TRI is an information processing system in real time);
4. It offers transparency in the firms' grouping, allowing for argument in the decisions.
5. It takes into account the preferences of the decision-maker (cf. Malecot, 1986).

The approach with DA is totally different than the ELECTRE TRI. With DA, the model is constructed once and it is used without any changes, while with ELECTRE TRI, the model is constructed taking into account the preferences of the decision maker and it can be modified in real time if the preferences of the decision-maker change or if new information is provided by the environment. Finally, ELECTRE TRI can be considered to be an effective operational tool for the prediction of business failure. It can be incorporated in the models' base of multicriteria decision support systems as those proposed by Siskos et al. (1994) and Zopounidis et al. (1992) and Zopounidis et al. (1995).

REFERENCES

Altman, E.I. (1968), "Financial ratios, discriminant analysis and the prediction of corporate bankruptcy", *The Journal of Finance 23*, 589-609.

Altman, E.I. (1984), "The success of business failure prediction models: An international survey", *Journal of Banking and Finance*, Vol. 8, No 2, 171-198.

Altman, E.I., Marco, G. and Varetto, F., (1994), "Corporate distress diagnosis : Comparisons using linear discriminant analysis and neural networks (the Italian experience)", *Journal of Banking and Finance*, Vol. 18, 505-529.
Alves, J.R., (1978), "The prediction of small business failure utilizing financial and nonfinancial data", Unpublished doctoral dissertation, University of Massachusetts., Massachussetts, U.S.A.
Andenmatten, A. (1995), *Evaluation de risque de défaillance des émetteurs d'obligations : Une approche par l'aide multicritère à la décision*, Lausanne, Presses Polytechniques et Universitaires Romandes.
Dimitras, A.I., Zanakis, S.H. and Zopounidis, C. (1996), "A survey of business failures with an emphasis on prediction methods and industrial applications", *European Journal of Operational Research* 90, 487-513.
Dimitras, A., Zopounidis, C. and Hurson, Ch. (1995), "A multicriteria decision aid method for the assessment of business failure risk", *Foundations of Computing and Decision Sciences*, Vol. 20., No 2, 99-112.
Eisenbeis, R.A. (1977), "Pitfalls in the applications discriminant analysis in business and economics", *The Journal of Finance 32*, 875-900.
Frydman, H., Altman, E.I. and Kao, D-L, (1985), "Introducing recursive partitioning for financial classification : The case of financial distress", *The Journal of Finance, Vol. XL, No 1*, 269-291.
Gupta, Y.P., Rao, R.P. and Bagghi, P.K. (1990), "Linear goal programming as an alternative to multivariate discriminant analysis : A Note", *Journal of Business Finance and Accounting*, 17 (4), 593-598.
Jones, F.L. (1987), "Current techniques in bankruptcy prediction", *Journal of Accounting Literature*, Vol. 6, 131-164.
Keasey, K. and Watson, R. (1991), "Financial distress prediction models : A review of their usefulness" *British Journal of Management*, Vol. 2, 89-102.
Le Rudulier, L. (1994), "L' approche multicritère en prévision de la faillite par la méthode ELECTRE TRI", Diplôme de fin d'Etudes, Technical University of Crete, Chania, Greece.
Luoma, M. and Laitinen, E.K., (1991), "Survival analysis as a tool for company failure prediction", *Omega*, Vol. 19, No 6, 673-678.
Malecot, J.-F. (1986), "Sait-on vraiment prévoir les défaillances d'entreprises ?", *ISMEA, Revue Sciences de Gestion* 9, 55-82.
Mareschal, B. and Brans, J.P. (1991), "BANKADVISER: An industrial evaluation system", *European Journal of Operational Research* 54, 318-324.
Massaglia, M. and Ostanello, A. (1991), "N-TOMIC: A decision support for multicriteria segmentation problems", in: P. Korhonen (ed.), *International Workshop on Multicriteria Decision Support*, Lecture Notes in Economics and Mathematical Systems 356, Berlin, Springer-Verlag, 167-174.
Messier, W.F. and Hansen, J.V. (1988), "Including rules for expert system development: An example using default and bankruptcy data", *Management Science*, Vol. 34, No 12, 1403-1415.
Ohlson, J.A. (1980), "Financial ratios and the probabilistic prediction of bankruptcy", *Journal of Accounting Research*, Spring, 109-131.
Roy, B. (1981), "A multicriteria analysis for trichotomic segmentation problems", in : P. Nijkamp and J. Spronk (eds), *Operational Methods*, Gower Press, 245-257.
Roy, B. (1991), "The outranking approach and the foundations of ELECTRE methods", *Theory and Decision* 31, 49-73.
Roy, B. and Bouyssou, D., (1993), *Aide multicritère à la décision : Méthodes et cas*, Paris, Economica.
Roy, B. et Moscarola, J. (1977), "Procédure automatique d'examen de dossiers fondée sur une segmentation trichotomique en présence de critères multiples", *RAIRO Recherche Opérationnelle*, Vol. 11, No 2, 145-173.
Scott, J. (1981), "The probability of bankruptcy: A comparison of empirical predictions and theoretical models", *Journal of Banking and Finance 5*, 317-344.
Siskos, Y., Zopounidis, C. and Pouliezos, A. (1994), "An integrated DSS for financing firms by an industrial development bank in Greece", *Decision Support Systems* 12, 151-168.
Vernimmen, P. (1978), "Panorama des recherches portant sur le risque du créancier", *Analyse Financière 1*, 54-61.
Yu, W. (1992), "ELECTRE TRI : Aspects méthodologiques et manuel d'utilisation", Document du LAMSADE no 74, Université de Paris-Dauphine, Paris, 100p.
Zavgren, C.V. (1983), "The prediction of corporate failure: The state of the art", *Journal of Financial Literature*, Vol. 2, 1-37.
Zmijewski, M.E. (1984), "Methodological issues related to the estimation of financial distress prediction models", *Studies on current Econometric Issues in Accounting Research*, 59-82.

Zollinger, M. (1982), "L' Analyse multicritère et le risque de crèdit aux enterprises", *Revue Française de Gestion*, 56-66.
Zopounidis, C. (1987), "A multicriteria decision making methodology for the evaluation of the risk of failure and an application", *Foundations of Control Engineering*, Vol. 12, No. 1, 45-67.
Zopounidis, C. (1995), *Evaluation du risque de défaillance de l'entreprise : Méthodes et cas d'application*, Paris, Economica .
Zopounidis, C., and Doumpos, M. (1997), "Preference disaggregation methodology in segmentation problems: The case of financial distress", *Working Paper* 97-01, Decision Support Systems Laboratory, Technical Univerisity of Crete.
Zopounidis, C., Godefroid, M. and Hurson, Ch. (1995), "Designing a multicriteria DSS for portfolio selection and management", in J. Janssen, C.H. Skiadas and C. Zopounidis (eds), *Advances in Stochastic Modelling and Data Analysis*, Dordrecht, Kluwer Academic Publishers, 261-292.
Zopounidis, C., Pouliezos, A. and Yannacopoulos, D. (1992), "Designing a DSS for the assessment of company performance and viability", *Computer Science in Economics and Management* 5, 41-56.

A NEW ROUGH SET APPROACH TO EVALUATION OF BANKRUPTCY RISK

Salvatore Greco[1], Benedetto Matarazzo[1], Roman Slowinski[2]

[1] Faculty of Economics, University of Catania, Corso Italia, 55, 95129 Catania, Italy
[2] Institute of Computing Science, Poznan University of Technology, 60-965 Poznan, Poland

Abstract: We present a new rough set method for evaluation of bankruptcy risk. This approach is based on approximations of a given partition of a set of firms into pre-defined and ordered categories of risk by means of dominance relations, instead of indiscernibility relations. This type of approximations enables us to take into account the ordinal properties of considered evaluation criteria. The new approach maintains the best properties of the original rough set analysis: it analyses only facts hidden in data, without requiring any additional information, and possible inconsistencies are not corrected. Moreover, the results obtained in terms of sorting rules are more understandable for the user than the rules obtained by the original approach, due to the possibility of dealing with ordered domains of criteria instead of non-ordered domains of attributes. The rules based on dominance are also better adapted to sort new actions than the rules based on indiscernibility. One real application illustrates the new approach and shows its advantages with respect to the original rough set analysis.

Keywords: Bankruptcy risk evaluation, Rough set approach, Approximation by dominance relations, Decision rules.

1 Introduction

Various methods have been proposed in the specialised literature for evaluation of the bankruptcy risk. According to Dimitras, Zanakis and Zopounidis (1995), the set of existing methods includes: univariate statistical methods, survival methods, discriminant analysis, linear probability model, logit and probit analysis, recursive partitioning algorithm, mathematical programming, multicriteria decision aid/support methods, expert systems.

Recently, a new method based on the *rough set approach* has been proposed for evaluation of bankruptcy risk (Slowinski and Zopounidis, 1995). The concept of rough set, introduced by Pawlak (1982), proved to be an effective tool for the analysis of an information table (financial information table) describing a set of objects (firms) by a set of multi-valued attributes (financial ratios and qualitative variables).

The major *results* obtained from the rough set approach are twofold:

- evaluation of the relevance of the considered attributes (criteria);

- generation of a set of decision rules from the information table in view of explaining a decision policy of the expert.

The main *advantages* of the rough set approach are the following:

- the set of decision rules derived by the rough set approach gives a generalised description of knowledge contained in the financial information table, eliminating any redundancy typical for original data;

- the rough set analysis is based on the original data only and it does not need any additional information, like probability in statistics or grade of membership in fuzzy set theory (for a thorough comparison of the rough set theory with discriminant analysis, fuzzy set theory and evidence theory see Krusinska, Slowinski and Stefanowski (1992), Dubois and Prade (1992) and Skowron and Grzymala-Busse (1993));

- rough set approach is a tool specifically suitable for analysing not only quantitative attributes but also qualitative ones (for a discussion about the importance of qualitative attributes in bankruptcy evaluation see Zopounidis (1987), Shaw and Gentry (1988), Peel, Peel and Pope (1986));

- the decision rules obtained from the rough set approach are based on facts, because each decision rule is supported by a set of real examples;

- the results of the rough set approach are easily understandable, while the results from other methods (credit scoring, utility function, outranking relation) need an interpretation of some technical parameters, with which the user is generally not familiar (for a quite extensive discussion on this subject see Roy, 1993).

As pointed out by Greco, Matarazzo and Slowinski (1996), the original rough set approach, however, does not consider the attributes with ordered domains. Nevertheless, in many real problems the *ordering properties* of the considered attributes may play an important role. In bankruptcy evaluation this problem occurs too. E.g. if firm A has a low value of the indebtment ratio (Total debt/Total assets) and firm B has a large value of the same ratio, within the original rough set approach, the two firms are discernible, but no preference is established between them two with respect to the attribute "indebtment ratio". Instead, from a decisional point of view, it would be better to consider firm A as preferred to firm B, and not simply "discernible", with respect to the attribute in question.

Motivated by the previous considerations, we propose a new approach to evaluation of bankruptcy risk based on the rough set philosophy. Similarly to the original rough set analysis, the proposed approach is based on approximations of a partition of the firms into some pre-defined risk categories analysing data from the financial information table. However, differently from the original rough set approach, the approximations are built using dominance relations instead of indiscernibility relations. This enables us to take into account the ordering properties of considered attributes.

The paper is organised in the following way. In the next section, basic ideas of the rough set theory are recalled. In section 3, the main concepts of the rough approximation by dominance relations are introduced. Then, in section 4, the rough set analysis by indiscernibility relations and by dominance relations are compared with respect to a real problem of bankruptcy evaluation. Final section groups conclusions.

2 Introductory remarks about the rough set theory

2.1 The general idea

The rough set concept proposed by Pawlak (1982, 1991) is founded on the assumption that with every object of the universe of discourse there is associated some information (data, knowledge). For example, if objects are firms submitted to a bankruptcy evaluation, their financial, economic and technical characteristics form information (description) about the firms. Objects characterised by the same description are indiscernible (similar) in view of available information about them. The *indiscernibility relation* generated in this way is the mathematical basis of the rough set theory.

Any set of indiscernible objects is called elementary set and forms a basic granule of knowledge (atom) about the universe. Any subset Y of the universe can either be expressed precisely in terms of the granules or roughly only. In the latter case, subset Y can be characterised by two ordinary sets, called *lower and upper approximations*. The two approximations define the *rough set*. The lower approximation of Y consists of all elementary sets included in Y, whereas the upper approximation of Y consists of all elementary sets having a non-empty intersection with Y. Obviously, the difference between the upper and the lower approximation constitutes the *boundary region* including objects which cannot be properly classified as belonging or not to Y, using the available information. Cardinality of the boundary region says, moreover, how exactly we can describe Y in terms of available data.

2.2 Information table

For algorithmic reasons, knowledge about objects will be represented in the form of an information table. The rows of the table are labelled by *objects*, whereas columns are labelled by *attributes* and entries of the table are *attribute-values*. In general, the notion of attribute differs from that of criterion, because the domain (scale) of a criterion has to be ordered according to a decreasing or increasing preference, while the domain of the attribute does not have to be ordered. We will use the notion of criterion only when the preferential ordering of the attribute domain will be important in a given context. Formally, by an *information table* we understand the 4-tuple S=<U,Q,V,f>, where U is a finite set of objects, Q is a finite set of *attributes*, $V = \bigcup_{q \in Q} V_q$ and V_q is a domain of the attribute q, and f:UQ→V is a total function such that $f(x,q) \in V_q$ for every q∈Q, x∈U, called an *information function* (cf. Pawlak , 1991).

The concepts of reduct and core are important in the rough set analysis of an information table. A *reduct* consists of a minimal subset of independent attributes ensuring the same quality of sorting as the whole set. There can be more than one reduct. The intersection of all the reducts is the *core*. It represents a collection of the most important attributes, i.e. the set of all the attributes which can not be eliminated without decreasing the quality of sorting.

2.3 Decision rules

An information table can be seen as *decision table* assuming the set of attributes $Q=C \cup D$ and $C \cap D = \emptyset$, where set C contains so called *condition attributes*, and D, *decision attributes*.

From the decision table, a set of *decision rules* can be derived and expressed as logical statements "if ...then ..." relating condition and decision attribute values. The decision rules are *exact* or *approximate* depending whether the condition attribute values correspond to unique decision attribute value or not. Different procedures for derivation of decision rules have been presented (e.g. by Slowinski and Stefanowski, 1992, Grzymala-Busse, 1992, Skowron, 1993, Mienko, Stefanowski, Toumi and Vanderpooten, 1996, Ziarko, Golan and Edwards, 1993).

3 Rough approximation by dominance relations

3.1 Basic concepts

$\forall q \in C$ let S_q be an outranking relation (Roy 1985) on U with respect to attribute q such that xS_qy means "x is at least as good as y with respect to attribute q". We suppose that S_q is a total preorder, i.e. a strongly complete and transitive binary relation, defined on V_q. Furthermore let $\mathbf{Cl}=\{Cl_t, t \in T\}$, T={1,...,n}, be a set of classes of U, such that each x∈U belongs to one and only one $Cl_t \in \mathbf{Cl}$.

We suppose that ∀r,s∈T such that r>s the elements of Cl_r are preferred (strictly or weakly (Roy, 1985)) to the elements of Cl_s. More formally, if S is a comprehensive outranking relation on U, i.e. if ∀x,y∈U xSy means "x is at least as good as y", we suppose

$$[x \in Cl_r,\ y \in Cl_s,\ r>s] \Rightarrow [xSy \text{ and not } ySx].$$

The following sets are also considered:

$$Cl_t^{\geq} = \bigcup_{s \geq t} Cl_s,$$

$$Cl_t^{\leq} = \bigcup_{s \leq t} Cl_s .$$

Let us remark that $Cl_1^{\geq} = Cl_n^{\leq} = U$, $Cl_n^{\geq} = Cl_n$ and $Cl_1^{\leq} = Cl_1$.
We say that x dominates y with respect to $P \subseteq C$, denoted xD_Py, if $x\, S_q\, y \;\; \forall q \in P$.
Given $P \subseteq C$ and $x \in U$ let be

$$D_P^+(x) = \{y \in U : yD_Px\},$$
$$D_P^-(x) = \{y \in U : xD_Py\}.$$

$\forall t \in T$ and $\forall P \subseteq C$ we define the lower approximation of $Cl_t^{\geq}$ with respect to P, denoted by $\underline{P}Cl_t^{\geq}$, and the upper approximation of $Cl_t^{\geq}$ with respect to P, denoted by $\overline{P}Cl_t^{\geq}$, as:

$$\underline{P}Cl_t^{\geq} = \{x \in U: D_P^+(x) \subseteq Cl_t^{\geq}\},$$
$$\overline{P}Cl_t^{\geq} = \bigcup_{x \in Cl_t^{\geq}} D_P^+(x).$$

Analogously, $\forall t \in T$ and $\forall P \subseteq C$ we define the lower approximation of $Cl_t^{\leq}$ with respect to P, denoted by $\underline{P}Cl_t^{\leq}$, and the upper approximation of $Cl_t^{\leq}$ with respect to P, denoted by $\overline{P}Cl_t^{\leq}$, as:

$$\underline{P}Cl_t^{\leq} = \{x \in U : D_P^-(x) \subseteq Cl_t^{\leq}\},$$
$$\overline{P}Cl_t^{\leq} = \bigcup_{x \in Cl_t^{\leq}} D_P^-(x).$$

Since the dominance relation D_P is reflexive, our definitions of lower and upper approximation are the same as proposed by Slowinski and Vanderpooten (1995,1996) with respect to approximations by similarity relations.
The P-boundary (doubtful region) of $Cl_t^{\geq}$ and $Cl_t^{\leq}$ are respectively defined as:

$$Bn_P(Cl_t^{\geq}) = \overline{P}Cl_t^{\geq} - \underline{P}Cl_t^{\geq},$$
$$Bn_P(Cl_t^{\leq}) = \overline{P}Cl_t^{\leq} - \underline{P}Cl_t^{\leq}.$$

$\forall t \in T$ and $\forall P \subseteq C$ we define the accuracy of the approximation of $Cl_t^{\geq}$ and $Cl_t^{\leq}$ as the ratios

$$\alpha_P(Cl_t^{\geq}) = \frac{card(\underline{P}Cl_t^{\geq})}{card(\overline{P}Cl_t^{\geq})},$$
$$\alpha_P(Cl_t^{\leq}) = \frac{card(\underline{P}Cl_t^{\leq})}{card(\overline{P}Cl_t^{\leq})},$$

respectively. The coefficient

$$\gamma_P(\mathbf{Cl}) = \frac{card\left(U - \left(\left(\bigcup_{t \in T} Bn_P(Cl_t^{\leq})\right) \cup \left(\bigcup_{t \in T} Bn_P(Cl_t^{\geq})\right)\right)\right)}{card\,(U)}$$

is called the *quality of approximation of partition* **Cl** by set of attributes P, or in short, *quality of classification*. It expresses the ratio of all P-correctly classified objects to all objects in the table.

Each minimal subset $P \subseteq C$ such that $\gamma_P(\mathbf{Cl}) = \gamma_C(\mathbf{Cl})$ is called a *reduct* of **Cl** and denoted by $RED_{\mathbf{Cl}}$. Let us remark that an information table can have more than one reduct. The intersection of all the reducts is called the *core* and denoted by $CORE_{\mathbf{Cl}}$.

3.2 Decision rules

We can derive a generalised description of the preferential information contained in a given decision table in terms of decision rules.

We will consider the following three types of decision rules:

1) $D_{\geq}$-decision rule, being a statement of the type:

$$[f(x,q_1) \geq r_{q1} \text{ and } f(x,q_2) \geq r_{q2} \text{ and} \ldots f(x,q_p) \geq r_{qp}] \Rightarrow x \in Cl_t^{\geq},$$

where $\{q_1, q_2, \ldots q_p\} \subseteq C$, $r_{q1} \in V_{q1}$, $r_{q2} \in V_{q2}, \ldots, r_{qp} \in V_{qp}$ and $t \in T$;

2) $D_{\leq}$-decision rule, being a statement of the type:

$$[f(x,q_1) \leq r_{q1} \text{ and } f(x,q_2) \leq r_{q2} \text{ and} \ldots f(x,q_p) \leq r_{qp}] \Rightarrow x \in Cl_t^{\leq},$$

where $\{q_1, q_2, \ldots q_p\} \subseteq C$, $r_{q1} \in V_{q1}$, $r_{q2} \in V_{q2}, \ldots, r_{qp} \in V_{qp}$ and $t \in T$;

3) $D_{\geq\leq}$-decision rule, being a statement of the type:

$[f(x,q_1) \geq r_{q1}$ and $f(x,q_2) \geq r_{q2}$ and... $f(x,q_k) \geq r_{qk}$ and $f(x,q_{qk+1}) \leq r_{qk+1}$ and ... $f(x,q_p) \leq r_{qp},]$ $\Rightarrow x \in Cl_t^{\leq}$ or $x \in Cl_s^{\geq}$,

where $\{q_1, q_2, \ldots q_k\} \subseteq C$, $\{q_{k+1}, q_{k+2}, \ldots q_p\} \subseteq C$, $r_{q1} \in V_{q1}$, $r_{q2} \in V_{q2}, \ldots, r_{qp} \in V_{qp}$, $s,t \in T$ such that $t<s$.

Let us observe that we can have $\{q_1, q_2, \ldots q_k\} \cap \{q_{k+1}, q_{k+2}, \ldots q_p\} \neq \emptyset$. If in the condition part of a $D_{\geq\leq}$-decision rule, for some $q \in C$, we have "$f(x,q) \geq r_q$" and "$f(x,q) \leq r_q$", then we can simply write "$f(x,q) = r_q$".

When speaking about decision rules, we will understand the three kinds of decision rules together.

A statement "$[f(x,q_1) \geq r_{q1}$ and $f(x,q_2) \geq r_{q2}$ and... $f(x,q_p) \geq r_{qp}] \Rightarrow x \in Cl_t^{\geq}$" is accepted as a $D_{\geq}$-decision rule if there exists at least one $w \in \underline{C}Cl_t^{\geq}$ such that $f(w,q_1) \geq r_{q1}$ and $f(w,q_2) \geq r_{q2}$ and... $f(w,q_p) \geq r_{qp}$ and there is no $u \in U\text{-}\underline{C}Cl_t^{\geq}$ such that $f(u,q_1) \geq r_{q1}$ and $f(u,q_2) \geq r_{q2}$ and... $f(u,q_p) \geq r_{qp}$.

Analogously, a statement "$[f(x,q_1) \leq r_{q1}$ and $f(x,q_2) \leq r_{q2}$ and...$f(x,q_p) \leq r_{qp}] \Rightarrow x \in Cl_t^{\leq}$" is accepted as a $D_{\leq}$-decision rule if there exists at least one $w \in \underline{C}Cl_t^{\leq}$ such $f(w,q_1) \leq r_{q1}$ and $f(w,q_2) \leq r_{q2}$ and... $f(w,q_p) \leq r_{qp}$ and there is no $u \in U\text{-}\underline{C}Cl_t^{\leq}$ such that $f(u,q_1) \leq r_{q1}$ and $f(u,q_2) \leq r_{q2}$ and... $f(u,q_p) \leq r_{qp}$.

Finally a statement "$[f(x,q_1) \geq r_{q1}$ and $f(x,q_2) \geq r_{q2}$ and... $f(x,q_k) \geq r_k$ and $f(x,q_{k+1}) \leq r_{k+1}$ and ... $f(x,q_p) \leq r_{qp}] \Rightarrow x \in Cl_t^{\leq}$ or $x \in Cl_s^{\geq}$", with $t<s$, is accepted as a $D_{\geq\leq}$-decision rule if there exists at least one $w \in Bn_C(Cl_t^{\leq}) \cap Bn_C(Cl_s^{\geq})$ such that $f(w,q_1) \geq r_{q1}$ and $f(w,q_2) \geq r_{q2}$ and ... $f(w,q_k) \geq r_{qk}$ and $f(w,q_{qk+1}) \leq r_{qk+1}$ and ... $f(w,q_p) \leq r_{qp}$ and there is no

$u \in U-(Bn_C(Cl_t^{\leq}) \cap Bn_C(Cl_s^{\geq}))$ such that $f(u,q_1) \geq r_{q1}$ and $f(u,q_2) \geq r_{q2}$ and ... $f(u,q_k) \geq r_{qk}$ and $f(u,q_{k+1}) \leq r_{qk+1}$ and ... $f(u,q_p) \leq r_{qp}$.

$D_{\geq}$-decision rules and $D_{\leq}$-decision rules are called *certain rules*, because they are obtained from lower approximations, while $D_{\geq\leq}$-decision rules are called *approximate rules*, because they are obtained from the boundaries.

A decision rule "$[f(x,q_1) \geq r_{q1}$ and $f(x,q_2) \geq r_{q2}$ and... $f(x,q_p) \geq r_{qp}] \Rightarrow x \in Cl_t^{\geq}$" is called *minimal* if there is no other decision rule "$[f(x,h_1) \geq e_{h1}$ and $f(x,h_2) \geq e_{h2}$ and... $f(x,h_k) \geq e_{hk}] \Rightarrow x \in Cl_s^{\geq}$" such that $\{h_1, h_2,...,h_k\} \subseteq \{q_1, q_2,...,q_p\}$, $e_j \leq r_j$ $\forall j \in \{h_1, h_2,...,h_k\}$ and $s \geq t$. A decision rule "$[f(x,q_1) \leq r_{q1}$ and $f(x,q_2) \leq r_{q2}$ and ...$f(x,q_p) \leq r_{qp}] \Rightarrow x \in Cl_t^{\leq}$" is called *minimal* if there is no other decision rule "$[f(x,h_1) \leq e_{h1}$ and $f(x,h_2) \leq e_{h2}$ and... $f(x,h_k) \leq e_{hk}] \Rightarrow x \in Cl_s^{\leq}$" such that $\{h_1, h_2,...,h_k\} \subseteq \{q_1, q_2,...,q_p\}$, $e_j \geq r_j$ $\forall j \in \{h_1, h_2,...,h_k\}$ and $s \leq t$. A decision rule "$[f(x,q_1) \geq r_{q1}$ and $f(x,q_2) \geq r_{q2}$ and... $f(x,q_k) \geq r_{qk}$ and $f(x,q_{qk+1}) \leq r_{qk+1}$ and ... $f(x,q_p) \leq r_{qp}] \Rightarrow x \in Cl_t^{\leq}$ or $x \in Cl_s^{\geq}$", is called *minimal* if there is no other decision rule "$[f(x,h_1) \geq e_{h1}$ and $f(x,h_2) \geq e_{h2}$ and ... $f(x,h_g) \geq e_{hg}$ and $f(x,h_{g+1}) \leq e_{hg+1}$ and ... $f(x,h_o) \leq e_{ho}] \Rightarrow x \in Cl_a^{\leq}$ or $x \in Cl_b^{\geq}$", such that $\{h_1, h_2,...,h_g\} \subseteq \{q_1, q_2,...,q_k\}$, $\{h_{g+1},...,h_o\} \subseteq \{q_{k+1},...,q_p\}$, $e_j \leq r_j$ $\forall j \in \{h_1, h_2,...,h_g\}$, $e_j \geq r_j$ $\forall j \in \{h_{g+1},..., h_o\}$, $a \leq t$ and $b \geq s$.

Let us observe that, since each decision rule is an implication, the minimal decision rules represent the implications such that there is no other implication with an antecedent at most of the same weakness and a consequent of at least the same strength.

We say that $y \in U$ supports the $D_{\geq}$-decision rule "$[f(x,q_1) \geq r_{q1}$ and $f(x,q_2) \geq r_{q2}$ and... $f(x,q_p) \geq r_{qp}] \Rightarrow x \in Cl_t^{\geq}$" if $f(y,q_1) \geq r_{q1}$ and $f(y,q_2) \geq r_{q2}$ and... $f(y,q_p) \geq r_{qp}$ and $y \in Cl_t^{\geq}$. Analogously, $y \in U$ supports the $D_{\leq}$-decision rule "$[f(x,q_1) \leq r_{q1}$ and $f(x,q_2) \leq r_{q2}$ and...$f(x,q_p) \leq r_{qp}] \Rightarrow x \in Cl_t^{\leq}$" if $f(y,q_1) \leq r_{q1}$ and $f(y,q_2) \leq r_{q2}$ and...$f(y,q_p) \leq r_{qp}$ and $y \in Cl_t^{\leq}$. Lastly $y \in U$ supports the $D_{\geq\leq}$-decision rules "$[f(x,q_1) \geq r_{q1}$ and $f(x,q_2) \geq r_{q2}$ and ... $f(x,q_k) \geq r_{qk}$ and $f(x,q_{qk+1}) \leq r_{qk+1}$ and ... $f(x,q_p) \leq r_{qp}] \Rightarrow x \in Cl_t^{\leq}$ or $x \in Cl_s^{\geq}$", if $f(y,q_1) \geq r_{q1}$ and $f(y,q_2) \geq r_{q2}$ and ... $f(y,q_k) \geq r_{qk}$ and $f(y,q_{qk+1}) \leq r_{qk+1}$ and ... $f(y,q_p) \leq r_{qp}$ and $y \in Cl_t^{\leq}$ or $y \in Cl_s^{\geq}$.

We call *complete* a *set* of decision rules such that

1) each $x \in \underline{C}Cl_t^{\geq}$ supports at least a $D_{\geq}$-decision rule obtained from $\underline{C}Cl_s^{\geq}$ with $s,t \in \{2,...,n\}$ and $s \geq t$,

2) each $x \in \underline{C}Cl_t^{\leq}$ supports at least a $D_{\leq}$-decision rule obtained from $\underline{C}Cl_s^{\leq}$ with $s,t \in \{1,...,n-1\}$ and $s \leq t$,

3) each $x \in (Bn_C(Cl_t^{\leq}) \cap Bn_C(Cl_s^{\geq}))$ supports at least a $D_{\geq\leq}$-decision rule obtained from $(Bn_C(Cl_v^{\leq}) \cap Bn_C(Cl_z^{\geq}))$ such that $t,s,v,z \in T$ and $t \leq v \leq z \leq s$.

We call *minimal* each *set* of minimal decision rules which is complete and such that there is no other complete set of minimal decision rules which has a smaller number of rules.

4 A real problem: evaluation of bankruptcy risk

4.1 Statement of the problem

The problem (Slowinski and Zopounidis, 1995) has been considered by a Greek industrial development bank, called ETEVA, which finances industrial and commercial firms in Greece. A sample of 39 firms was chosen. With the co-operation of the ETEVA's financial manager, the selected firms were classified into three pre-defined categories of risk for the year 1988. The result of the classification is represented by decision attribute d making a trichotomic partition of the firms: d=1 means "unacceptable", d=2 means "uncertainty", d=3 means "acceptable". The firms were evaluated according to the following 12 condition attributes: A1=earnings before interests and taxes/total assets, A2=net income/net worth, A3=total liabilities/total assets, A4=total liabilities/cash flow, A5=interest expenses/sales, A6=general and administrative expense/sales, A7=managers' work experience, A8=firm's market niche-position, A9=technical structure-facilities, A10=organization-personnel, A11=special competitive advantage of firms, A12=market flexibility. The first six attributes are quantitative (financial ratios) and the last six are qualitative. The six qualitative attributes were modelled according to an ordinal scale (4 better than 3, 3 better then 2 and so on). The evaluations of the six quantitative attributes were transformed into ordinal scales by means of some norms, following from the financial manager's experience and some standards of the corporate financial analysis. Therefore, the rough set analysis was performed on the coded decision table presented in Table 1.

Table 1. Coded decision table

	A1	A2	A3	A4	A5	A6	A7	A8	A9	A10	A11	A12	d
F1	2	2	2	2	1	3	5	3	5	4	2	4	3
F2	4	5	2	3	3	3	5	4	5	5	4	5	3
F3	3	5	1	1	2	2	5	3	5	5	3	5	3
F4	2	3	2	1	2	4	5	2	5	4	3	4	3
F5	3	4	3	2	2	2	5	3	5	5	3	5	3
F6	3	5	3	3	3	2	5	3	4	4	3	4	3
F7	3	5	2	3	4	4	5	4	4	5	3	5	3
F8	1	1	4	1	2	3	5	2	4	4	1	4	3
F9	3	4	3	3	2	4	4	2	4	3	1	3	3
F10	3	4	2	1	2	2	4	2	4	4	1	4	3
F11	2	5	1	1	3	4	4	3	4	4	3	4	3
F12	3	3	4	4	3	4	4	2	4	4	1	3	3
F13	1	1	2	1	1	3	4	2	4	4	1	4	3
F14	2	1	1	1	4	3	4	2	4	4	3	3	3
F15	2	3	2	1	1	2	4	4	4	4	2	5	3
F16	2	3	4	3	1	5	4	2	4	3	2	3	3
F17	2	2	2	1	1	4	4	4	4	4	2	4	3

Table 1. Coded decision table (continued)

	A1	A2	A3	A4	A5	A6	A7	A8	A9	A10	A11	A12	d
F18	2	1	3	1	1	3	5	2	4	2	1	3	3
F19	2	1	2	1	1	3	4	2	4	4	2	4	3
F20	2	1	2	1	1	5	4	2	4	4	2	4	3
F21	2	1	1	1	1	3	2	2	4	4	2	3	2
F22	1	1	3	1	2	1	3	4	4	4	3	4	2
F23	2	1	2	1	1	2	4	3	3	2	1	2	2
F24	1	1	1	1	1	1	3	2	4	4	2	3	2
F25	2	2	2	1	1	3	3	2	4	4	2	3	2
F26	2	2	1	1	1	3	2	2	4	4	2	3	2
F27	2	1	2	1	1	3	2	2	4	4	2	4	2
F28	1	1	4	1	3	1	2	2	3	3	1	2	2
F29	3	4	4	3	2	3	3	4	4	4	3	4	2
F30	3	1	3	3	1	2	2	3	4	4	2	3	2
F31	1	1	2	1	1	1	3	3	4	4	2	3	1
F32	3	5	2	1	1	1	3	2	3	4	1	3	1
F33	2	2	1	1	1	1	3	3	3	4	3	4	1
F34	2	1	1	1	1	1	2	2	3	4	3	4	1
F35	1	1	2	1	1	1	2	1	4	3	1	2	1
F36	1	1	3	1	2	1	2	1	3	3	2	3	1
F37	1	1	1	1	1	1	2	2	4	4	2	3	1
F38	1	1	3	1	1	1	1	1	4	3	1	3	1
F39	2	1	1	1	1	1	1	1	2	1	1	2	1

4.2 The results from classical rough set approach

The accuracy of all approximations is perfect, i.e. equal to one. Therefore the quality of sorting is also equal to one. There are 26 reducts. The core is composed of attribute A_7. Table 2 presents the quality of sorting by single attributes.

Table 2. Quality of sorting by single attributes

Attribute	A_1	A_2	A_3	A_4	A_5	A_6	A_7	A_8	A_9	A_{10}	A_{11}	A_{12}
Quality of sorting	.026	.103	0.0	.051	.051	.205	.282	.103	.154	.128	.026	.128

To select a reduct, the following procedure (Slowinski, Slowinski, Stefanowski, 1988) was used. The core $\{A_7\}$ was augmented by one of the remaining attributes and the pair that gave the highest quality of sorting was chosen. Then, the chosen pair was augmented by one of the remaining attributes and the triple that gave the highest quality of sorting was selected, and so on, until the quality was equal to one. The subsets of attributes that have simultaneously attained the quality equal to one were suggested as the best reducts. According to this procedure, the best reducts are $\{A_6, A_7, A_8, A_{11}\}$ and $\{A_1, A_7, A_8, A_{11}\}$.

A set of decision rules describing the decision table was constituted by a set of 15 exact rules using attributes from the reduct $\{A_6, A_7, A_8, A_{11}\}$. Only 31 descriptors are used in this set of decision rules. It is less than 7% of all descriptors (468) appearing in the initial decision table. The most compact set of decision rules was also calculated. This second set of decision rules using nine attributes was composed of 11 exact rules and 23 descriptors only, i.e. less than 5% of all original descriptors.

4.3 The results from approximations by dominance relations

With this approach we approximate the following sets of objects:

$Cl_1^{\leq}$={F31, F32, F33, F34, F35, F36, F37, F38, F39},

$Cl_2^{\leq}$={F21, F22, F23, F24, F25, F26, F27, F28, F29, F30, F31, F32, F33, F34, F35, F36, F37, F38, F39},

$Cl_2^{\geq}$={F1, F2, F3, F4, F5, F6, F7, F8, F9, F10, F11, F12, F13, F14, F15, F16, F17, F18, F19, F20, F21, F22, F23, F24, F25, F26, F27, F28, F29, F30},

$Cl_3^{\geq}$={F1, F2, F3, F4, F5, F6, F7, F8, F9, F10, F11, F12, F13, F14, F15, F16, F17, F18, F19, F20}.

Let us point out that

- $x \in Cl_1^{\leq}$ means "x is unacceptable",

- $x \in Cl_2^{\leq}$ means "x is at most uncertain", i.e. "x is uncertain or unacceptable",

- $x \in Cl_2^{\geq}$ means "x is at least uncertain", i.e. "x is uncertain or acceptable",

- $x \in Cl_3^{\geq}$ means "x is acceptable".

The lower approximation and upper approximation of $Cl_1^{\leq}$, $C_2^{\leq}$, $C_2^{\geq}$ and $Cl_3^{\geq}$ are equal, respectively, to $\underline{C}Cl_3^{\geq}=Cl_3^{\geq}$, $\overline{C}Cl_3^{\geq}=Cl_3^{\geq}$, $\underline{C}Cl_2^{\geq}=Cl_2^{\geq}-\{F24\}$, $\overline{C}Cl_2^{\geq}=Cl_2^{\geq}\cup\{F31\}$, $\underline{C}Cl_1^{\leq}=Cl_1^{\leq}-\{F31\}$, $\overline{C}Cl_1^{\leq}=Cl_1^{\geq}\cup\{F24\}$.

Let us observe that the C-doubtful region of the decision is composed of two firms: F24 and F31.

There are only the following four reducts for **Cl**:

$$RED_{Cl}^{1}=\{A_1, A_3, A_7, A_9\},$$

$$RED_{Cl}^{2}=\{A_1, A_5, A_7, A_9\},$$

$$RED_{Cl}^{3}=\{A_3, A_6, A_7, A_9\},$$

$$RED_{Cl}^{4}=\{A_5, A_6, A_7, A_9\}.$$

The core of **Cl** is $CORE_{Cl} = \{A_7, A_9\}$.

Table 3 presents the quality of sorting by single attributes.

Table 3. Quality of sorting by single attributes

Attribute	A_1	A_2	A_3	A_4	A_5	A_6	A_7	A_8	A_9	A_{10}	A_{11}	A_{12}
Quality of sorting	.026	0.0	0.0	.026	.051	.205	.282	.103	.154	.128	.026	.128

The best reduct was calculated with the same procedure considered in the previous subsections as summarised in Tables 4 and 5.

Table 4. Quality of sorting by triples of attributes including the core $\{A_7, A_9\}$

Attributes A_7, A_9 +	A_1	A_2	A_3	A_4	A_5	A_6	A_8	A_{10}	A_{11}	A_{12}
Quality of sorting	.744	.667	.795	.641	.769	.744	.692	.590	.641	.667

Table 5. Quality of sorting by quadruples of attributes including the triple $\{A_3, A_7, A_9\}$

Attributes A_3, A_7, A_9 +	A_1	A_2	A_4	A_5	A_6	A_8	A_{10}	A_{10}	A_{12}
Quality of sorting	0.949	0.846	0.795	0.795	0.949	0.821	0.821	0.821	0.846

The best quadruples are therefore $\{A_1, A_3, A_7, A_9\}$ and $\{A_3, A_6, A_7, A_9\}$.

With respect to $\{A_1, A_3, A_7, A_9\}$ we can extract the following minimal set of decision rules (within parentheses there are the examples supporting the corresponding rule):

1) if $f(x,7)\geq 4$ and $f(x,9)\geq 4$ then $x \in Cl_3^{\geq}$ (F1, F2, F3, F4, F5, F6, F7, F8, F9, F10, F11, F12, F13, F14, F15, F16, F17, F18, F19, F20),

2) if $f(x,3)\geq 4$ then $x \in Cl_2^{\geq}$ (F8, F12, F16, F28, F29),

3) if $f(x,7)\geq 4$ then $x \in Cl_2^{\geq}$ (F1, F2, F3, F4, F5, F6, F7, F8, F9, F10, F11, F12, F13, F14, F15, F16, F17, F18, F19, F20, F23),

4) if $f(x,3)\geq 3$ and $f(x,7)\geq 3$ then $x \in Cl_2^{\geq}$ (F5, F6, F8, F9, F12, F16, F18, F22, F29),

5) if $f(x,1)\geq 2$ and $f(x,9)\geq 4$ then $x \in Cl_2^{\geq}$ (F1, F2, F3, F4, F5, F6, F7, F9, F10, F11, F12, F14, F15, F16, F17, F18, F19, F20, F21, F25, F26, F27, F29, F30),

6) if f(x,7)≤3 then $x \in Cl_2^{\leq}$ (F21, F22, F24, F25, F26, F27, F28, F29, F30, F31, F32, F33, F34, F35, F36, F37, F38, F39),

7) if f(x,9)≤3 then $x \in Cl_2^{\leq}$ (F23, F28, F32, F33, F34, F36, F39),

8) if f(x,1) ≤1, f(x,3)≤3 and f(x,7)≤2 then $x \in Cl_1^{\leq}$ (F35, F36, F37, F38),

9) if f(x,3) ≤3, f(x,7)≤3 and f(x,9)≤3 then $x \in Cl_1^{\leq}$ (F32, F33, F34, F36, F39),

10) if f(x,1)≤3, f(x,3)≤2, f(x,7)=3 and f(x,9)≥4 then $x \in Cl_1^{\leq}$ **or** $x \in Cl_2^{\geq}$ (F24, F31).

With respect to $\{A_3, A_6, A_7, A_9\}$ we can extract the following minimal set of decision rules:

1) if f(x,7)≥4 and f(x,9)≥4 then $x \in Cl_3^{\geq}$ (F1, F2, F3, F4, F5, F6, F7, F8, F9, F10, F11, F12, F13, F14, F15, F16, F17, F18, F19, F20),

2) if f(x,3)≥4 then $x \in Cl_2^{\geq}$ (F8, F12, F16, F28, F29),

3) if f(x,6)≥2 then $x \in Cl_2^{\geq}$ (F1, F2, F3, F4, F5, F6, F7, F8, F9, F10, F11, F12, F13, F14, F15, F16, F17, F18, F19, F20, F21, F23, F25, F26, F27, F29, F30),

4) if f(x,3)≥3 and f(x,7)≥2 and f(x,9) ≥4 then $x \in Cl_2^{\geq}$ (F5, F6, F8, F9, F12, F16, F18, F22, F29, F30),

5) if f(x,7)≤3 then $x \in Cl_2^{\leq}$ (F21, F22, F24, F25, F26, F27, F28, F29, F30, F31, F32, F33, F34, F35, F36, F37, F38, F39),

6) if f(x,9)≤3 then $x \in Cl_2^{\leq}$ (F23, F28, F32, F33, F34, F36, F39),

7) if f(x,3) ≤3, f(x,6)≤1 and f(x,7) ≤2 then $x \in Cl_1^{\leq}$ (F34, F35, F36, F37, F38, F39),

8) if f(x,3) ≤3, f(x,6)≤1 and f(x,9) ≤3 then $x \in Cl_1^{\leq}$ (F32, F33, F34, F36, F39).

9) if f(x,3)≤2, f(x,6)≤1, f(x,7)≥4 and f(x,9)≥4 then $x \in Cl_1^{\leq}$ **or** $x \in Cl_2^{\geq}$ (F24, F31).

Several minimal sets of decision rules can be extracted from the whole decision table presented in Table 1. One of them is the following:

1) if f(x,7)≥4 and f(x,9)≥4 then $x \in Cl_3^{\geq}$ (F1, F2, F3, F4, F5, F6, F7, F8, F9, F10, F11, F12, F13, F14, F15, F16, F17, F18, F19, F20),

2) if f(x,6)≥2 then $x \in Cl_2^{\geq}$ (F1, F2, F3, F4, F5, F6, F7, F8, F9, F10, F11, F12, F13, F14, F15, F16, F17, F18, F19, F20, F21, F23, F25, F26, F27, F29, F30),

3) if f(x,3)≥3 and f(x,8)≥2 then $x \in Cl_2^{\geq}$ (F5, F6, F8, F9, F12, F16, F18, F22, F28, F29, F30),

4) if f(x,7)≤3 then x∈ $Cl_2^{\leq}$ (F21, F22, F24, F25, F26, F27, F28, F29, F30, F31, F32, F33, F34, F35, F36, F37, F38, F39),

5) if f(x,9)≤3 then x∈ $Cl_2^{\leq}$ (F23, F28, F32, F33, F34, F36, F39),

6) if f(x,1) ≤1, f(x,3)≤3 and f(x,7) ≤2 then x∈ $Cl_1^{\leq}$ (F35, F36, F37, F38),

7) if f(x,3) ≤3, f(x,7)≤3 and f(x,9)≤3 then x∈ $Cl_1^{\leq}$ (F32, F33, F34, F36, F39),

8) if f(x,1)≤3, f(x,3)≤2, f(x,7)=3 and f(x,9)≥4 then x∈ $Cl_1^{\leq}$ or x∈ $Cl_2^{\geq}$ (F24, F31).

The minimal set of decision rules extracted from the decision table reduced to {A_1, A_3, A_7, A_9} uses 20 descriptors, which represent 4.27% of all descriptors appearing in the initial decision table, while the minimal set of decision rules extracted from the decision table reduced to {A_6, A_7, A_8, A_{11}} uses 19 descriptors, which represent 4.06% of all descriptors appearing in the initial decision table. Lastly, the minimal set of decision rules extracted from the whole decision table, uses six attributes but 17 descriptors only, i.e. 3.63% of all original descriptors.

4.4 Comparison of the results

The advantages of the rough set approach based on dominance relations over the original rough set analysis based on the indiscernibility relation can be summarised in the following points.

The results of the approximation are more satisfactory. This improvement is represented by a smaller number of reducts (only 4 from the approximation by dominance against 26 from the approximation by indiscernibility) and by a larger core ({A_7, A_9} against {A_7}). These two features are generally recognised as desirable properties of a good approximation (Pawlak, 1991, Slowinski and Stefanowski, 1996). Let us observe that even if the quality of the approximation obtained by indiscernibility is equal to 1, while the quality of approximation by dominance is equal to 0.949, this is another point in favour of the new approach. In fact, this difference is due to the firms F24 and F31. Let us notice that with respect to the evaluations (condition attributes) shown in the coded decision table, F31 dominates F24; however F31 has a comprehensive evaluation (decision attribute) worse than F24. Therefore, this can be interpreted as an inconsistency revealed by the approximation by dominance that cannot be pointed out when the approximation is done by indiscernibility.

From the viewpoint of the quality of the set of decision rules extracted from the decision table with the two approaches, let us remark that the decision rules obtained from the approximation by dominance relations give a more synthetic representation of information contained in the decision table. All the three minimal sets of decision rules obtained from the new approach have a smaller number of rules and use a smaller number of attributes and descriptors than the set of the decision rules obtained from the classical rough set approach.

Furthermore, the decision rules obtained from the approximation by dominance relations generally perform better when applied to new objects. E.g. let us consider a firm x having the following evaluations: f(x,7)=4 and f(x,9)=5. Using the two sets of decision rules obtained from the original rough set approach we are not able to classify the firm x. On the contrary, the decision rule

r1: "if f(x,7)≥4 and f(x,9)≥4 then $x \in Cl_3^{\geq}$"

enables us to classify x as "acceptable" on the basis of all the three minimal sets of decision rules obtained from the approximation by dominance. Let us remark that in one of the two sorting algorithms obtained from the approximation by indiscernibility, there is one decision rule which is very similar to r1:

r2: "if f(x,7)=4 and f(x,9)=4 then $x \in Cl_3$".

From comparison of decision rules r1 and r2, it is clear that rule r1 has an application wider than rule r2.

5 Conclusions

We presented a new rough set method for bankruptcy evaluation. From a theoretical viewpoint this new method enables us to consider attributes with ordered domains and sets of objects divided in ordered pre-defined categories. The main idea of the proposed approach is an approximation of some sets of firms which comprehensively belong to the same class of risk by means of dominance relations. Furthermore we showed that the basic concepts of rough set theory can be restored in the new context. We also applied the approach to a real problem of bankruptcy risk evaluation already solved with the classical rough set approach. The comparison of the results proved the usefulness of the new method.

Acknowledgement: The research of the first two authors has been supported by grant No. 96.01658.CT10 from Italian National Council for Scientific Research (CNR). The research of the third author has been supported by KBN research grant from the State Committee for Scientific Research (Komitet Badan Naukowych).

References

Dimitras, A. I., Zanakis, S. H., Zopounidis, C., "A survey of business failures with an emphasis on prediction methods and industrial applications", *European Journal of Operational Research*, 90, 1996, 487-513.

Dubois, D., Prade, H., "Putting rough sets and fuzzy sets together", in Slowinski, R. (ed.), *Intelligent Decision Support: Handbook of Applications and Advances of the Rough Sets Theory,* Kluwer Academic, Dordrecht, 1992, 203-232.

Greco, S., Matarazzo, B., Slowinski, R., *Rough Approximation of Preference Relation by Dominance Relations*, ICS Research Report 16/96, Warsaw University of Technology, Warsaw, 1996.

Grzymala-Busse, J.W, "LERS - a system for learning from examples based on rough sets", in Slowinski, R., (ed.) *Intelligent Decision Support. Handbook of Applications and Advances of the Rough Sets Theory*, Kluwer Academic Publishers, Dordrecht, 1992, 3-18.

Mienko, R., Stefanowski, J., Toumi, K., Vanderpooten, D., *Discovery-oriented induction of decision rules*, Cahier du LAMSADE 141, Université de Paris Dauphine, Paris, 1996.

Krusinska, E. Slowinski, R., Stefanowski, J., "Discriminant versus rough set approach to vague data analysis", *Applied Stochastic Models and Data Analysis*, 8, 1992, 43-56.

Pawlak, Z., Rough sets, *International Journal of Information & Computer Sciences*, 11, 1982, 341-356.

Pawlak, Z., *Rough Sets. Theoretical Aspects of Reasoning about Data*, Kluwer Academic Publishers, Dordrecht, 1991.

Peel, M.J., Peel, D.A., and Pope, P.F., Predicting corporate failure - Some results for the UK corporate sector", *OMEGA*, 14, 1, 1986, 5-12.

Roy, B., *Méthodologie Multicritère d'Aide à la Décision*, Economica, Paris, 1985.

Roy, B., "Decision science or decision aid science", *European Journal of Operational Research*, Special Issue on Model Validation in Operations Research, 66, 1993, 184-203.

Shaw, M.J., Gentry, J.A., "Using an expert system with inductive learning to evaluate business loans", *Financial Management*, 17, 3, 1988, 45-56.

Skowron, A., "Boolean reasoning for decision rules generation", in Komorowski, J., Ras, Z. W., (eds.), *Methodologies for Intelligent Systems*, (Lecture Notes in Artificial Intelligence, Vol. 689), Springer -Verlag, Berlin, 1993, 295-305.

Skowron, A., Grzymala-Busse, J.W:, "From the rough set theory to the evidence theory", in Fedrizzi, M., Kacprzyk, J., Yager, R.R., (eds.), *Advances in the Dempster-Shafer Theory of Evidence*, John Wiley, New York, 1993, 193-236.

Slowinski, K., Slowinski, R., Stefanowski, J., "Rough sets approach to analysis of data from peritoneal lavage in acute pancreatitis", *Medical Informatics*, 13, 1988, 145-159.

Slowinski, K., Stefanowski, J., "On Limitations of Using Rough Set Approach to Analyse Non-Trivial Medical Information Systems", in Shusaku Tsumoto and al. (eds.), *Proceedings of the Fourth International Workshop on Rough Sets, Fuzzy Sets and Machine Discovery*, November 1996, Tokyo, 176-183.

Slowinski, R., Stefanowski, J., "RoughDAS and RoughClass software implementations of the rough sets approach", in Slowinski, R. (ed.), *Intelligent*

Decision Support. Handbook of Application and Advances of the rough Sets Decision Theory, Kluwer Academic Publishers, Dordrecht, 1992, 445-456.

Slowinski, R., Vanderpooten, D., *Similarity relation as a basis for rough approximations*, ICS Research Report 53/95, Warsaw University of Technology, Warsaw, 1995.

Slowinski, R., Vanderpooten, D., *A generalised definition of rough approximations*, ICS Research Report 4/96, Warsaw University of Technology, Warsaw, 1996.

Slowinski, R., Zopounidis, C., "Application of the Rough Set Approach to Evaluation of Bankruptcy Risk", *Intelligent Systems in Accounting, Finance and Management*, 4, 1995, 27-41.

Ziarko, W., Golan, D., Edwards, D., "An application of DATALOGIC/R knowledge discovery tool to identify strong predictive rules in stock market data", in *Proc. AAAI Workshop on Knowledge Discovery in Databases*, Washington D.C., 1993, 89-101.

Zopounidis, C., "A multicriteria decision making methodology for the evaluation of the risk of failure and an application", *Foundations of Control Engineering*, 12/1, 45-67, 1987.

FINCLAS: A MULTICRITERIA DECISION SUPPORT SYSTEM FOR FINANCIAL CLASSIFICATION PROBLEMS

Constantin Zopounidis, Michael Doumpos

Technical University of Crete
Department of Production Engineering and Management
Decision Support Systems Laboratory
University Campus, 73100 Chania, Greece

Abstract: A significant portion of financial decision problems concerns the sorting of the alternatives into a set of predefined classes. Such financial decision problems include the assessment of bankruptcy risk, credit granting problems, venture capital investments and financing decisions in general, country risk assessment, portfolio selection and management, etc. Several techniques and methods have been proposed in the past for the study of financial classification problems, originating from different scientific fields, including statistical analysis, mathematical programming, multicriteria decision aid and artificial intelligence. The application of these methods in real world problems where the decisions have to be taken in real-time, calls upon a powerful and efficient tool to support practitioners financial analysts in implementing these techniques according to the available data of each specific problem. This paper presents the FINCLAS (FINancial CLASsification) decision support system for financial classification problems. The basic characteristic of this system is its enriched financial modelling capabilities. The system using both financial data and qualitative information concerning the operation of the firms, provides a sorting of the firms in classes of risk. The classification is achieved through the use of the UTADIS (UTilités Additives DIScriminantes) multicriteria decision aid method. The several parts of the FINCLAS system are discussed in detail, and an application of the system is presented.

Keywords: Classification, Financial problems, DSS, Multicriteria decision aid

1. Introduction and review

Financial decision problems constitute a significant part of real world decisions which are characterized by their high complexity, the plethora of factors which are involved, both of quantitative and qualitative nature, and the difficulty of determining a specific decision making process. In a daily basis many practitioners, including financial and credit analysts, managers of firms and credit institutions, individual investors, etc., have to deal with a vast amount of information and data, which must be examined and analyzed in order to take the appropriate decisions.

In some financial decision problems such as the assessment of bankruptcy risk, credit granting, country risk assessment, venture capital investments, portfolio selection and management, etc., ranking a set of alternatives (i.e. firms, credit applications, countries, investment projects, among others) from the best one to the worst one, does not necessarily provide a solution to the examined problem. In such cases, it would be more appropriate to sort the alternatives into homogenous predefined classes in order to derive suitable decisions concerning the financing of a firm or a country, the granting of a credit application, the implementation of an investment project, etc.

The techniques which have already been applied in financial classification problems include statistical analysis methods, rough sets, techniques originated from the field of artificial intelligence (i.e. expert systems, and neural networks), multicriteria decision aid (MCDA) methods, and multicriteria decision support systems (MCDSSs), among others (Table 1).

Table 1: Techniques already applied in financial classification problems

Statistical techniques	***Rough sets***	***Expert systems***	***Neural networks***	***MCDA methods***	***MCDSSs***
Altman, 1968	Slowinski and Zopounidis, 1995	Bouwman, 1983	Altman et al., 1994	Zopounidis, 1987	Mareschal and Brans, 1991
Jensen, 1971	Dimitras et al., 1996a	Ben-David and Sterling, 1986	Wilson and Sharda, 1994	Khoury and Martel, 1990	Siskos et al., 1994
Gupta and Huefner, 1972	Slowinski et al., 1997	Elmer and Borowski, 1988	Back et al., 1995	Jablonský, 1993	Zopounidis et al., 1995
Martin, 1977		Messier and Hansen, 1988	Boritz and Kennedy, 1995	Dimitras et al., 1995	Zopounidis et al., 1996
Peel, 1987		Shaw and Gentry, 1988	Serrano-Cinca, 1996	Jacquet-Lagrèze, 1995	
Casey et al., 1986		Cronan et al., 1991		Zopounidis, 1995	
Keasey et al., 1990		Michalopoulos and Zopounidis, 1993		Zopounidis and Doumpos, 1996	
Skogsvik, 1990		Matsatsinis et al., 1997			

The statistical techniques constitute the first and the most popular approach in the study of financial classification problems. They were the first to take into consideration the multidimensional nature of financial decisions combining several decision variables in the same classification model. However, soon after their first applications in finance they were criticized mainly because of their statistical restrictions such as the distribution, the multicollinearity, and the role of the decision variables, the difficulty in the explanation of the error rates, the reductions in dimensionality, the definition of the classes (groups), and the selection of the a priori probabilities or costs of misclassification (Eisenbeis, 1977).

The rough sets theory constitutes a promising tool for the study of financial classification problems. Based on examples of past decisions, the aim of the rough set approach is to develop a set of decision rules representing the preferences of the decision maker in an easily understandable form, using only the significant attributes (decision variables). Moreover, rough sets are able to handle both quantitative and qualitative criteria, and even non-monotone preferences. On the contrary, they are unable to take into account continuous variables in a direct manner. The continuous variables are transformed into discrete ones through a discretization process, which should be performed prudently to represent the preferences of the decision maker.

The expert systems technology has attracted in the past the interest of many researchers in the field of financial management. Their basic advantages concern their ability to provide estimations using an understandable form of reasoning based on the knowledge of human experts, as well as their capability to explain their estimations using natural language. Despite these advantages and the initial enthusiasm on the expert systems technology, their implementation revealed several limitations. The major drawbacks in the implementation of expert systems in the study of financial decision problems concern the significant amount of time which is needed to elicit precise knowledge from experts, the accuracy needed during the modeling of the decision variables, and their inflexibility to the changes of the decision environment (Tafti and Nikbakht, 1993; Klein and Methlie, 1995).

Neural networks have started to gain an increasing interest among the financial and operational researchers during the last decade. Neural networks simulate the structure of human brain in order to derive estimations. Once the neural network has been trained, it can be easily used in new situations. Taking advantage of several parallel processing units (neurons), neural networks are able to provide real time estimations, but most important of all they can be easily adapted to the changes of the decision environment. The criticism on neural networks mainly focuses on three basic aspects: (i) the determination of their structural parameters (e.g. number of layers, number of processing units in each layer, etc.) is rather arbitrary based usually on large number of experimental tests, (ii) their operation which has been characterized as a "black box", does not provide the decision maker with information regarding the justification of the obtained results, and (iii) a significant amount of time may be needed during the training phase of the neural network (Altman et al., 1994).

Multicriteria decision aid methods (MCDA) constitute a significant tool for the study of financial classification problems. MCDA methods are free of restrictive statistical assumptions, they incorporate the preferences of the decision maker (financial/credit analysts, managers of banks or firms, investors, etc.) into the analysis of financial decision problems, they are capable of handling qualitative criteria, and they are easily updated taking into account the dynamic nature of the decision environment as well as the changing preferences of the decision maker.

Multicriteria decision support systems (MCDSSs) constitute a significant category of decision support systems. MCDSSs provide the necessary means for implementing several MCDA techniques in order to support individuals and managers of firms, credit institutions and banks in making and implementing effective decisions in real time. MCDSSs' interactive structure and operation enables them to integrate data base management with MCDA methods, to be flexible and adaptable to the changes in the decision environment as well as to the cognitive style and the preferences of different decision makers.

Closing this brief review of the techniques which have been applied in financial classification problems, it is worth noting the comprehensive survey of all the methods applied in the prediction of business failure, presented by Dimitras et al. (1996b).

This paper presents the FINCLAS (FINancial CLASsification), a multicriteria decision support system for financial classification problems. The FINCLAS system, in its present form, aims at classifying a set of firms in classes of risk. The basic inputs to the system include both financial data, and qualitative information such as the quality of management, the organization of the firm, its market position, etc. The system using the UTADIS method (UTilités Additives DIScriminantes, Devaud et al., 1980; Jacquet-Lagrèze, 1995), and a variant of the UTADIS method, classifies the firms in classes of risk. Moreover, the system incorporates an enriched financial model base module, including the differential balance sheet, the table of sources and uses of funds, and financial forecasting methods such as the linear regression and the sales percentage methods.

Initially, the structure, the basic characteristics, and the modules of the FINCLAS system are presented in detail (section 2), followed by an application of the system in real world data (section 3). Finally, in section 4, the conclusions and the future perspectives are described.

2. Structure of the FINCLAS multicriteria decision support system

The FINCLAS system operates on any IBM compatible personal computer using the operating system MS Windows 3.1 or higher. Microsoft's Visual Basic 3.0 Professional Edition was used as the programming environment for the development of the system, taking advantage of the graphical user interface capabilities of the MS Windows operating system, as well as its object-oriented programming features.

The basic modules of the FINCLAS system as well as their integration are presented in Figure 1. The user interface is designed to be user friendly. Through the friendly graphical user interface, the decision maker can easily communicate with the system, and the smooth transfer of data between the data base and the model base is achieved. The data base of the system includes all the necessary information for the evaluation of the firms. This information, as mentioned above, includes the financial data of the firms, and qualitative information relevant to their internal operation, as well as their relation to the market. The model base incorporates the UTADIS method, a variant of the UTADIS method, and several financial models which can provide the necessary support to the decision makers in identifying the basic financial characteristics of the firms. The financial module includes financial ratios, several graphical presentations of the information derived by the financial statements, the differential balance sheet, the table of sources and uses of funds, and financial forecasting methods, such as the linear regression and the sales percentage methods.

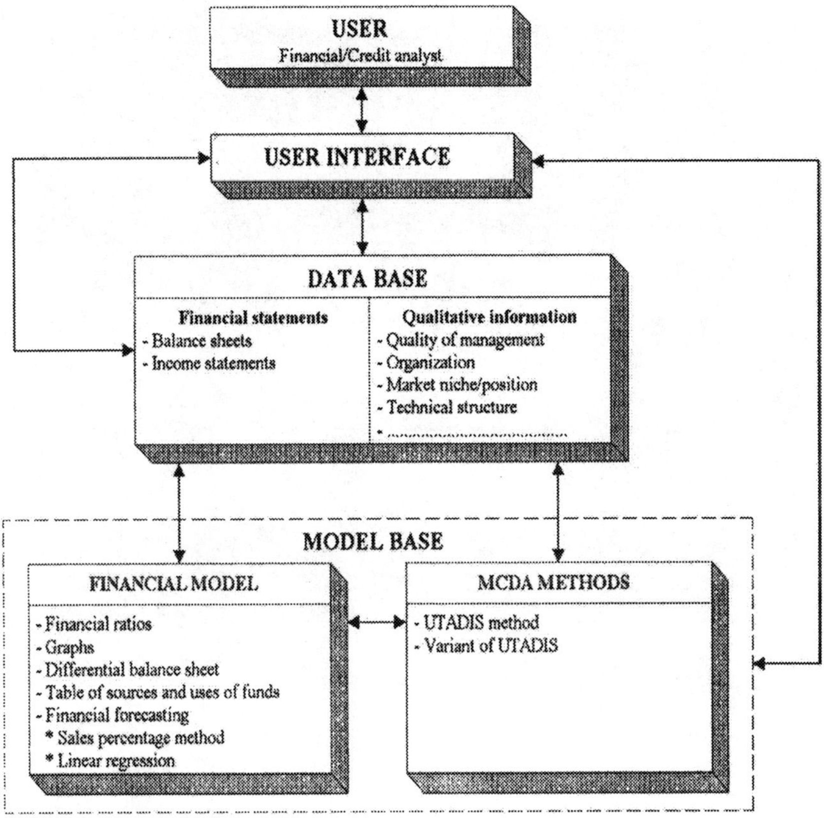

Figure 1: Structure of the FINCLAS system

2.1. Data base

The data base includes two types of information. The first one concerns the financial data of the firms. These data can be drawn from the basic financial statements of the firms (i.e. the balance sheet and the income statement). The financial data are further used to calculate some financial ratios (profitability ratios, solvency and liquidity ratios, managerial performance ratios) which are used as evaluation criteria for the classification of the firms. In order to examine the firms, taking into consideration the dynamic nature of the environment in which they operate, the analysis should not be bounded on the static information of a single balance sheet and income statement. Therefore, the financial data used by the system concern a five years period which is considered as an adequate time period for the inference of reliable estimations concerning the viability of firms. Using historical data the financial/credit analyst can examine the trend of specific items of the balance sheet and the income statement, as well as the trend of financial ratios.

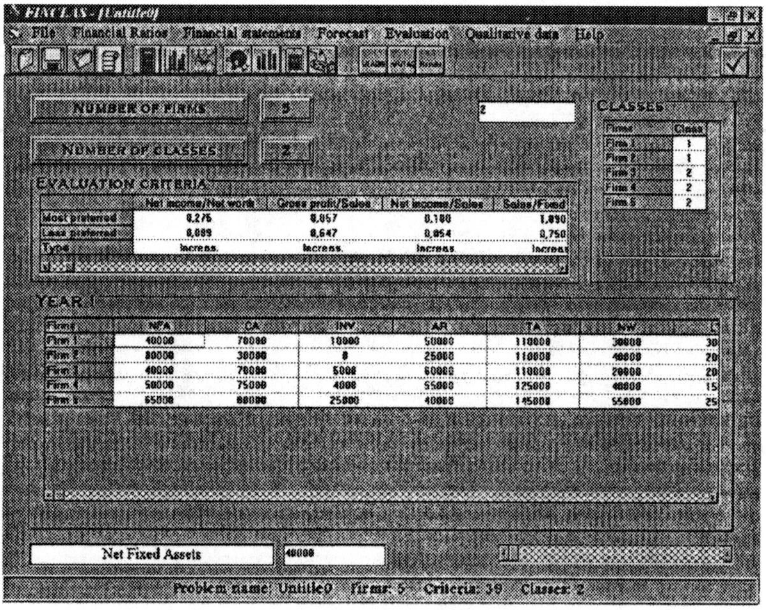

Figure 2: Basic screen of the system (data base)

Figure 2 presents the basic screen of the FINCLAS system. Through this screen the user/decision maker inserts the financial data of the firms, using several tables. This type of communication is already used by all the commercial spreadsheet software packages (Excel, 1-2-3, etc.). Therefore, no specialized knowledge or experience is required by the decision maker in order to insert the necessary data in the system's data base. The user/decision maker can easily communicate with the system through an input device such as the keyboard or the mouse. Furthermore, on the top of the screen there are several menus and tools which facilitate the use of the different modules of the system.

In addition to the financial information which is necessary for the estimation of the financial status of a firm, as mentioned above, some qualitative information is also significant. Thus, the second type of information that is included in the data base of the FINCLAS system concerns some significant strategic variables which can describe the general internal operation of a firm, as well as its relation with the market. Such strategic variables include the quality of management, the technical structure of the firm, its market position, its organization, the general performance and the perspectives of its business sector, the special know-how that the firm possesses concerning its production methods, etc. (Figure 3). These variables are mainly of qualitative nature affecting both the long term and short term operation of the firm, being sometimes of even greater significance than the quantitative measures and criteria (i.e. financial ratios) which are commonly used in financial decision problems.

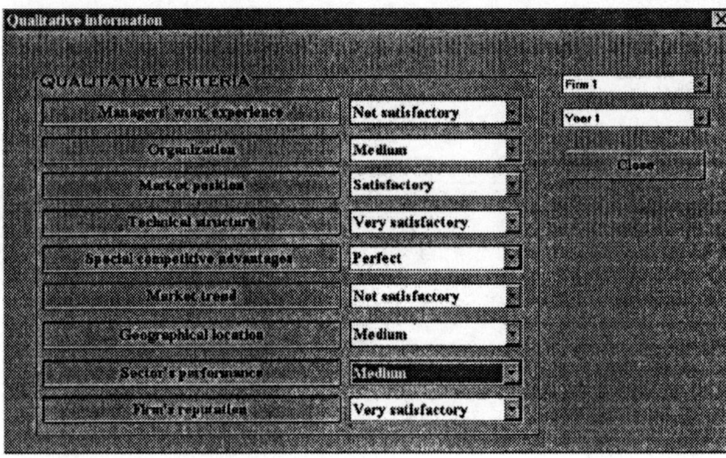

Figure 3: Qualitative evaluation criteria

2.2. Financial model base

Before any decision is taken by the financial/credit analyst of a credit institution or by the manager of a firm, he/she should be fully aware of the characteristics of the alternatives (i.e. firms) under consideration. The financial model base module of the FINCLAS system provides some necessary tools which can help the decision maker to get a clearer view concerning the financial position and operation of the firms. In its present form, the financial model base module of the FINCLAS system includes the following modelling techniques:

1. financial ratios,
2. graphical presentations,
3. the differential balance sheet,
4. the table of sources and uses of funds,

5. two financial forecasting methods: the linear regression method and the sales percentage method.

Financial ratios and graphical presentations

The basic tool for the financial analysis and examination of the firms are the financial ratios, which represent relations between the accounts of the balance sheet and the income statement. The financial model base of the FINCLAS system includes 30 financial ratios concerning the profitability, the solvency, and the managerial performance of firms (Courtis, 1978). A small sample of the financial ratios which are included in the financial model of the FINCLAS system is presented below:

1. *Profitability ratios:*
   - $\frac{\text{Net income}}{\text{Net worth}}$ (i.e. financial profitability)
   - $\frac{\text{Earnings before interest and taxes}}{\text{Total assets}}$ (i.e. industrial profitability)
   - $\frac{\text{Net income}}{\text{Sales}}$ (i.e. net profit margin)
2. *Solvency ratios:*
   - $\frac{\text{Total liabilities}}{\text{Total assets}}$ (i.e. total debt capacity)
   - $\frac{\text{Net worth}}{\text{Long term debt + Net worth}}$ (i.e. long term debt capacity)
   - $\frac{\text{Current assets}}{\text{Current liabilities}}$ (i.e. general liquidity)
3. *Managerial performance ratios:*
   - $\frac{\text{General and administrative expenses}}{\text{Sales}}$ (i.e. importance of general and administrative expenses)
   - $\frac{\text{Accounts receivable} \cdot 365}{\text{Sales}}$ (i.e. credit policy)
   - $\frac{\text{Inventory}}{\text{Current assets}}$ (i.e. stock importance)

The financial ratios are calculated for a five years period in order to provide the decision maker with useful information concerning the evolution of the financial status of a firm.

The appropriate presentation of the information that can be derived from the basic financial statements of the firms, as well as from the financial ratios, are essential for any financial/credit analyst. This information is presented through several graphs, illustrating the structure of the balance sheet and the income statement (Figures 4 and 5), as well as the trend of the financial ratios (Figure 6), and the trend of some important accounts of the financial statements, such as sales, net income, total assets, total liabilities, etc., for the five years period for which the firms are examined.

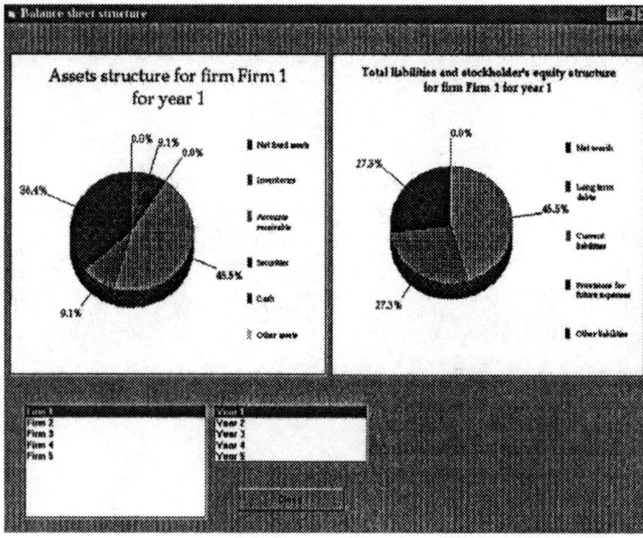

Figure 4: Structure of the balance sheet

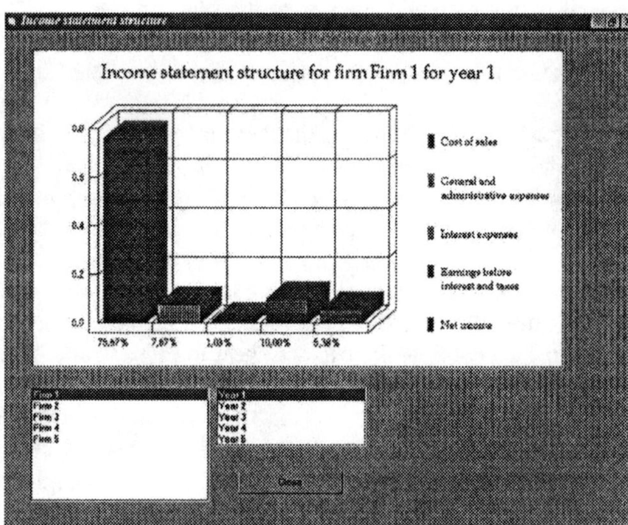

Figure 5: Structure of the income statement

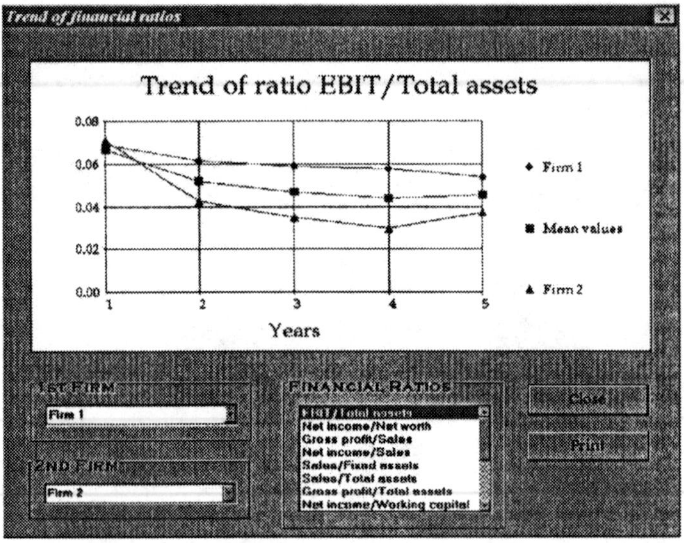

Figure 6: Trend of financial ratios

The differential balance sheet and the table of sources and uses of funds

The balance sheet of a firm provides static information concerning the assets of a firm as well as its funds (total liabilities and stockholders' equity) which have been used to finance the investments in these assets. This static information should be further enriched with a dynamic analysis of the flows (inflows and outflows) of the firm and their uses that affect its financial position. The changes of the assets and liabilities of firms can be adequately presented through the table of sources and uses of funds.

A source of funds can be defined as the decrease in any asset, or the increase in any account of total liabilities and stockholders' equity. Similarly, a use of funds can be defined as the increase in any asset, or the decrease in any account of total liabilities and stockholders' equity. Therefore, the first step for the construction of the table of sources and uses of funds is the distinction of the flows in those which constitute sources of funds and in those which constitute uses of funds. This task can be accomplished through the differential balance sheet. The differential balance sheet distinguishes the accounts of the balance sheet in those which have increased and in those which have decreased in the period of two successive years.

Using the differential balance sheet as the basis and together with some additional information, the construction of the table of sources and uses of funds is accomplished. The additional information that is required concerns the possessions and concessions of fixed assets (purchases and sales of fixed assets), the depreciation policy, possible increase in capital stock by incorporation of retained earnings, etc. Through the table of sources and uses of funds the investment and financing policy of a firm can be examined and analyzed. Figure 7 gives an example of the information that the table of sources and uses of funds provides:

- the amount of capital that was used for investments, the portion of this amount concerning financial investments (i.e. participations), and physical investments,
- the amount of cash flow,
- the amount of dividends,
- the change of working capital, the change of working capital required, the change of treasury, etc.

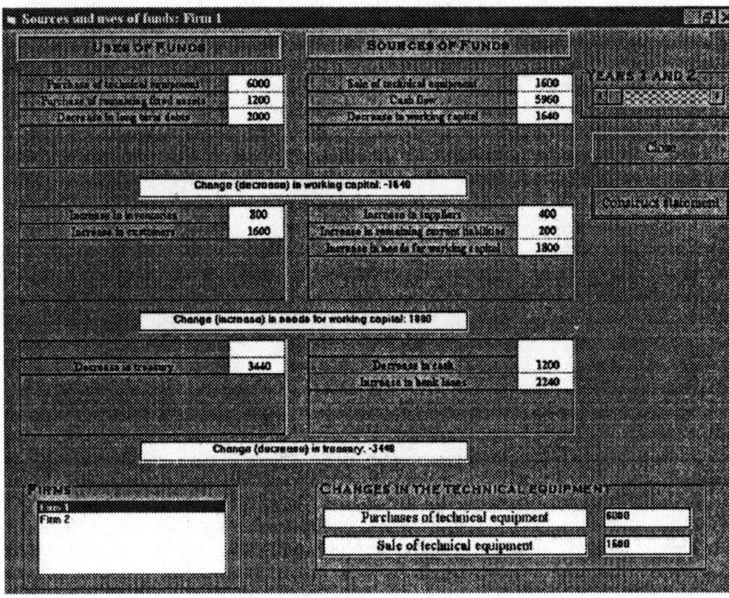

Figure 7: Table of sources and uses of funds

Furthermore, through the table of sources and uses of funds, some additional important financial ratios can be calculated, such as the ratios "cash flow/investments in fixed assets", and "dividends/cash flow". The table of sources and uses of funds can also be used as a tool for supervising the investments of a firm, examining a posteriori the financial planning strategy of the firm in relation to the expected investment results.

Financial forecasting methods

The financial model base of the FINCLAS system also incorporates two financial forecasting techniques: (i) the sales percentage method, and (ii) the linear regression method.

The aim of the sales percentage method is to forecast the external financing required for a given increase in sales, as well as the percentage of the increase in sales that should be financed by external funds, according to the changes of the accounts of the balance sheet with regard to the increase in sales.

Figure 8, presents the sales percentage method. The necessary inputs include the increase in sales, the net profit margin, the dividend payout ratio, and the changes of the accounts of the balance sheet expressed as a percentage to the increase in sales. According to this information the system computes the amount of external financing required (EFR) for a specific increase in sales which is defined by the firm (economic and marketing department), as well as the percentage of the increase in sales that should be financed by external funds (percentage of external funds required-PEFR), using the following formulas:

$$EFR = \frac{A}{S} \cdot \Delta S - \frac{B}{S} \cdot \Delta S - m \cdot b \cdot S_1$$

$$PEFR = \frac{EFR}{\Delta S}$$

where,

S : current amount of sales

S_1 : predicted amount of sales,

$\Delta S = S_1 - S$: increase in sales,

$\frac{A}{S}$: assets which are increased with regard to sales,

$\frac{B}{S}$: liabilities which are increased with regard to sales,

m : net profit margin,

b : retained earnings percentage.

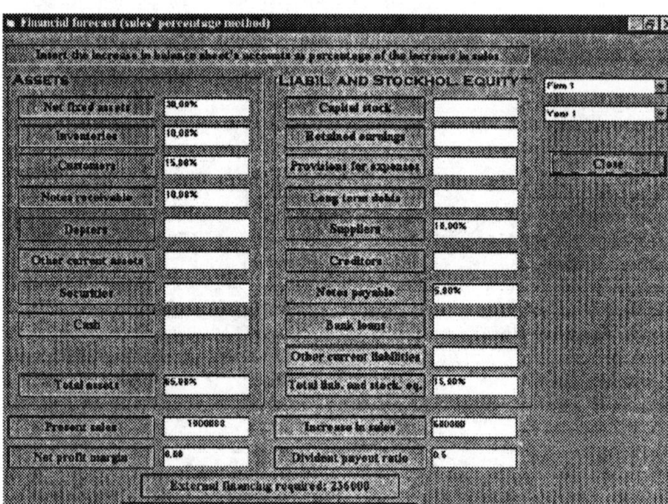

Figure 8: Sales percentage method

The linear regression method is a well known statistical technique for the study of forecasting problems. Using the linear regression method the decision maker can forecast the sales (or other important accounts of the balance sheet and the income statement) of a firm. The linear regression module of the FINCLAS system, and more specifically the simple linear regression, is presented in Figure 9. The decision maker initially, has to determine the independent and the dependent variables. In the case of Figure 9, the decision maker wants to forecast the inventories (dependent variable) according to the sales (independent variable). The time period of the regression analysis and the corresponding historical data, are also required. According to these historical data, the system develops a linear regression model, which in the case of Figure 9 is used to forecast the inventories of the firm, with regard to the sales. The correlation coefficient of these two variables is also computed.

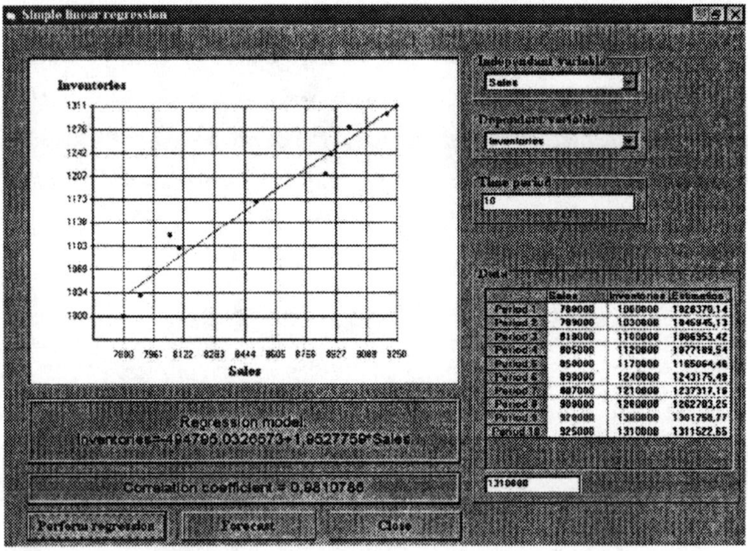

Figure 9: Linear regression method

2.3. The UTADIS method

In the present analysis of the system, the classification of the firms in classes of risk is achieved through the UTADIS multicriteria method. The UTADIS method is an ordinal regression method, based on the preference disaggregation approach of MCDA (Zopounidis, 1997). Given a predefined classification of the alternatives (i.e. firms), the objective of the UTADIS method is to estimate an additive utility function and the utility thresholds that classify the alternatives in their original classes with the minimum misclassification error. The estimation of both the additive utility function and the utility thresholds is accomplished through linear programming techniques. A brief description of the UTADIS method and its variation that is incorporated in the FINCLAS system is presented below.

Let $g_1, g_2, ..., g_m$ be a consistent family of m evaluation criteria, and $A=\{a_1, a_2, ..., a_n\}$ a set of n alternatives to be classified in Q ordered classes $C_1, C_2, ..., C_Q$ which are defined *a priori*:

$$C_1 \; \boldsymbol{P} C_2 \ldots C_{Q\text{-}1} \; \boldsymbol{P} C_Q$$

where, $\boldsymbol{P}$ denotes the strict preference relation, between the classes.

The global utility $U(a)$ of an alternative $a \in A$ is of an additive form:

$$U(a) = \sum_{i=1}^{m} u_i[g_i(a)]$$

where $u_i[g_i(a)]$ is the marginal utility of the alternative a for the criterion g_i. The marginal utilities represent the relative importance of the evaluation criteria in the classification model.

For each evaluation criterion i the interval $G_i = [g_{i\bullet}, g_i^{\bullet}]$ of the values is defined. $g_{i\bullet}$ and $g_i^{\bullet}$ are the less and the most preferred values, respectively, of the criterion i for all the alternatives belonging to A. The interval G_i is divided into a_i-1 equal intervals $[g_i^j, g_i^{j+1}]$, j=1, 2, ..., a_i-1. a_i, is defined by the decision maker as the number of estimated points for every marginal utility u_i. Each point g_i^j can be calculated using linear interpolation:

$$g_i^j = g_{i\bullet} + \frac{j-1}{a_i - 1}\left(g_i^{\bullet} - g_{i\bullet}\right)$$

The aim is to estimate the marginal utilities in each of these points. Suppose that the evaluation of an alternative a on criterion i is $g_i(a) \in [g_i^j, g_i^{j+1}]$. The marginal utility of the alternative action a, $u_i[g_i(a)]$, can be approximated through linear interpolation:

$$u_i[g_i(a)] = u_i(g_i^j) + \frac{g_i(a) - g_i^j}{g_i^{j+1} - g_i^j}\left[u_i(g_i^{j+1}) - u_i(g_i^j)\right] \qquad (1)$$

Supposing that the preferences of the decision maker on each one of the evaluation criteria are monotone, the following constraint must be satisfied:

$$u_i(g_i^{j+1}) - u_i(g_i^j) \geq 0, \qquad \forall i$$

The monoticity constraints are converted into non-negativity constraints through the following transformations:

$$\left.\begin{aligned}
&w_{ij}=u_i\left(g_i^{j+1}\right)-u_i\left(g_i^j\right)\geq 0 \quad \forall i,j\\
&u_i\left(g_{i*}\right)=0\\
&u_i\left(g_i^*\right)=\sum_{k=1}^{a_i-1} w_{ik}\\
&u_i\left(g_i^j\right)=\sum_{k=1}^{j-1} w_{ik}
\end{aligned}\right\}$$

Using these transformations, (1) can be written as:

$$u_i[g_i(\alpha)]=\sum_{k=1}^{j-1} w_{ik}+\frac{g_i(\alpha)-g_i^j}{g_i^{j+1}-g_i^j}\left[\sum_{k=1}^{j} w_{ik}-\sum_{k=1}^{j-1} w_{ik}\right]$$

There are two possible misclassification errors relative to the global utility $U(a)$. The over-estimation error $\sigma^+(a)$ and the under-estimation error $\sigma^-(a)$. The over-estimation error exists in cases when an alternative according to its utility is classified to a lower class than the class that it really belongs. On the other hand, the under-estimation error exists in cases when an alternative according to its utility is classified to a higher class than the class that it really belongs.

The classification of the alternatives is achieved through the comparison of each utility with the corresponding utility thresholds u_i ($u_1 > u_2 > \ldots > u_{Q-1}$):

$$\begin{aligned}
&U(a)\geq u_1 && \Rightarrow a\in C_1\\
&\ldots\ldots\ldots\ldots &&\\
&u_k\leq U(a)< u_{k-1} && \Rightarrow a\in C_k\\
&\ldots\ldots\ldots\ldots &&\\
&U(a)< u_{Q-1} && \Rightarrow a\in C_Q
\end{aligned}$$

The assessment of both, the marginal utilities $u_i\,[g_i\,(a)]$ and the utility thresholds u_k, is achieved through the following linear program:

$$\text{Minimize } F=\sum_{a\in C_1}\sigma^+(\alpha)+\ldots+\sum_{a\in C_k}\left[\sigma^+(\alpha)+\sigma^-(\alpha)\right]+\ldots+\sum_{a\in C_Q}\sigma^-(\alpha)$$

subject to the constraints:

$$\sum_{i=1}^{m} u_i[g_i(\alpha)]-u_1+\sigma^+(a)\geq 0 \qquad \forall a\in C_1$$

$$\left.\begin{aligned}
&\sum_{i=1}^{m} u_i[g_i(\alpha)]-u_{k-1}-\sigma^-(\alpha)\leq-\delta\\
&\sum_{i=1}^{m} u_i[g_i(\alpha)]-u_k+\sigma^+(\alpha)\geq 0
\end{aligned}\right\} \qquad \forall a\in C_k$$

$$\sum_{i=1}^{m} u_i[g_i(\alpha)]-u_{Q-1}-\sigma^-(a)\leq-\delta \qquad \forall a\in C_Q$$

$$\sum_{i=1}^{m}\sum_{j=1}^{a_i-1} w_{ij}=1$$

$$u_{k-1}-u_k\geq s \qquad k=2, 3, \ldots, Q-1$$

$w_{ij} \geq 0$, $\sigma^{+}(a) \geq 0$, $\sigma^{-}(a) \geq 0$,

where δ is a small positive real number, used to ensure the strict inequality of $U(a)$ to u_{k-1} ($\forall a \in C_k$, $k > 1$) and u_{Q-1} ($\forall a \in C_Q$). The threshold s is used to denote the strict preference between the utility thresholds that distinguish the classes ($s>\delta>0$).

A variant of the UTADIS method which is incorporated in the FINCLAS system, is to minimize the total number of misclassifications instead of minimizing the total misclassification error. This is achieved by solving the following mixed-integer linear program:

Minimize $F = \sum_{a \in A} M^{+}(a) + M^{-}(a)$

subject to the constraints:

$$\sum_{i=1}^{m} u_i[g_i(a)] - u_1 + M^{+}(a) \geq 0 \qquad \forall a \in C_1 \qquad (2)$$

$$\left.\begin{aligned} &\sum_{i=1}^{m} u_i[g_i(a)] - u_{k-1} - M^{-}(a) \leq -\delta \qquad (3)\\ &\sum_{i=1}^{m} u_i[g_i(a)] - u_k + M^{+}(a) \geq 0 \qquad (4) \end{aligned}\right\} \forall a \in C_k$$

$$\sum_{i=1}^{m} u_i[g_i(a)] - u_{Q-1} - M^{-}(a) \leq -\delta \qquad \forall a \in C_Q \qquad (5)$$

$$\sum_{i=1}^{m} \sum_{j=1}^{a_i - 1} w_{ij} = 1$$

$u_{k-1} - u_k \geq s \qquad k = 2, 3, \ldots, Q-1$

$w_{ij} \geq 0$, $M^{+}(a), M^{-}(a) \in \{0,1\}$

Jacquet-Lagrèze and Siskos (1982) proposed a similar approach for maximizing the Kendall's τ rank correlation coefficient in order to develop an additive utility model as consistently as possible with the preordering of the alternatives defined by the decision maker.

In the proposed variant of the UTADIS method, $M^{+}(a)$ and $M^{-}(a)$ are boolean numbers. Taking into consideration the fact that $U[g(a)] \in [0,1]$ and $u_k \in (0,1]$ ($\forall a \in A$, $k=1, 2, \ldots, Q-1$), the difference between the global utility of an alternative and the utility thresholds is always lower than 1. Consequently, the three following cases hold.

(i) If an alternative is classified by the decision maker in class C_k while its global utility is lower than the utility threshold u_k, then adding to its global utility an amount equal to $M^{+}(a)=1$ will certainly lead to the correct classification of this alternative and the constraints (2) and (4) are satisfied.

(ii) If an alternative is classified by the decision maker in class C_k while its global utility is lower than the utility threshold u_{k-1}, then subtracting from the global utility an amount equal to $M^{-}(a)=1$ will certainly lead to the correct classification of this alternative and the constraints (3) and (5) are satisfied.

(iii) On the other hand, in cases when an alternative is correctly classified according to its global utility, then $M^{+}(a)$ and $M^{-}(a)$ will be set equal to 0 since all the constraints (2)–(5) are satisfied.

According to these remarks if $M^{+}(a)=0$ and $M^{-}(a)=0$ then the alternative a is correctly classified by the additive utility model to its original class, otherwise if $M^{+}(a)=1$ or $M^{-}(a)=1$ then the alternative a is misclassified. A drawback of this formulation is that although the boolean variables $M^{+}(a)$ and $M^{-}(a)$ indicate whether an alternative is correctly classified or not, in cases of misclassification the magnitude of the misclassification error is not taken under consideration (e.g. if an alternative is misclassified by one or more classes). For instance, according to this formulation, two misclassifications of the type $C_k \rightarrow C_{k+1}$ and $C_k \rightarrow C_{k+2}$ will have the same impact on the objective function of minimizing the number of misclassifications. This issue should be studied further in the near future in order to develop a formulation which considers the magnitude of the misclassification errors. Moreover, it should be noted that the computational difficulties of this mixed-integer linear program make it applicable only for small scale problems.

It is important to note that apart from the classification of the firms, the FINCLAS system using the UTADIS method and its variant, provides the competitive level between the firms of the same class (i.e. which ones are the best and the worst firms of a specific class), according to their global utilities. The results of the UTADIS method are presented through the screen of Figure 10.

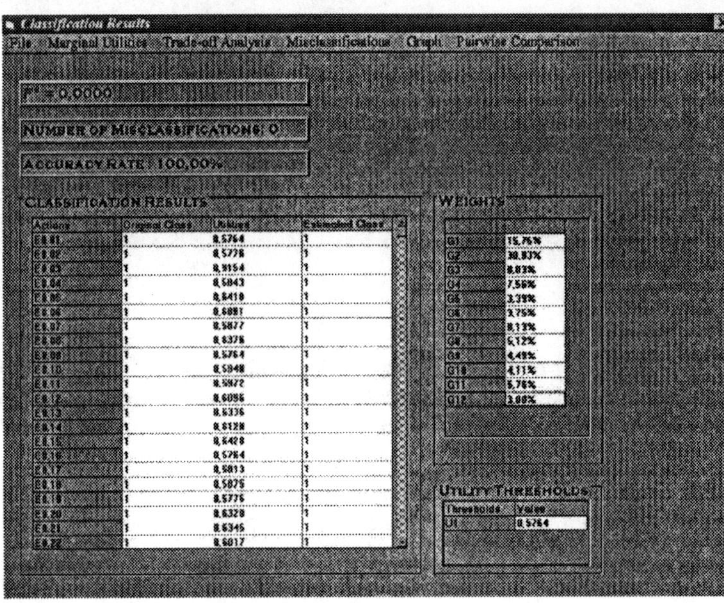

Figure 10: Results of the UTADIS method

Through the screen of Figure 10, the original and the estimated class in which the firms belong are presented, as well as the global utilities of the firms, the utility thresholds which distinguish the classes, the weights of the evaluation criteria, the total number of misclassifications, and the accuracy rate. The developed additive utility model can be stored so that it can be used to evaluate new firms which are inserted in the data base of the system.

Figure 11 presents the marginal utilities of the evaluation criteria estimated by the UTADIS method. The values of the marginal utilities show the relative weight of the evaluation criteria (the relative importance of each criterion in the classification model). The decision maker selects the criterion, for which he is interested to see the corresponding marginal utilities, from a list of all the evaluation criteria which were used to develop the additive utility model. Marginal utilities of the selected criterion are represented in a graph. The horizontal axis of the graph represents the possible values of the criterion and the vertical axis represents the utilities of the selected criterion.

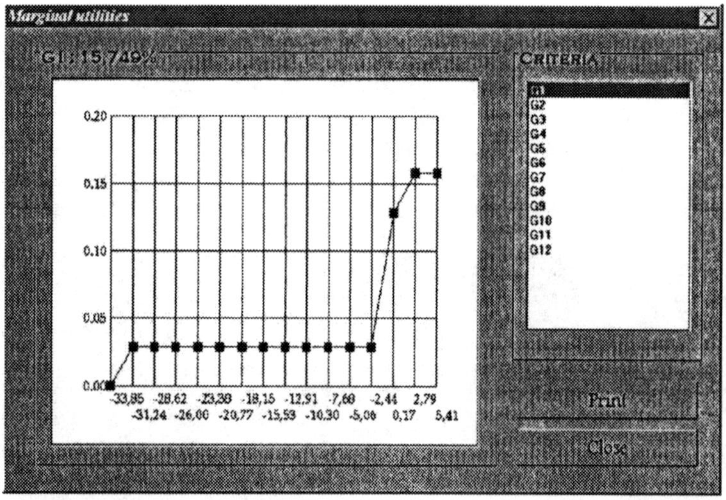

Figure 11: Marginal utilities

Furthermore, the system according to the marginal utilities of the evaluation criteria can provide a pairwise comparison of firms according to their utilities on each criterion (Figure 12). The decision maker selects the firms that he wants to compare. The pairwise comparison is presented through a graph illustrating the utility of the firms under consideration on each of the evaluation criteria. Moreover, the results of the comparison are presented through natural language to the decision maker.

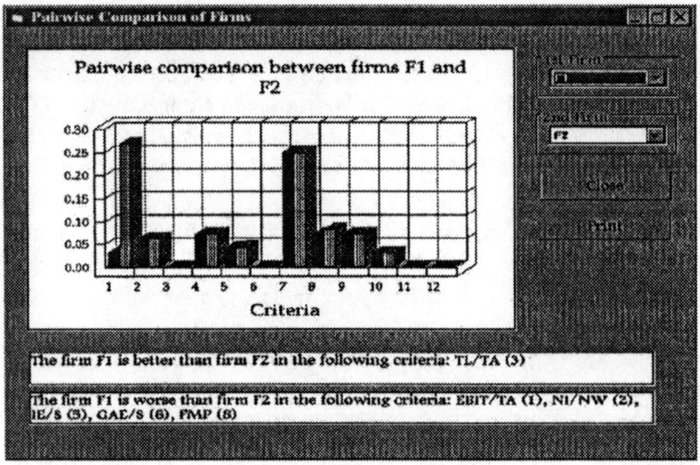

Figure 12: Pairwise comparison of firms

Although, the ranking of firms is not the objective of the UTADIS method (i.e. the principal objective is to classify the firms in their original classes), the firms could be ranked according to their global utilities (scores) from the best ones (dynamic firms) to the most risky ones. The ranking of the firms is presented through the graph of Figure 13. In the same graph the utility thresholds are also presented, as well as all the possible misclassifications (in the case of Figure 13, there are no misclassifications), using three dimensional bars of different colors.

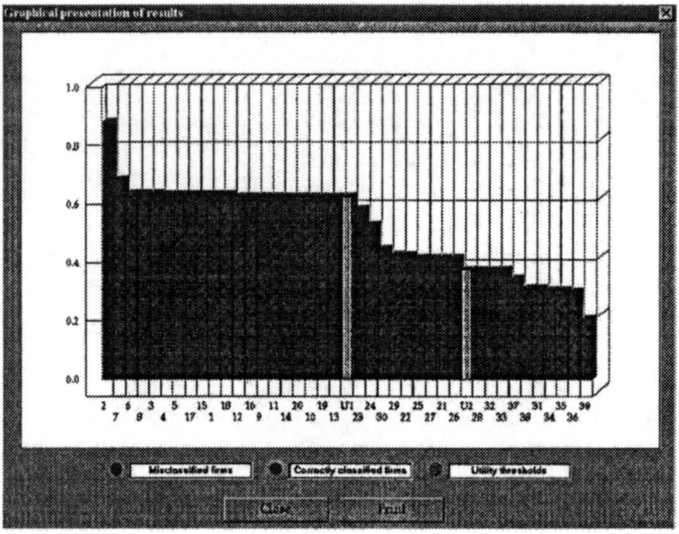

Figure 13: Ranking of firms according to their global utilities

3. An application

The FINCLAS system has been applied in a real world problem concerning the evaluation of bankruptcy risk, which is originated from the study of Slowinski and Zopounidis (1995). The application involves 39 firms which were classified by the financial manager of the ETEVA industrial and development bank in Greece in three classes:

- The acceptable firms, including firms that the financial manager would recommend for financing (class C_1).
- The uncertain firms, including firms for which further study is needed (class C_2).
- The unacceptable firms, including firms that the financial manager would not recommend to be financed by the bank (class C_3).

The sample of the 39 firms includes 20 firms which are considered as acceptable firms (healthy firms), belonging in class C_1, 10 firms for which a further study is needed, belonging in class C_2, and finally, 9 firms which are considered as bankrupt, belonging in class C_3.

The firms are evaluated along 12 criteria (Table 2). The evaluation criteria include six quantitative criteria (financial ratios) and six qualitative criteria (Siskos et al., 1994; Slowinski and Zopounidis, 1995).

Table 2: Evaluation criteria (Source: Slowinski and Zopounidis, 1995)

Code	Evaluation criteria
G_1	Earnings before interest and taxes/Total assets
G_2	Net income/Net worth
G_3	Total liabilities/Total assets
G_4	Total liabilities/Cash flow
G_5	Interest expenses/Sales
G_6	General and administrative expenses/Sales
G_7	Managers' work experience
G_8	Firm's market niche/position
G_9	Technical structure-facilities
G_{10}	Organization-personnel
G_{11}	Special competitive advantages of firms
G_{12}	Market flexibility

The classification of the firms according to their utility and the utility thresholds u_1 and u_2 which are calculated by the UTADIS method, are presented in Table 3. Figure 14 presents the marginal utilities of the evaluation criteria.

Table 3: Classification results by the UTADIS method

Firms	Original class	Utility	Estimated class
F1	C_1	0.6451	C_1
F2	C_1	0.9796	C_1
F3	C_1	0.8777	C_1
F4	C_1	0.6527	C_1
F5	C_1	0.6443	C_1
F6	C_1	0.6467	C_1
F7	C_1	0.6600	C_1
F8	C_1	0.6604	C_1
F9	C_1	0.6308	C_1
F10	C_1	0.6227	C_1
F11	C_1	0.6351	C_1
F12	C_1	0.6452	C_1
F13	C_1	0.6229	C_1
F14	C_1	0.6314	C_1
F15	C_1	0.6230	C_1
F16	C_1	0.6436	C_1
F17	C_1	0.6277	C_1
F18	C_1	0.6435	C_1
F19	C_1	0.6248	C_1
F20	C_1	0.6321	C_1
Utility threshold u_1		0.6226	
F21	C_2	0.3836	C_2
F22	C_2	0.3847	C_2
F23	C_2	0.6102	C_2
F24	C_2	0.3727	C_2
F25	C_2	0.3859	C_2
F26	C_2	0.3851	C_2
F27	C_2	0.3862	C_2
F28	C_2	0.3871	C_2
F29	C_2	0.4001	C_2
F30	C_2	0.3861	C_2
Utility threshold u_2		0.3726	
F31	C_3	0.3096	C_3
F32	C_3	0.3717	C_3
F33	C_3	0.3717	C_3
F34	C_3	0.3657	C_3
F35	C_3	0.2004	C_3
F36	C_3	0.3303	C_3
F37	C_3	0.3382	C_3
F38	C_3	0.2970	C_3
F39	C_3	0.2286	C_3

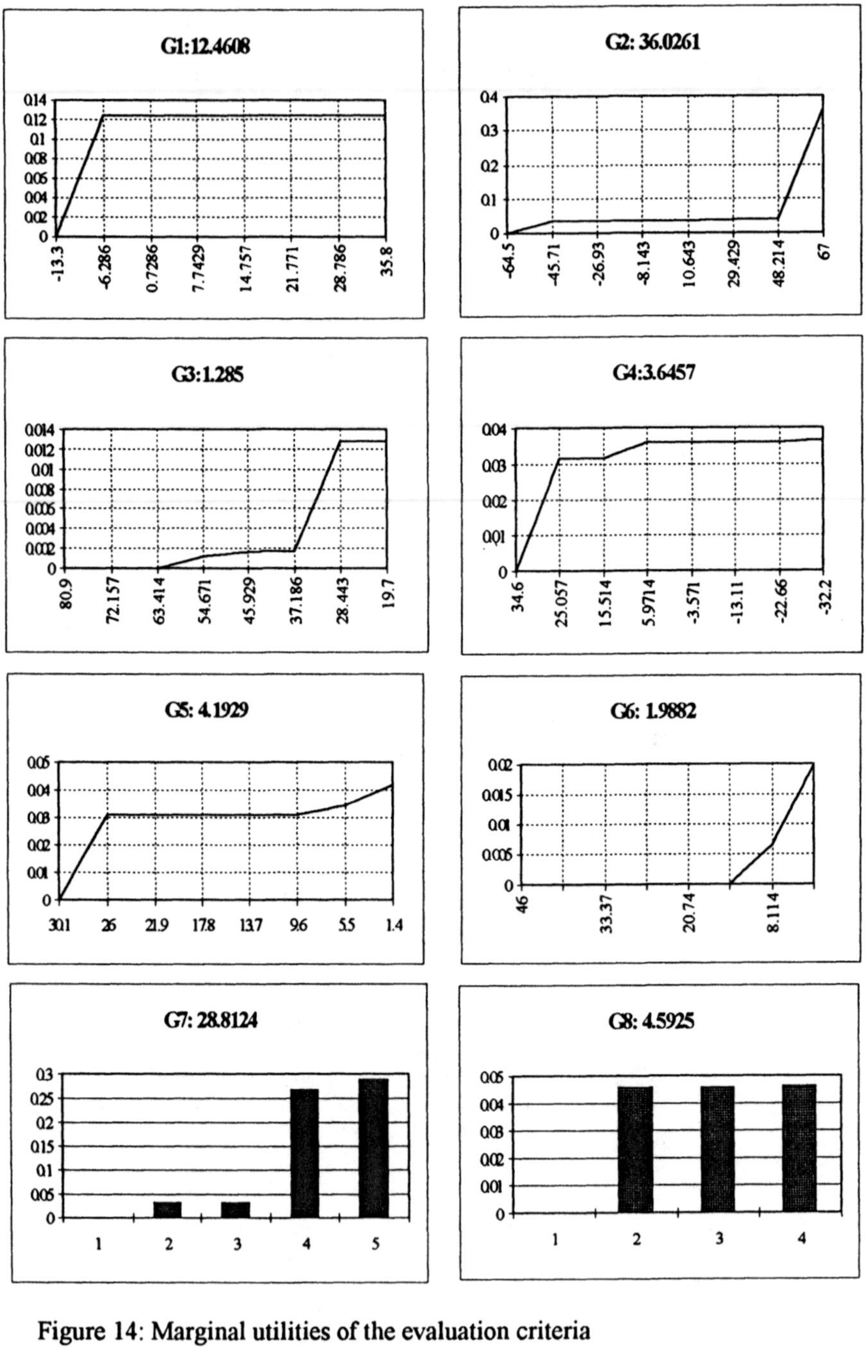

Figure 14: Marginal utilities of the evaluation criteria

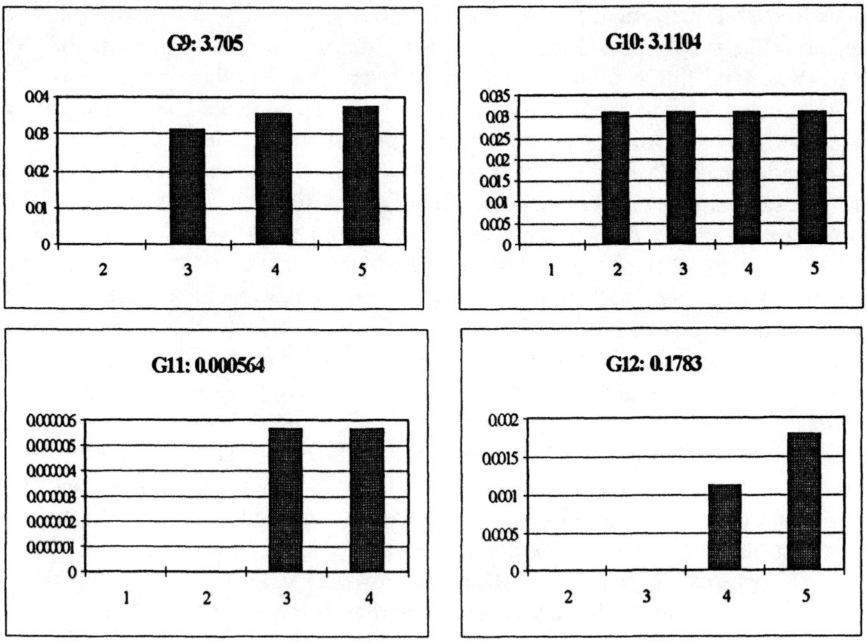

Figure 14: Marginal utilities of the evaluation criteria (continued)

According to the achieved results there are no misclassifications, obtaining a classification accuracy of 100%. This result is comparable with the results derived by the application of the rough sets approach in the same problem (Slowinski and Zopounidis, 1995). The most significant criterion in the sorting model is G_2 (financial profitability) with a weight of 36.02%. Two other important criteria are G_7 (managers' work experience) and G_1 (industrial profitability) which have a weight of 28.81% and 12.46% respectively. These two criteria were also found to be the most important for the rough sets approach (they were included in the core, cf. Slowinski and Zopounidis, 1995).

4. Conclusions and future perspectives

The FINCLAS system is a multicriteria decision support system for financial classification problems. In its present form, the system is basically oriented in the field of corporate failure risk, including the assessment of bankruptcy risk, and credit granting problems. The aim of the FINCLAS system is not to derive final decisions, but to support the financial/credit analysts in their financial decisions. The decision maker has an active role on the decision making process and on the operation of the system. The system provides several modelling tools that the decision maker can use to derive integrated estimations concerning the financial characteristics and behavior of the firms under consideration, as well as their overall performance and viability. The enriched financial model base module of the system provides a clear view of the current financial position of the firm, and of its future development perspectives. On the other hand, using the UTADIS multicriteria sorting method, the decision maker can develop a classification model which represents his preferences and decision policy. Once this classification model has been developed, it can be used to evaluate every new firm that is inserted in the data base of the system, in real time. The basic features of the UTADIS method and its variant which enable them to be a powerful and promising tool in the study of financial classification problems include: (i) the fact that they are free of restrictive statistical assumptions, (ii) their capability to handle both quantitative and qualitative criteria, (iii) their flexibility which enables the decision maker to develop models which can be easily updated considering the dynamic nature of the decision environment as well as his/her changing preferences, (iv) the minimal information that they require by the decision maker, since only a classification of the alternatives in predefined homogeneous classes is needed, and (v) their interactive operation enables the decision maker to play a significant role in the decision analysis process and provides valuable insight information regarding his/her preferences and decision policy.

The current development of the system involves:

- The enrichment of its model base with some multivariate statistical tools, such as principal components analysis, factor analysis, etc., which could help the financial/credit analyst to get a better idea concerning the financial data of the examined firms (identification of significant financial ratios and the correlation between them, determination of the financial characteristics of the firms, etc.). The model base could also be enriched to include several other analytical techniques such as credit scoring models based on discriminant analysis, and/or MCDA methods for classification problems.
- Furthermore, additional financial models could be incorporated in the current structure of the system to provide the capability of analyzing various financial classification problems including portfolio selection and management, country risk assessment, and financial planning, among others.
- Finally, a significant improvement to the current structure of the system would be the incorporation of an expert system in it. The expert system could be used to provide expert advice on the problem under consideration, assistance to the use of the several modules of the system, explanations concerning the results of the statistical, MCDA, or financial models, support on structuring the decision making process, as well as recommendations and further guidance for the future actions that the decision maker should take in order to implement successfully his/her decisions.

References

Altman, E.I. (1968), "Financial ratios, discriminant analysis and the prediction of corporate bankruptcy", *The Journal of Finance* 23, 589-609.

Altman, E.I., Marco, G. and Varetto, F. (1994), "Corporate distress diagnosis: Comparisons using linear discriminant analysis and neural networks (the Italian experience)", *Journal of Banking and Finance* 18, 505-529.

Back, B., Oosterom, G., Sere, K. and Van Wezel, M. (1995), "Intelligent information systems within business: Bankruptcy predictions using neural networks", in: G. Doukidis, B. Galliers, T. Jelassi, H. Kremar and F. Land (eds.), *Proceedings of the 3rd European Conference on Information Systems*, 99-111.

Ben-David, A. and Sterling, L. (1986), "A prototype expert system for credit evaluation", in: L.F. Pau (ed.), *Artificial Intelligence in Economics and Management*, Elsevier Science Publisher, North-Holland, 121-128.

Boritz, J.E. and Kennedy, D.B. (1995), "Effectiveness of neural network types for prediction of business failure", *Expert Systems with Applications: An International Journal* 9, 4, 503-512.

Bouwman, M.J. (1983), "Human diagnostic reasoning by computer: An illustration from financial analysis", *Management Science* 29, 6, 653-672.

Casey, M., McGee, V. and Stinkey, C. (1986), "Discriminating between reorganized and liquidated firms in bankruptcy", *The Accounting Review*, April, 249-262.

Cronan, T.P., Glorfeld, L.W. and Perry, L.G. (1991), "Production system development for expert systems using a recursive partitioning induction approach: An application to mortgage, commercial and consumer lending", *Decision Sciences* 22, 812-845.

Courtis, J.K. (1978), "Modelling a financial ratios categoric framework", *Journal of Business Finance & Accounting* 5, 4, 371-387.

Devaud, J.M., Groussaud, G. and Jacquet-Lagrèze, E. (1980), "UTADIS: Une méthode de construction de fonctions d'utilité additives rendant compte de jugements globaux", *European Working Group on Multicriteria Decision Aid*, Bochum.

Dimitras, A.I., Slowinski, R. and Zopounidis, C. (1996a), "Business failure prediction using rough sets", *Paper presented at the IFORS '96, 14th Triennial Conference*, July 8-12, 1996.

Dimitras, A.I., Zanakis, S.H. and Zopounidis, C. (1996b), "A survey of business failures with an emphasis on prediction methods and industrial applications", *European Journal of Operational Research* 90, 487-513.

Dimitras, A.I., Zopounidis, C. and Hurson, C. (1995), "A multicriteria decision aid method for the assessment of business failure risk", *Foundations of Computing and Decision Sciences* 20, 2, 99-112.

Duchessi, P. and Belardo, S. (1987), "Lending analysis support system (LASS): An application of a knowledge-based system to support commercial loan analysis", *IEEE Transactions on Systems, Man, and Cybernetics* 17, 4, 608-616.

Eisenbeis, R. (1977), "The pitfalls in the application of discriminant analysis in business, finance and economics", *The Journal of Finance* 32, 723-739.

Elmer, P.J. and Borowski, D.M. (1988), "An expert system approach to financial analysis: The case of S&L bankruptcy", *Financial Management* 17, 66-76.

Gupta, M.C. and Huefner, R.J. (1972), "A cluster analysis study of financial ratios and industry characteristics", *Journal of Accounting Research*, Spring, 77-95.

Jablonský, J. (1993), "Multicriteria evaluation of clients in financial houses", *Central European Journal for Operations Research and Economics* 2, 3, 257-264.

Jacquet-Lagrèze, E. (1995), "An application of the UTA discriminant model for the evaluation of R & D projects", in: P.M. Pardalos, Y. Siskos, C. Zopounidis (eds.), *Advances in Multicriteria Analysis*, Kluwer Academic Publishers, Dordrecht, 203-211.

Jacquet-Lagrèze, E. and Siskos, Y. (1982), "Assessing a set of additive utility functions for multicriteria decision making: The UTA method", *European Journal of Operational Research* 10, 151-164.

Jensen, R.E. (1971), "A cluster analysis study of financial performance of selected firms", *The Accounting Review* XLVI, January, 36-56.

Keasey, K., McGuinness, P. and Short, H. (1990), "Multilogit approach to predicting corporate failure-Further analysis and the issue of signal consistency", *OMEGA* 18, 1, 85-94.

Khoury, N.T. and Martel, J.M. (1990), "The relationship between risk-return characteristics of mutual funds and their size", *Finance* 11, 2, 67-82.

Klein, M. and Methlie, L.B. (1995), *Knowledge based decision support systems with applications in business* (2nd ed.), John Wiley & Sons, Chichester.

Mareschal, B. and Brans, J.P. (1991), "BANKADVISER: An industrial evaluation system", *European Journal of Operational Research* 54, 318-324.

Martel, J.M. and Khoury N. (1994), "Une alternative à l'analyse discriminante en prévision de faillite: Un indice multicritère", *ASAC '94*, Halifax, Nouvelle Ecosse.

Martin, D. (1977), "Early warning of bank failure: A logit regression approach", *Journal of Banking and Finance* 1, 249-276.

Massaglia, R. and Ostanello, A. (1991), "N-TOMIC: A support system for multicriteria segmentation problems", in: P. Korhonen (ed.), *International Workshop on Multicriteria Decision Support*, Lecture Notes in Economics and Mathematics Systems 356, Springer-Verlag, Berlin, 167-174.

Matsatsinis, N.F., Doumpos M. and Zopounidis, C. (1997), "Knowledge acquisition and representation for expert systems in the field of financial analysis", *Expert Systems with Applications: An International Journal* 12, 2, (in press).

Messier, W.F. and Hansen, J.V. (1988), "Inducing rules for expert system development: An example using default and bankruptcy data", *Management Science* 34, 12, 1403-1415.

Michalopoulos, M. and Zopounidis, C. (1993), "An expert system for the assessment of bankruptcy risk", in: B. Papathanassiou and K. Paparrizos (eds.), *Proceedings of 2nd Balkan Conference on Operational Research*, 151-163.

Peel, M.J., (1987), "Timeliness of private company reports predicting corporate failure", *Investment Analysis* 83, 23-27.

Pomerol, J.C. and Ibp, L. (1993), "Multicriteria DSSs: State of the art and problems", *Central European Journal for Operations Research and Economics* 2, 3, 197-211.

Roy, B. (1981), "A multicriteria analysis for trichotomic segmentation problems", in: P. Nijkamp and J. Spronk (eds.), *Operational Methods*, Gower Press, 245-257.

Roy, B. (1996), *Multicriteria methodology for decision aiding*, Kluwer Academic Publishers, Dordrecht.

Roy, B. and Moscarola, J. (1977), "Procédure automatique d'examem de dossiers fondée sur une segmentation trichotomique en présence de critères multiples", *RAIRO Recherche Operationnele* 11, 2, 145-173.

Serrano-Cinca, C. (1996), "Self organizing neural networks for financial diagnosis", *Decision Support Systems* 17, 227-238.

Shaw, M. and Gentry, J.A. (1988), "Using an expert system with inductive learning to evaluate business loans", *Financial Management*, Autumn, 45-56.

Siskos, Y. and Yannacopoulos, D. (1985), "UTASTAR: An ordinal regression method for building additive value functions", *Investigação Operacional* 5, 1, 39-53.

Siskos, Y., Zopounidis, C. and Pouliezos, A. (1994), "An integrated DSS for financing firms by an industrial development bank in Greece", *Decision Support Systems* 12, 151-168.

Skogsvik, K. (1990), "Current cost accounting ratios as predictors of business failure: The Swedish case", *Journal of Business Finance and Accounting* 17, 1, 137-160.

Slowinski, R. and Zopounidis, C. (1995), "Application of the rough set approach to evaluation of bankruptcy risk", *International Journal of Intelligent Systems in Accounting, Finance and Management* 4, 27-41.

Slowinski, R., Zopounidis, C. and Dimitras, A.I. (1997), "Prediction of company acquisition in Greece by means of the rough set approach", *European Journal of Operational Research* (in press).

Tafti, M.H.A. and Nikbakht, E. (1993), "Neural networks and expert systems: New horizons in business finance applications", *Information Management & Computer Security* 1, 1, 22-28.

Wilson, R.L. and Sharda, R. (1994), "Bankruptcy prediction using neural networks", *Decision Support Systems* 11, 545-557.

Yu, W. (1992), "ELECTRE TRI: Aspects methodologiques et manuel d' utilisation", *Document du LAMSADE*, no 74, Univesité de Paris-Dauphine.

Zopounidis, C. (1987), "A multicriteria decision-making methodology for the evaluation of the risk of failure and an application", *Foundations of Control Engineering* 12, 1, 45-67.

Zopounidis, C. (1995), *Evaluation du risque de défaillance de l'entreprise: Méthodes et cas d'application*, Economica, Paris.

Zopounidis, C. (1997), "Multicriteria decision aid in financial management", *Proceedings of EURO XV-INFORMS XXXIV Joint International Meeting* (in press).

Zopounidis, C., and Doumpos, M. (1996), "Preference disaggregation methodology in segmentation problems: The case of financial distress", in: C. Zopounidis (ed.), *New Operational Approaches for Financial Modelling*, Springer-Verlag, Berlin-Heidelberg (to appear).

Zopounidis, C., Godefroid, M. and Hurson, Ch. (1995), "Designing a multicriteria decision support system for portfolio selection and management", in: J. Janssen, C.H. Skiadas and C. Zopounidis (Eds), *Advances in Stochastic Modelling and Data Analysis*, Kluwer Academic Publishers, Dordrecht, 261-292.

Zopounidis, C., Matsatsinis, N.F. and Doumpos, M. (1996), "Developing a multicriteria knowledge-based decision support system for the assessment of corporate performance and viability: The FINEVA system", *Fuzzy Economic Review* 1, 2, 35-53.

A MATHEMATICAL APPROACH OF DETERMINING BANK RISKS PREMIUM

Jyoti Gupta and Philippe Spieser

Département Finance, Groupe Ecole Supérieure de Commerce de Paris,
79 avenue de la république, 75011 Paris, France.

Abstract: The theory of options allows to establish a system of specific risk premiums designed to insure a bank against bankruptcy on a logic and rational basis because basically, paying a premium transforms a risky deposit into a riskless investment.The basic mechanism is that the payment of a premium is equivalent to the purchase of a put on the assets of the bank. The impact of different parameters on the value of that put is analyzed : level of debts, value of the market of the bank, volatility, pay-out ratio of dividend distribution .
Conclusions related to the management of the firm are drawn from the different simulations.

Keywords: Banking system, risk premium of deposits, bankruptcy, call, put.

INTRODUCTION

The growing number of banks bankruptcies since the middle of the 1970s, the growing volatilty of interest rates in the world, and the deregulation process in developed countries have pushed national monetary authorities and international control organisations onto making mandatory steps in bank management (see BRI-Bank for International Settlements reports)
It can be noticed that in the USA, the number of banks going bankrupt amounted to 6 per annum between 1945 and 1980 with a maximum 17 in 1976, but since then, this number has dramatically increased:

YEAR	1985	1986	1987	1988	1989	1990
N.1	14 400	14 200	13.700	13 600	12 700	12 878
N.2	120	138	184	200	206	169

(continued)

YEAR	1991	1992	1993	1994	1995
N.1	11.920	11200	10100	9850	9600
N.2	127	114	71	37	44

N.1 Number of banks insured by the FDIC
N.2 Number of bankrupcies

What's more, the number of the banks which may set a problem to the FDIC(Federal Deposit Insurance Corporation, which insures banks in the US) stands to about 10% of all the banks covered by this institution in 1991.
But one can nevertheless notice that after 1991 the situation has improved due to two factors:

-the government took measures to reorganize banks ans sometimes even save them from bankruptcy

-there has been a big movement of mergers and acquisitions within the banking industry at the beginning of the nineties .

In France, the agreements of 4 banks were canceled by the Bank Commission and a procedure of "redressement judiciaire" (legal bankruptcy regulation) was undertaken against the Banque de participation et de Placement, United Banking Corporation, Lebanese Arab Bank and the Société Anonyme de Crédit rural de Seine Maritime.

The Commission has pronounced the same sanction against the Industrial Bank of Monaco at the beginning of 1990. Other banks have gone through paramount difficulties, which are now almost got over with-the French and Portugese Bank, the International Bank for West Africa, the French Bank of Trade and Saudi European Bank ; moreover, during the first term of 1991, three banks were deprived of their agreement.
Since then, the number of bankruptcies has also decreased.

Eventually in Germany, although no bank has gone bankrupt since the Herstatt Bank, the new bank regulation of 1985 determined the rules of the business, under the controle of the Federal Office of Bank Controle.

The monetary authorities took two kinds of measures:

-on the one hand, the creation of a guarantee funds to which each bank-member contributes and a system of insurances that enables the members to be paid back whatever occurs, up to a determined threshold. For instance in the U.S, the F.D.I.C is warrant up to 100 000$ by deposit (1990).The premium paid by each bank is determined in proportion to deposits, regardless of the real value of their assets,

-on the other hand, the organization of a set of rules to increase the safety of the banks : indicators like the liquidity ratio, a minimum amount of capital, the division of risks etc. are now used.

The cornerstone of the system proposed is the prudential ratio adopted by the council of European Communities on the 18th of December 1989 and simultaneously by thr B.R.I by the name of Ratio Cooke.

In France, the supervision of credit institutions is on the responsability of the "Banking Commission" (Commission Bancaire") according to the law issued on the 24 th of January 89. The features of the French system is the existence of a collective mechanism of support to the banks when one of them has difficulties instead of a guarantee funds.

The premium system based on deposits and regardless of the quality of the assets seems risky, because two banks owning the same amount of deposits would pay the same premium whatever the nature of their engagements and their cash, the quality of the management of their assets and liabilities etc. Moreover, one can consider that a system of uniform rate premium is against the principle of free competition and is against the very logics of a deregulation process.

To raise that contradiction, two different approaches can be adopted by monetary Authorities :

-either the amount of the premium is set in fonction of the risk of the bank, which means the quality of its management has to be taken into account,
-or a system of control can be determined, which forces banks to reinforce their social capital according to their assets estimated both quantitatively and qualitatively.

The main argument against the first approach was to remind that a system of different premiums is unacceptable because it demands an accurate estimation of the risk of a bank in fonction of observed data. If such a criterium does not exist, the premiums are determined on a more or less subjective and therefore inacceptable basis.

The aim of this article is to show that the theory of options allows to establish a system of specific premiums on a logic and rational basis because basically, paying a premium transforms a risky deposit into a riskless investment.The basic mechanism is that the payment of a prime is equivalent to the purchase of a put on the assets of the bank.
If debts are superior to the net asset of the bank, the premium is a guarantee for the creditors of the bank. Whatever happens, individual depositors are anyway paid back and interbank credits are taken into account to avoid an overall reaction of the bank system.The model we develop here is based on the works of Merton (1973), Black and Scholes (1973), Ron and Verma (1986) (cf bibliography). But previous works did not simulate the different impact of the parameters on the premium level.

This article has two main parts :
-the first one describes the analytic environment of the model;
-the second deals with the study of the impact of different parameters : level of debts, value of the market of the bank, volatility, rate of dividend distribution.

1 ANALYTIC APPROACH OF THE MODEL

1.1 General approach

According to Merton (1977), insuring a bond issued by a company against the risk of no pay-back is equivalent to the purchase of a put option of an insurance institution. The maturity of the option must be, in that case, equal to the maturity of the bond, and the exercise price of the option is the same as the price at which the bond is paid back.
Still according to Merton, this hypothesis is justified even in the case of a bank that owns a portfolio of debts (a set of bonds with different maturities and even deposit accounts)
The arguments used to put together the vehicles of debts stem from the observation that audit made by control authorities equal the true maturity of those debts to the period between two missions of control.

We adopt following notation :
VA = market value of assets
D = total Debts
T = period between two audits
V= instantaneous standard deviation of the return of assets

The formula giving the insurance premium 'd' for a deposit of 1FF is the following :

$$d=N(y+\sigma V\sqrt{T}) - (1-\delta)^n VA/ D N(y) \quad \text{(Eq 1)}$$

N being the cumulted density function of the normal gaussian law

and

$$y= (Ln(D/(1-\delta)^n VA) - \sigma^2 VT/2) / \sigma V\sqrt{T}$$

Three remarks about this equation :

- The value of 'd' is independent of the riskfree rate, which can be accounted for by the fact that in the Black-Scholes (1973) model, the riskfree rate is resorted to update the striking price, whereas in our case D is the present value of debts. One must also keep in mind that interest rates influence the market value of assets and its volatility.
- The monetary authorities follow closely the market value of assets and will inject capital only when VA'<D<VA (VA' is the market value of assets after insurance). Thus, the premium value should be determined in relation with the value of VA and of σV after the premium payment. The link between the premium before and after the insurance procedure must take account of the profits due to the diminished risk and the losses due to an increase in competition. An estimation of these profits

is provided by the difference between the return on deposit greater than the riskfree rate.
So :
VA' = VA + G - P
with :
VA = market value before insurance
VA' = market value after insurance
G = profit or surplus due to the insurance process
P = loss resulting from competition.

But at the equilibrium, the competition between banks compels them to transfer the surplus to clients and the insurance procedure does not threat the market equilibrium. Under these circumstances, the premium must be estimated when G = P and then VA = VA'.

If Merton's approach is justified (estimate the premium that each bank must pay at the optimum), the first problem is the one related to the value of VA and of σA.

1.2. Determination of VA and of σA.

Merton (1977) puts forward an implicit assessment method of VA and of σA based on the Black-Scholes model.
The bank's market stock value can be regarded as an option call on its assets, option which maturity tallies with that of its debts.
If T stands for the maturity of debts, D is the value of debts at maturity,VC the market value of the capital, we get the following formula:

$$VC = VA\ N(x) - D\ N(x - \sigma V \sqrt{T})\text{(Eq.2)}$$

$$\text{with } x = (\text{Ln}\ (VA/D) + \sigma^2\ VT/2)\ /\ \sigma V \sqrt{T}$$

Besides we know that standard-deviations of capital and assets (respectively σC and σV) are linked by the following statement -which is typically an elasticity one :

$$\sigma C/\sigma V = (\ VA/VC\ )\ /\ (\ \delta VC/\delta VA\ ).$$

We finally also know that :

$$(\ \delta VC/\delta VA) = N\ (x)$$

and consequently :

$$\sigma C = [VA\ N(x)\ \sigma V]\ /\ VC \qquad \text{(Eq.3)}$$

We can now calculate σV and VA from equations 2 and 3 since VC and σV are known.

1.3. An improved version

We have modified the formulation of equation 2 by taking account of the fact that when the bank's net market value becomes negative, the control authorities do not systematically require the bank's dissolution.

The first goal of the authorities is to ascertain to what point the bank may be assisted to remain solvent through capital injection. The bank's dissolution is considered as a last resort if $VA<kD$ when $k<1$.
So, when

$kD \leq VA \leq D$,

the authorities inject capital into the bank and the amount of injected funds equals (1-k)D such that VA = D.
The value of k depends on the monetary policy, on the economic and sociological environment of banks. One may remember the example of the Banque Arabe Libanaise and the discussions between the Banque de France and the other banks.

The improved version becomes :

$$VC = VA\,N(x) - k\,D\,N(x - \sigma V\sqrt{T})$$
(Eq. 4)

with : $x = (\mathrm{Ln}\,(VA/kD) + \sigma^2\,VT/2) / \sigma V\sqrt{T}$

$$\sigma V = (\sigma C.VC) / VA\,N(x) \qquad \text{(Eq. 5)}$$

In a nutshell,we can contend that the guarantee given by the authorities is similar to a put option which striking price is the future value of D and of maturity T = 1 (conventionally audits are conducted once a year).
The option is bought by the bank in return for the payment of a premium.
Besides the bank's shareholders hold a call with striking price of
(k*future value of D ≤VA ≤ future value of D) and a maturity equal to 1.

Hence,when :

k*future value of D ≤ VA ≤future value of D

then monetary authorities make capital injections equal to the amount of
(FVD-VA).

In that case, shareholders retain ownership of the bank. However, if VA $\leq$ kFVD authorities inject (FVD-VA) into the bank but the shareholders lose control of it.

1.4 Estimation of the premium 'd'

The different steps of solving the model are the following :

Given observed variables :

VA = market value of the shareholder's equity
D = total debt
σC = volatility estimation of the share price

the first step is to estimate the the equations 4 and 5 to price VA and σV . The solving has led to a numerical convegence algorithm since σV depends on 'x' , this latter depending on VC and VA.. With initial estimations of VC and VA we calculated a first estimation of VC and we let VC converge to the observed values.- the convergence was fortunately rapid .

We used the following estimations for the initial solution :
VA = market capitalization and total debts
σ V = σC VC/D

1.5 Problem of minimum social capital

The model provides us with a solution to the following question: if the authorities choose a constant premium what should be the minimum social capital required? .
If α is a constant premium defined by the authorities, the quantity I of supplementary capital that the bank should raise to increase the value of 'd' is given by the following steps:

We know that (cf. Eq 1)

$$d = N(y + V\sqrt{T}) - (1 - \delta)^n \, VA/D \, N(y)$$

where $y = (\mathrm{Ln}\,[D / (1-\delta)^n\, VA] - \sigma^2\, VT/2) / V\sqrt{T}$

and T maturity equal to 1.

Moreover the input of supplementary capital will increase the market value of assets by a quantity of I and simultaneously decrease the volatility of the assets σV .

The market value will reach VA + I.

Under those conditions, the volatility will decrease by a rythm that Ron and Verma describe as linear:

$\sigma V' = \sigma V\,(VA / (VA + I))$

where $\sigma V'$ refers to the instantaneous standard error after input of new capital. This injection of supplementary capital has positive effects on the financial situation of the bank since it reduces the ratio Debts / Total assets and improves simultaneously the prudential ratio Capital/ Total assets.
Then the risk on the assets decreases immediately after that but when liquidities are reinjected, the volatility σV varies according to the risk of the new assets acquired by the bank.

If the risk on those new assets is identical to the former one, $\sigma V' = \sigma V$ and we get:

$$\alpha = N\,[y + \sigma V\, VA/(VA+1)] - (1-\delta)^2\,(VA + 1) / DN(y)$$

$$y = [\mathrm{Ln}\,(D/ (1 - \delta)^n\,(VA + 1))] - \sigma^2 V/2(VA+I))]/ \sigma V\,(VA/VA + I)$$

This system of equations allows us to calculate I.

We can go now to the simulation phase.

II.- Simulations

The first simulation that we will comment is the relation between the insurance premium ond the pay-out ratio('d' and δ).

We initialized the data at the following levels :

D = debts = 96

VA = market value of asset =99

VC= market value of the capital = 3
T= period between two audits = 1
σC= instant volatility of the market =0.3
K = maximum coefficient beyond which the bank is liquidated

The convegence process led to a value of the assets of 99.
We selected different values of δ and estimated the values of the premium by using the equation system 1 to 4.

Chart 1 gives the premium value for different levels of δ.

The study if the link between the pay-out ratio and the premium shows a high sensitity which can be mathematically explained by the convexity of the second derivative of the option price function and economically by the fact that the higher the dividends are, the less profits can increase social capital . We may say that there is an optimal trade-off between the pay-out ratio that satisfies tha shareholders and the acceptable level of the premium that the bank has to pay. If the financial situation is deteriorating, the mangers dispose of an efficient weapon to reduce the premium: to convince the shareholders to accept a lower dividend.

The graphical representation is the following one

Chart 1

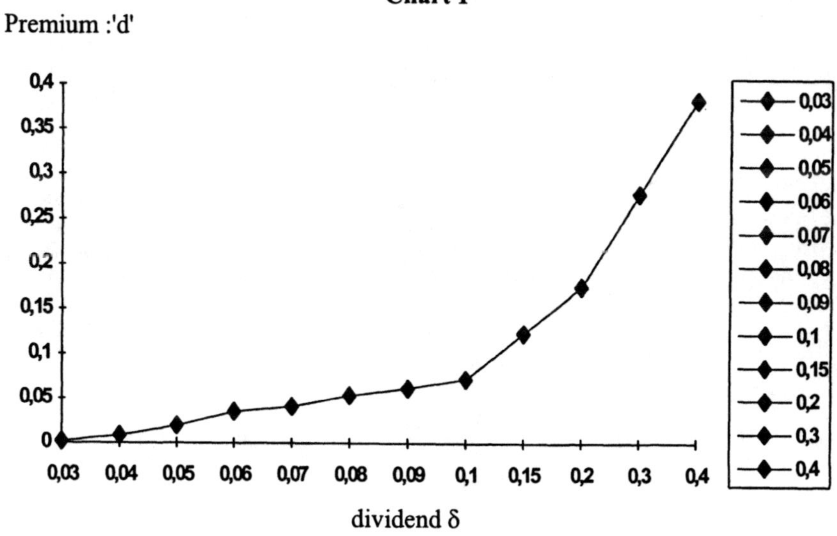

The other simulations we will comment are the following ones:

2.1 the premium 'd' for different levels of σC (standard deviation of the value of the capital)

The chart 2 below gives 'd' as a function of σC

with following inputs :
VC=3 K=0.95 VA=99 D=96
δ=0,01 T=1 VA=99

The figures not given here show that the premium growths exponentially such as for small values of σ and for a steady growth (by 0.05), the premium increases almost linearly in the beginning, then at a faster pace.This is quite logical since a high volatility shows that the banks assets are invested in risky portfolios.
An other conclusion seems to be of noticeable importance: since the premium is sensitive to volatility, we may assert that an efficient way to reduce the premium paid by the bank is to manage dynamically the share price. The behavior of this price must be regarded as a strategical parameter of the Finance Department. The management is partially responsible for the level of the premium and must take all decisions likely to reduce the premium prior any constrainig decison of the Central Bank.

Chart 2

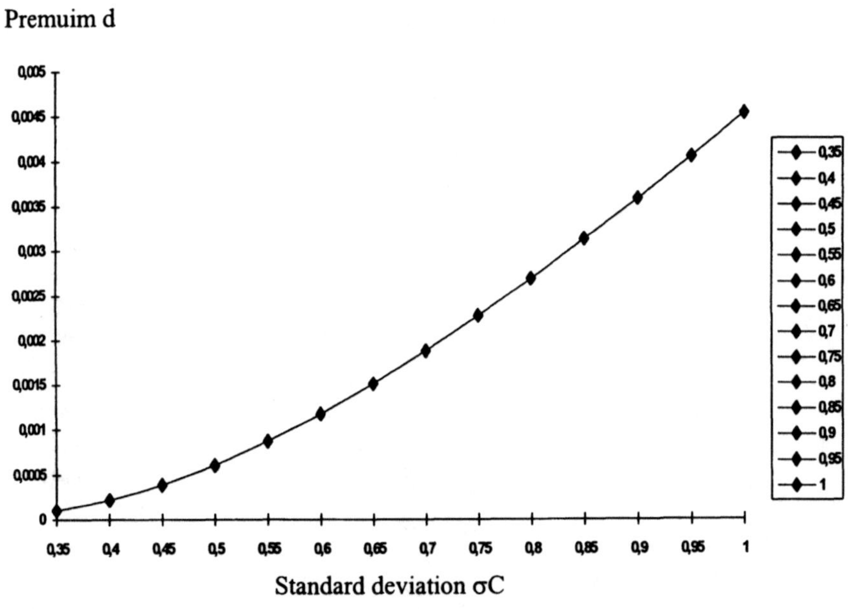

2.2 The influence due to the input of supplementary capital on the premium's value

Initial conditions are the following ones:
VC =5 VA = 100.2 D =95
σC =0.3 K =0.95 n=1

The following chart gives the α value for different values of I,incremented by 0.05 (cf chart3)

Chart 3

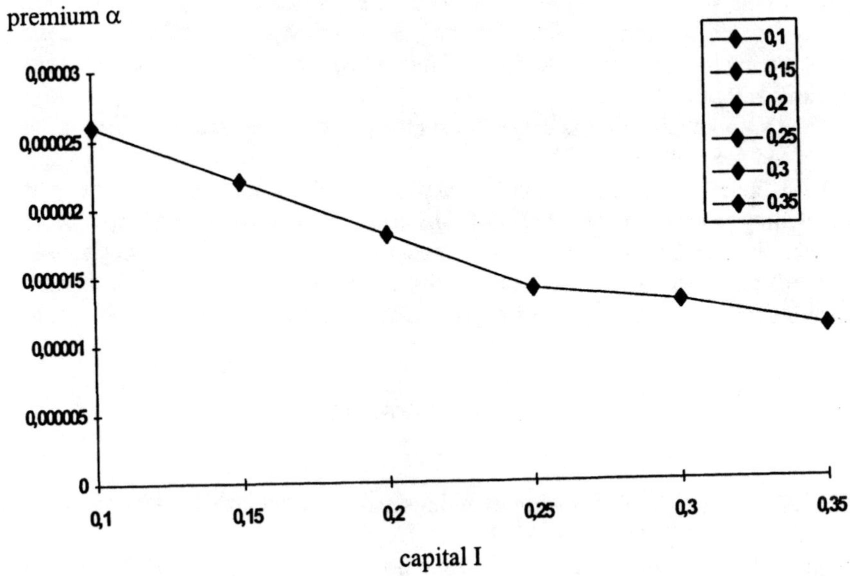

We also made a simulation by using other input data closer to realistic figures and above all offering a higher sensitivity of α related to I (increased by 0,5). with following inputs

VA = 1000 D = 991 VC = 9

K = 0.95 δ = 0.01

Chart 4

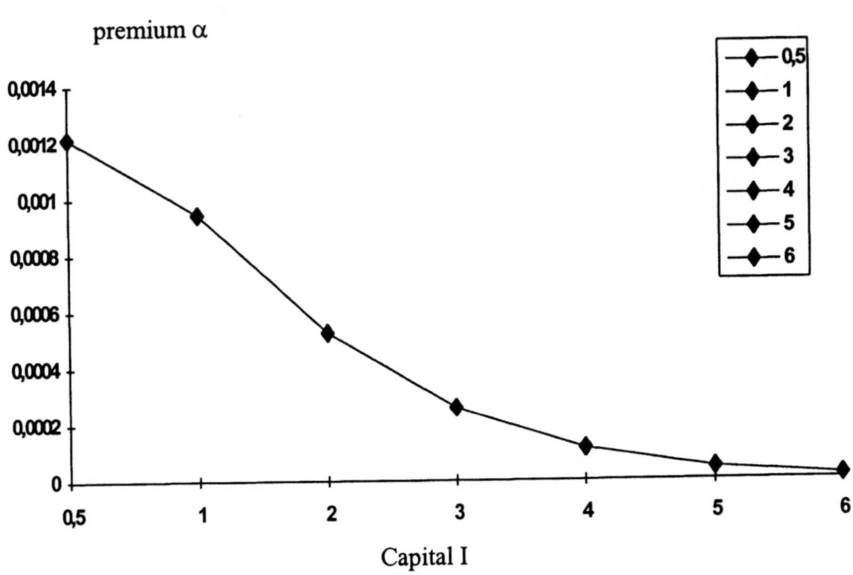

CONCLUSION

The conclusion we may draw from the tables are numereous:

-it is very useful to simulate the impacts of different parameters to rank the different parameters according tto thier relative efficiency. This has not been done by previous authors.

- the input of capital is very efficient to reduce the premium -we may even say more and more efficient.

Indeed the function between α and I is exponential. This reveals that, when the firm is willing to raise capital, it should raise as much as it can being aware of its strategical objectives in its shareholders "geography". Of course it is much more difficult to raise capital than to look for the volatility of its sharebut it is more efficient because more lasting. It manages a significant growth potential.

BIBLIOGRAPHY

1. Black F.Scholes M. "The pricing of options and corporate liabilities", *Journal of Political Economy*, May-June 1973.
2. Merton R.C., Theory of rational option pricing, *Bell Journal of Economics* 1973.
3. Merton R.C., Analytic derivation of cost of deposit insurance and loan deposit, *Journal of Banking and Finance*, June1977.
4. Merton, R.C., On the cost of deposit insurance when there are surveillance costs, *Journal of Business*, July 1978.
5. Merton, R.C., *Continuous Time Finance*, Blackwell 1990.
6. Ron and Verma, Pricing risk adjested deposit insurance and option based model, *Journal of Finance*, September 1986.

III. LINEAR AND STOCHASTIC PROGRAMMING IN PORTFOLIO MANAGEMENT

DESIGNING CALLABLE BONDS USING SIMULATED ANNEALING [1]

Martin R. Holmer [1], Dafeng Yang [2], Stavros A. Zenios [2]

[1] HR&A, Inc, Washington, DC 20005.

[2] Operations & Information Management Department, The Wharton School, University of Pennsylvania, Philadelphia, PA 19104.

Abstract: Where it is shown that the design of a new issue of callable bonds — its lockout period, the schedule of redemption prices, and its time to maturity — need not be an art, but can be a science.

Keywords: Simulated annealing, Optimization, Callable bonds

INTRODUCTION

Quantitative techniques for pricing callable bonds are today well understood and widely adopted in practice. Such techniques range from the estimation of the term structure of credit spreads for corporate debt [9], to the estimation of risk premia due to the call option, liquidity and default risk of these securities [8, 7], and the calculation of durations, convexities and the like [2].

These tools are serving well portfolio managers, investors and traders. For example, the manager of a portfolio with callable assets can apply these tools to conduct a rich/cheap analysis of available bonds to guide trading actions. Or she can estimate the duration and convexity of several bonds, and design a portfolio which is duration- and convexity-matched against her liabilities.

Unfortunately, existing tools are not of much assistance to institutions who *issue* callable debt. In particular, the design of a particular issue of callable bonds — i.e., specification of the *lockout period* during which the bonds can not be called, a schedule of *redemption prices* thereafter, and the time to *maturity* — still remains an art. Nevertheless, bonds with embedded options — callable or putable bonds, and

[1] Research partially supported by NSF grant CCR-91-04042. This work was completed while S.A. Zenios was visiting the Universities of Urbino and Bergamo, Italy, under a fellowship from the GNAFA and GNIM groups of Consiglio Nazionale delle Richerche (CNR). He is currently visiting Professor of Management Science, Department of Public and Business Administration, University of Cyprus, Nicosia, CYPRUS.

bonds with sinking fund provisions — are major borrowing instrument for corporations, utilities, government agencies, and other financial institutions.

For example, agencies like Fannie Mae and Freddie Mac fund more than 95% of their mortgage asset portfolios via debt issuance. Their liabilities exceed today $150 billion. Much of this debt is in the form of callable bonds. The rationale for this choice is simple. Callable bonds can be retired in tandem with the mortgage assets: as interest rates drop and mortgages prepay, the agencies may call the bonds.

Herein lies an important observation: Bonds are issued to finance a specific collection of assets. Hence, they should be "designed for a purpose"! It is not sufficient to examine only the structure of the liabilities. The characteristics of the funded assets should be integrated in the design of the bond issue. This is an example of *integrated product management* [5]: The design of the financial product should be integrated with the asset side of the balance sheet. Then we can define the *quality* of the product in conjunction with the purpose for which it was designed in the first place. With a measure of quality at hand we can then fine tune the design specifications of the product to improve its quality.

Let us draw a simple analogy from manufacturing. Is it a sign of a "good quality" car that it has high mileage? Or that it goes from 0 to 60 mph in 6 seconds? Or that passengers can open the door and walk away after a 40 mph frontal impact? High mileage and safety are characteristics of a good family sedan. Superior acceleration and stability are the predominant characteristics of good racing cars. The car designer does not start the design until the potential market of the new vehicle is first established.

Returning to the world of financial products, the usual goal of issuing callable bonds with the lowest average coupon is short-sighted. It ignores the purpose of the product. Instead, the product should be designed so that its return — during the institution's portfolio planning period — moves in tandem with the return of the assets. Such co-movements should take place under a wide variety of economic scenarios. The decision maker's attitude towards risk — i.e., the level of risk aversion — should also be taken into consideration.

This paper develops a systematic framework for *designing* callable bonds that can be retired in tandem with the assets, even under extreme interest rate scenarios. We first define a measure of the *quality* of the financial product, that incorporates the decision maker's risk preference. The bond design problem is then parametrized using a few simple, and widely accepted, parameters. A methodology is then developed to fine-tune these parameters — either heuristically or using a *simulated annealing* optimization method — to design a product of the highest quality.

A SYSTEMATIC FRAMEWORK FOR DESIGNING CALLABLE BONDS

The procedure for designing a callable bond — or any other financial product for that matter — consists of two major components. First, we need a quantifiable measure of the *quality* of the product. Second, we need to parametrize the design characteristics of the financial product. For a callable bond these characteristics are the *lockout period* during which the bond can not be called, a schedule of *redemption prices* after the end of the lockout period, and the time to *maturity* of the bond. Once the design characteristics are parametrized our framework specifies procedures to adjust these parameters so that the measure of quality is maximized.

In this paper we consider the issuance of callable debt to fund a mortgage asset. This is the problem faced, primarily, by the secondary mortgage market agencies. Focusing on a single problem does not diminish the generality of the proposed framework but it simplifies our discussion. A similar framework can be adopted by corporations and utilities that issue callable debt to fund their general assets. Few modifications are needed in designing, for example, insurance policies (like SPDAs, GICs etc) to be held against the general assets of an insurer.

Measuring the Quality of a Callable Bond

Consider a financial institution that raises \$100 in funds by issuing callable bonds at par prices. In addition to these borrowed funds, it invests E dollars from the shareholder's equity to purchase \$$(100+E)$ in assets. Assume that at the end of the holding period the liabilities realize a return of $1 + R_L$ and the assets a return of $1 + R_A$. The position of the institution at the end of this holding period is given by the terminal wealth WT:

$$\begin{aligned} WT &= (100 + E)(1 + R_A) - 100(1 + R_L) \\ &= 100(R_A - R_L) + E(1 + R_A) \end{aligned} \tag{1}$$

The return to the shareholders is given by the *Return on Equity* (ROE):

$$ROE = \frac{WT}{E}$$

$$= \quad 1 + R_A + 100\frac{R_A - R_L}{E} \tag{2}$$

Note that if the return on the assets is greater than the return on the liabilities (i.e., $R_A - R_L > 0$) the shareholders may realize infinite ROE just by investing zero equity in the first place. What is wrong? We assumed known returns on both assets and liabilities during the holding period. Unfortunately, returns are uncertain and may take several values — which we call *scenarios* — during the holding period. Furthermore, the return on the assets can not exceed the return of the liabilities for all scenarios. This would imply the presence of riskless arbitrage profits in the financial markets.

To complete the analysis we specify a set of *scenarios* $\Omega = \{1, 2, 3, \ldots, S\}$. For simplicity we assume that all scenarios are equally likely, and they can occur with probability $1/S$. With each element of this set $s \in \Omega$ we associate a return for the assets R^s_A and a return for the liabilities R^s_L. These returns are driven by realizations of interest rates [12]. Figure 1 illustrates the holding period returns of a mortgage-backed security and three different bonds under alternative interest rate scenarios, during a 36-month holding period.

Terminal wealth and *ROE* are now scenario dependent. They are given by:

$$WT^s = 100(R^s_A - R^s_L) + E(1 + R^s_A) \tag{3}$$

$$ROE^s = 1 + R^s_A + 100\frac{R^s_A - R^s_L}{E}. \tag{4}$$

We mention, as an aside, that the market-value solvency of the institution — i.e., having positive terminal wealth — can be guaranteed if sufficient equity is invested. It is easy to confirm that the terminal wealth WT^s is non-negative if $E > 100\frac{R^s_L - R^s_A}{1+R^s_A}$, for all scenarios $s \in \Omega$. (Under the extreme case scenario, however, the *ROE* is zero.)

Risk Aversion and Certainty Equivalent Return on Equity

The problem of designing a callable bond, that maximizes return on equity, is now complicated by the fact that no single bond is likely to have superior ROE^s for all plausible scenarios $s \in \Omega$. For example, it is obvious from Figure 1 that bonds B and C are superior to bond A. But the comparison between bonds B and C is not so easy.

To compare alternative bond designs we need to incorporate the decision maker's risk aversion for different levels of *ROE*. Of particular

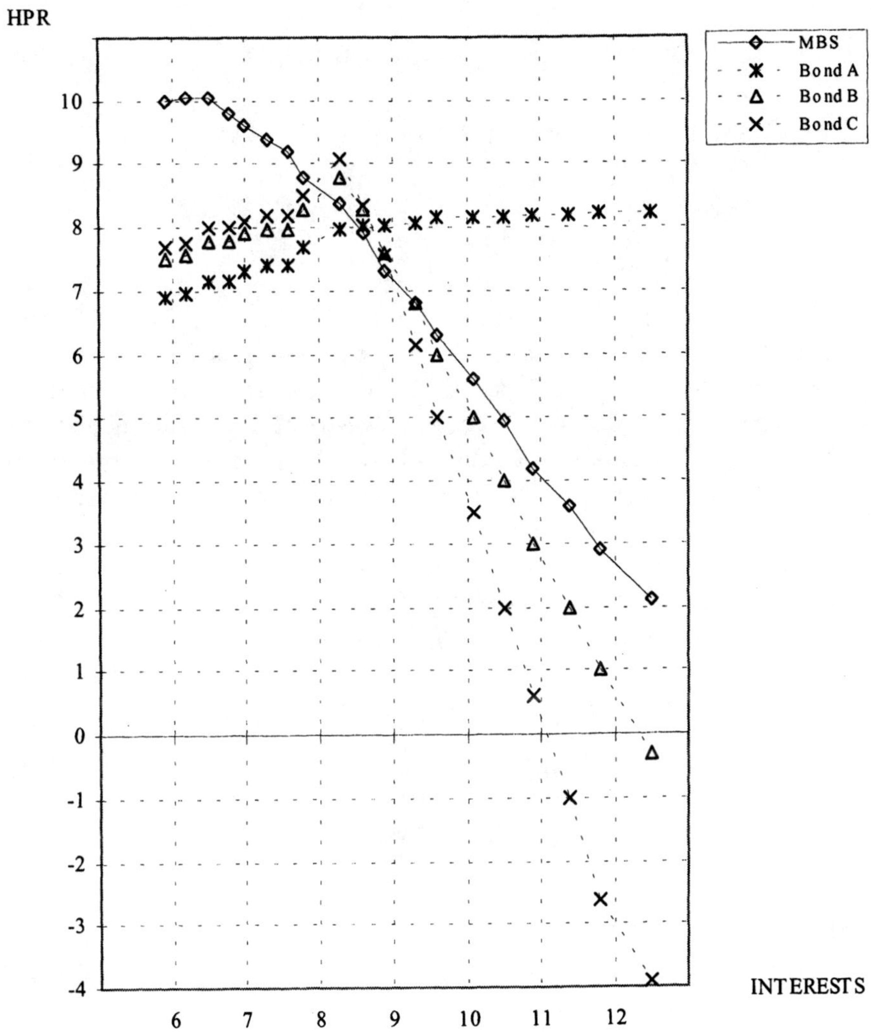

Figure 1: Holding period returns of a mortgage-backed security (MBS) and three alternative callable bonds, under different interest rate scenarios. In this example it is obvious that bonds B and C are preferable liabilities to hold against the MBS asset, as opposed to bond A.

interest, in this regard, are scenarios under which the *ROE* is less than one. Let $\mathcal{U}\{\cdot\}$ denote the utility function of the decision maker. Then — under the expected utility maximization hypothesis — the issuing institution will prefer the bond that maximizes the expected utility

$$\frac{1}{S}\sum_{s\in\Omega}\mathcal{U}\{ROE^s\}. \tag{5}$$

With each bond we can associate its *Certainty Equivalent Return on Equity* (*CEROE*) defined by

$$\mathcal{U}\{CEROE\} = \frac{1}{S}\sum_{s\in\Omega}\mathcal{U}\{ROE^s\}. \tag{6}$$

The certainty equivalent return is simply that level of return (known with complete certainty) that makes the decision maker indifferent between accepting this level of certain return, versus taking her chances with returns ROE^s. For a *risk neutral* investor the certainty equivalent is the expected value of the S values of ROE^s. A *risk averse* individual is willing to accept a lower certain return than the expected value, in order to avoid playing the odds.

Assuming non-decreasing utility functions (i.e., under the popular maxim that "more is better") we can simply rank bonds by their *CEROE*. The bond with the highest value of CEROE provides the best design. A typical choice of a utility function, which we used in our numerical experiments, is the *logarithmic* utility function

$$\mathcal{U}\{ROE^s\} = \log(ROE^s), \tag{7}$$

where log denotes the natural logarithm.

Designing a Callable Bond

We have, so far, developed a measure of the *quality* of a bond (i.e., its *CEROE*) that incorporates the characteristics of the funded assets, the decision maker's risk attitudes and the stochastic nature of interest rates. In this section we define the bond design parameters and then outline the design procedure.

Design Parameters

A callable bond is unambiguously specified by the following four parameters. These are the *design parameters* of the problem; see Figure 2.

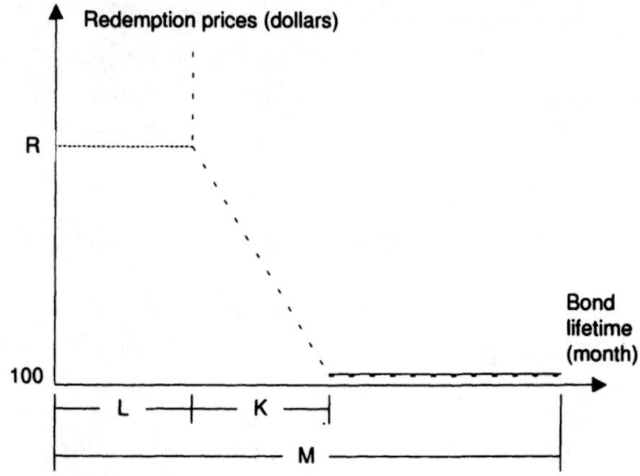

Figure 2: The four parameters of the bond design problem: (1) L lockout period, (2) R redemption price at first call date, (3) M time to maturity, and (4) K time after which the security can be called at par.

Lockout period (L): This is the period following issuance during which the bond can not be called.

Redemption price at first call date (R): This is the price at which the bond can be called at the first call date. This price is usually at a premium above par.

Time to maturity (M): This has the usual meaning for all fixed-income securities.

Schedule of redemption prices: The redemption price at the first call date is at a premium above par. This premium declines as the security approaches maturity. A schedule of redemption prices during the term of the bond are specified by the issuer. We assume, for simplicity, that the redemption price declines linearly from a starting value of R a the end of the lockout period, to par at period K. The term during which the security can be called at a premium (i.e., K) is the design parameter. Other schedules of redemption prices — instead of the linear decline — can be easily specified as well.

Design Procedure

We now develop a systematic procedure for generating bond designs that have as large *CEROE* as possible. Figure 3 gives a schematic representation of this framework.

Initialization: Parametrize the design characteristics of the bond and assume some initial values. The specification of the holding period, generation of the interest rate scenarios, and estimation of the holding period returns for the funded assets, are also part of the initialization step.

Step 1: Estimate the coupon rate (C) that prices the bond at par [12].

Step 2: Generate returns for the bond, using the same interest rate scenarios and holding period used to estimate the returns of the funded assets. Scenarios of holding period returns are estimated [12].

Step 3: Compute the *CEROE* of the specified bond design using equation (6). If some suitable termination criteria are satisfied (i.e., the *CEROE* is "sufficiently" large) the procedure terminates. Otherwise proceed to the next step.

Step 4: Adjust the bond design parameters using some appropriate rule to improve its *CEROE* and return to Step 1.

This iterative procedure seeks designs that are optimal, that is designs that have the largest possible (i.e., maximum) *CEROE*. Of particular interest are the rules by which the bond's design parameters are adjusted in Step 4 in order to maximize the *CEROE*.

We point out that there exist multiple, locally optimal, solutions to the bond design problem. It may be possible to find a design that is optimal only in a small neighborhood of the design parameter. That is, if any of the parameters are adjusted by a small amount the *CEROE* will get worse. However, changing some of the parameters by a larger amount may actually improve the *CEROE* of the design. In this respect we found standard optimization methods inadequate. Superior results were obtained using *simulated annealing*, a method well suited for problems with multiple local optima [6].

The next two sections develop a series of rules for adjusting the bond design parameters in Step 4. As the rules become more complex, they result in improved bond designs. First, we develop rules that identify locally optimal solutions, and then develop the simulated annealing method for identifying an overall best.

All experiments were carried out on the problem of designing a callable bond to fund a mortgage-backed security with weighted average coupon (WAC) 9.5% and weighted average maturity (WAM) of 360 months. The holding period return of the mortgage security, during a 36-month holding period, is shown in Figure 1.

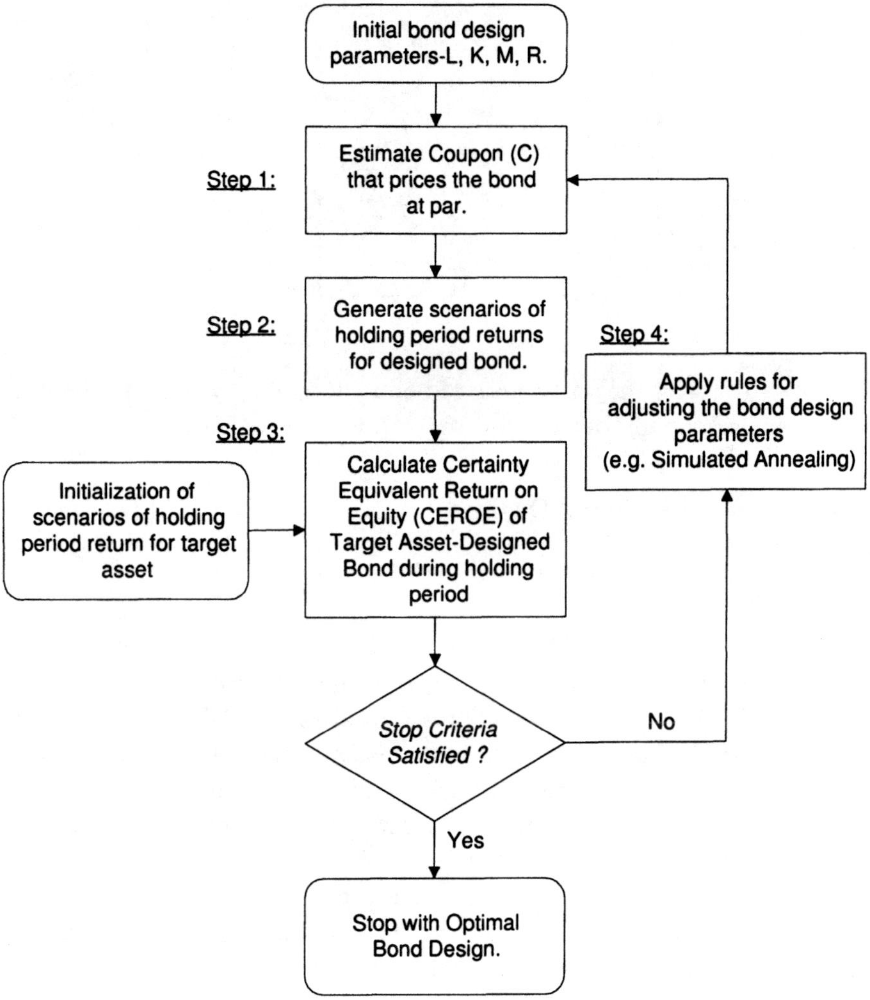

Figure 3: A framework for the optimal design of callable bonds.

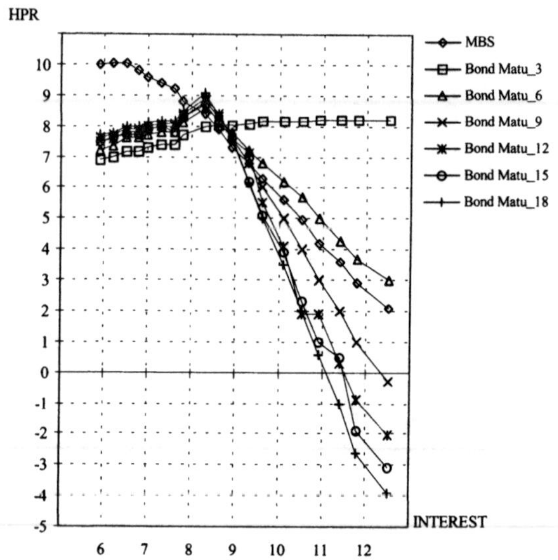

Figure 4: Holding period returns of bonds designed by varying the maturity, while keeping lockout and redemption prices fixed.

Locally Optimal Bond Designs

Single-parameter Optimizations

The first rule makes one-dimensional changes of the bond design parameters. One parameter is adjusted at a time, while all others are kept constant. Figure 4 shows the holding period returns of bonds designed with the lockout fixed at $L = 12$, redemption price starting at $R = 100$ and remaining constant until maturity, and with varying maturity. The $CEROE$ of increasing maturities is calculated at 12-month steps. The best $CEROE$ achieved with this rule is 1.4085.

Similarly, Figure 5 shows the holding period returns of bonds designed with maturity fixed at $M = 336$, redemption price starting at $R = 100$ and remaining constant until maturity, and with varying lockout. The best $CEROE$ is 1.3726.

Double-parameter Optimization

Once some preliminary experimentation with the single parameter optimizations is completed we may proceed to adjust two parameters simultaneously. We already have a good understanding of what range of values of maturity and lockout will produce high $CEROE$. Figure 6

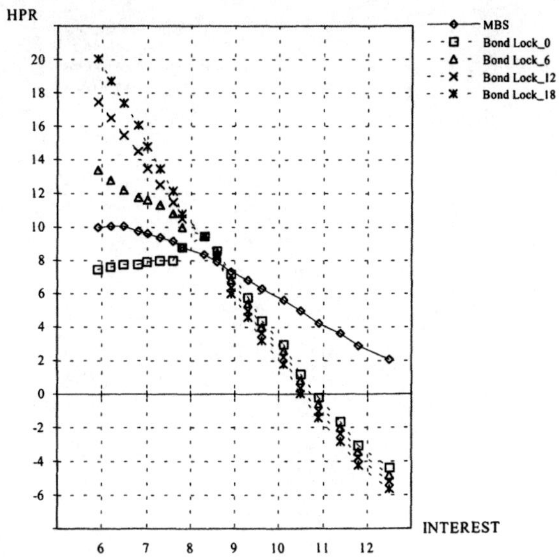

Figure 5: Holding period returns of bonds designed by varying the lock-out period, while keeping maturity and redemption prices fixed.

illustrates the *CEROE* of bonds designed using this approach. The best bond thus designed has $L = 0$ and $M = 288$, while its redemption prices were kept fixed at $R = 100$. Its *CEROE* is 1.4099.

Multi-parameter Optimization

The single- and double-parameter optimizations of the previous sections were done by trial-and-error. A fair amount of user intervention is required in order to determine the next trial point of parameter setting. We now automate this process, and in doing so we permit the optimization of the design by varying more than two parameters at a time.

The objective of the design is to achieve the maximum *CEROE*, which is a function of the four design variables. Hence, we can express the problem as the optimization program:

$$\text{Maximize}_{\,L,R,M,K}\ CEROE = \Phi(L, R, M, K). \qquad (8)$$

The *CEROE*, which is a function of the design parameters, is maximized over the search space of the four parameters. Note, however, that the function Φ is not explicitly specified. Instead, the value of Φ for a given setting of its arguments L, R, M, K is obtained from Steps 1 to 3

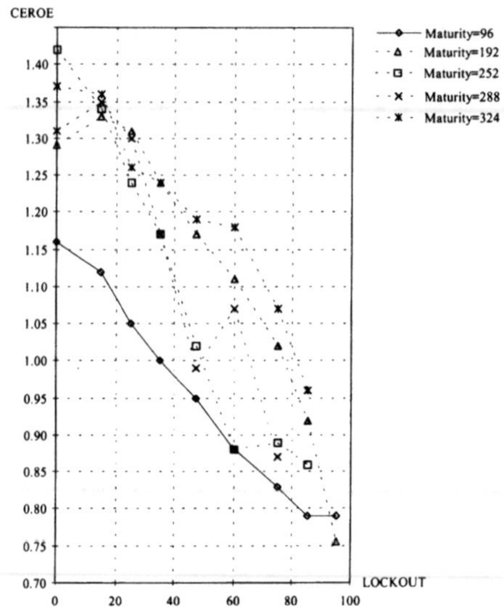

Figure 6: Certainty Equivalent Return on Equity (*CEROE*) of bonds designed by varying simultaneously the lockout period and the maturity of the bond.

of the design procedure (Figure 3). We can use any unconstrained optimization algorithm to search the space of the four design variables to solve this maximization program. In our work we used Powell's method [11]. Figure 7 illustrates the holding period return for bonds designed with this method. Note that the algorithm converged to different "optimal" designs depending on the initial value of the parameters! This was anticipated, as the function Φ does not necessarily have a unique global maximum. Instead, alternative local solutions are obtained depending on the starting point. In our experiments we obtained two alternative solutions with CEROE 1.2316 and 1.4056.

Benchmark: Duration-matched Bond Design

It is common in fixed-income portfolio management to match assets and liabilities on a *duration* basis. A duration matched portfolio implies that the asset-liability gap will remain invariant, for small changes of interest rates. While the shortcomings of duration matching are well understood — and we do not stop to repeat them here — this approach remains quite popular.

Using the double-parameter optimization rule we designed several bonds and for each one we calculated its *option adjusted duration* [12].

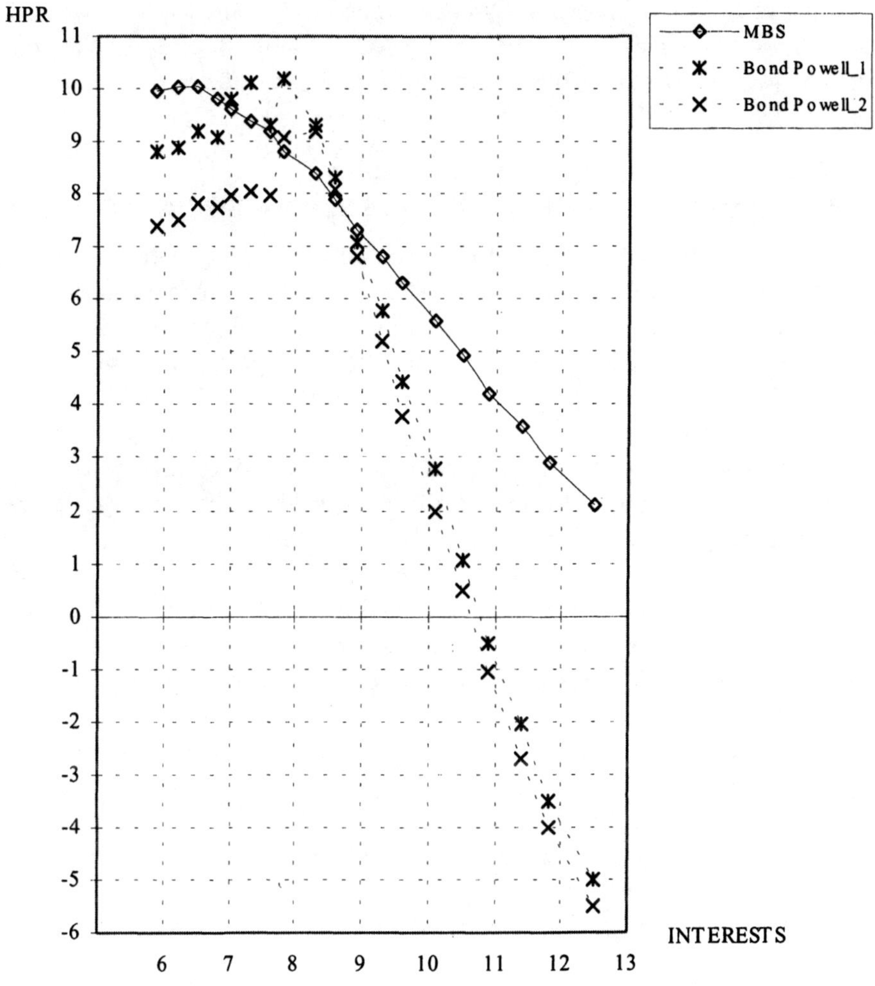

Figure 7: Holding period returns of two bonds designed by varying simultaneously their parameters, using an optimization algorithm. These bonds are examples of local optimal solutions. Their *CEROE* are 1.2316 and 1.4056.

Figure 8 illustrates the duration of bonds with varying lockout and maturity parameters. From this figure we can easily identify those bonds that have the same duration like the funded mortgage assets. For each such bond we can also calculate its *CEROE*. The best *CEROE* thus obtained (1.3313) is for a bond with $L = 3$, $M = 120$, $R = 100.00$ and $K = 0$.

Globally Optimal Bond Designs through Simulated Annealing

We have already mentioned that the bond design problem does not have a unique, global optimal, solution. To solve such optimization problems we resorted to the method of *simulated annealing*.

Simulated annealing is a method for difficult optimization problems, developed based on the physics of *metal annealing* [6]. The analogy with physics is simple. Scientists noticed that when a metal is cooled from a liquid to a solid state, its molecules may reach different equilibrium states of low energy. The lowest the energy level the stronger is the metal. Fast quenching — that is, cooling the metal very quickly — may lead to a solid with relatively high energy, and the metal is brittle. If, instead, the metal is reheated — so that its molecules reach a higher level of energy than the current equilibrium state — and then it is re-cooled, then most likely the new state will have lower energy. The process is repeated several times.

A simulated annealing algorithm, applied to the generic optimization problem

$$\text{Minimize}_{\,x} \quad = F(x),$$

proceeds as follows:

Initialization: Choose the initial temperature $T = T_0$ and the speed of annealing α, with $0 \leq \alpha \leq 1$. Large values of α correspond to a slow annealing process which has the highest probability of finding a global solution. $\alpha = 0$ corresponds to fast quenching, and the solution reached is unlikely to be globally optimal. Choose a starting point $x = x^0$ with corresponding "energy" value $= F(x^0)$.

Step Calculation Rule: Calculate a new trial point x' and evaluate its energy $' = F(x')$.

Step Acceptance Rule: Let $\Delta = -'$. If $\Delta > 0$ then the trial point x' is accepted. Set $x \leftarrow x'$, $\leftarrow'$ and proceed with a new Step Calculation.

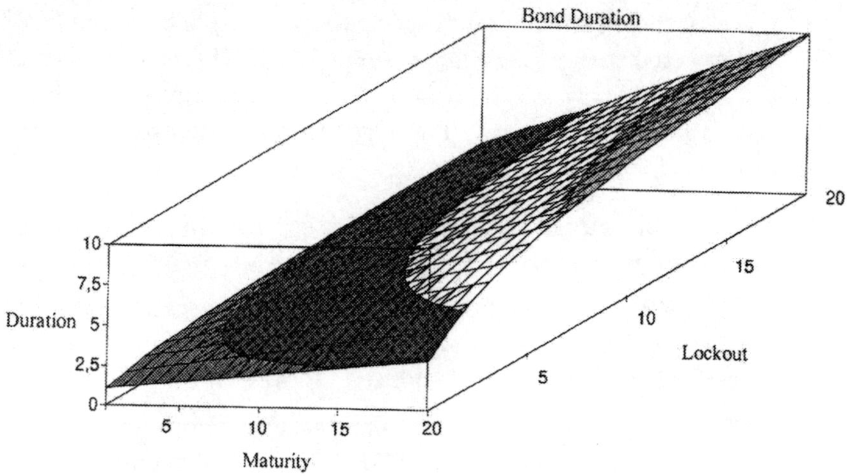

(1)

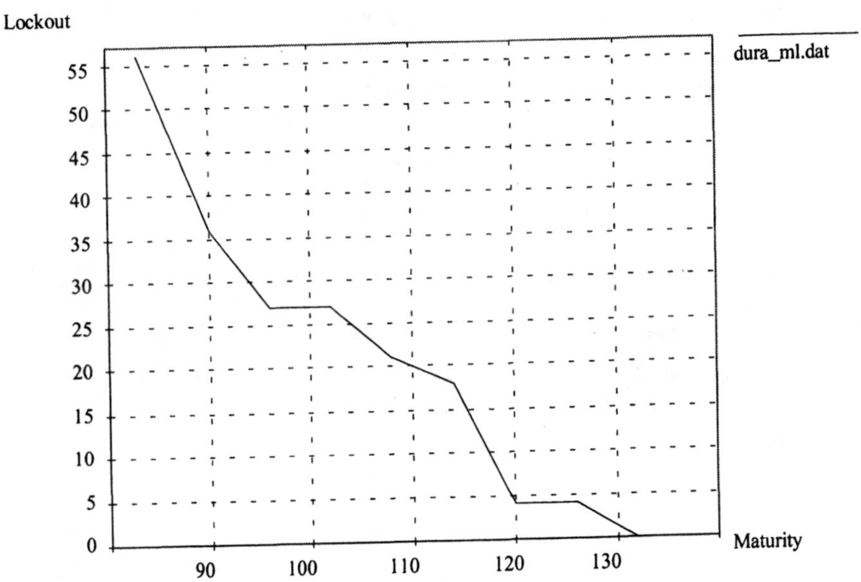

(2)

Figure 8: (1) Option adjusted duration of bonds with varying lockout and maturity parameters. (2) The combination of maturity and lockout periods that result into callable bonds with identical option adjusted duration like the target mortgage asset.

If $\Delta < 0$ then the trial point is accepted with probability $\exp(\Delta/T)$, and rejected with probability $1-\exp(\Delta/T)$. This rule implies that a step may be taken even if it *increases* the energy value . Such a step is known as a *jump*. The algorithm uses jumps to get away from local solutions.

Termination Check: If the number of steps has not exceeded some maximum number (NSTEPS) proceed with a Step Calculation.

If NSTEPS steps have been taken at this temperature, and some jumps have been made, reduce the temperature to $T \leftarrow \alpha T$ and proceed with a Step Calculation.

If NSTEPS steps have been taken at this temperature but no jumps have been made then STOP. The current point is optimal.

We applied the simulated annealing algorithm to the bond design problem specified as the *CEROE* maximization in eqn. (8). The energy function we want to minimize is the negative CEROE, i.e., $= -CEROE = -\Phi(L, R, M, K)$. The algorithm applies a Step Calculation Rule to the parameters L, R, K, M in order to get a bond design. It then calculates the *CEROE* of the design specified in eqn. 6 and applies the Step Acceptance Rule. The Step Calculation Rule for the optimal bond design problem is as follows:

Step Calculation Rule for Optimal Bond Design:

1. Select, at random, one of the four design parameters L, R, K, M.

2. If L has been selected then choose a random, integer, value for L such that $0 \leq L \leq M - K$. Keep the remaining parameters unchanged. Evaluate the energy as the negative *CEROE* of this design.

3. If R has been selected choose a random, real, value for R in the range [100,120]. (This range is arbitrary; redemption prices are expected to remain close to par.) Keep the remaining parameters unchanged. Evaluate the energy as the negative *CEROE* of this design.

4. If M has been selected then choose a random, integer, value for M such that $K + L \leq M \leq 360$. Keep the remaining parameters unchanged. Evaluate the energy as the negative *CEROE* of this design.

Bond design technique	*CEROE* of best bond design
Single-parameter optimization (varying maturity)	1.4085
Single-parameter optimization (varying lockout)	1.3726
Double-parameter optimization (varying maturity and lockout)	1.4099
Duration-matched bond	1.3313
Multi-dimensional optimization (local optimization algorithm)	1.4056
Multi-dimensional optimization (simulated annealing)	**1.4117**

Table 1: The quality of callable bonds — measured as certainty equivalent return on equity (CEROE) against the target mortgage assets — designed using different techniques.

5. If K has been selected then choose a random, integer, value for K such that $0 \leq K \leq M - L$. Keep the remaining parameters unchanged. Evaluate the energy as the negative *CEROE* of this design.

This simulated annealing algorithm was applied to our test problem. Figure 9 illustrates the holding period return of the optimal bond design obtained with this method. The *CEROE* is 1.4117, and it was achieved for a bond with $L = 8$, $M = 334$, $R = 104.00$ and $K = 27$.

SUMMARY AND CONCLUSIONS

We have presented details of a general framework for designing callable bonds in order to fund any specified collection of assets. Table 1 summarizes the results of the five techniques when applied to a specific problem of designing a callable bond to fund a mortgage-backed security. The superiority of a simulated annealing algorithm, developed expressly for this problem, is illustrated from the results of this table.

The developed framework is based on the ideas of *holding period returns.* It analyzes these returns for both the asset and the liability side of the balance sheet in an integrated fashion. The decision maker's risk preference is taken explicitly into consideration through the use of a utility function and the concept of *certainty equivalent of return.* In this respect our results have also shown that bond designs based on the

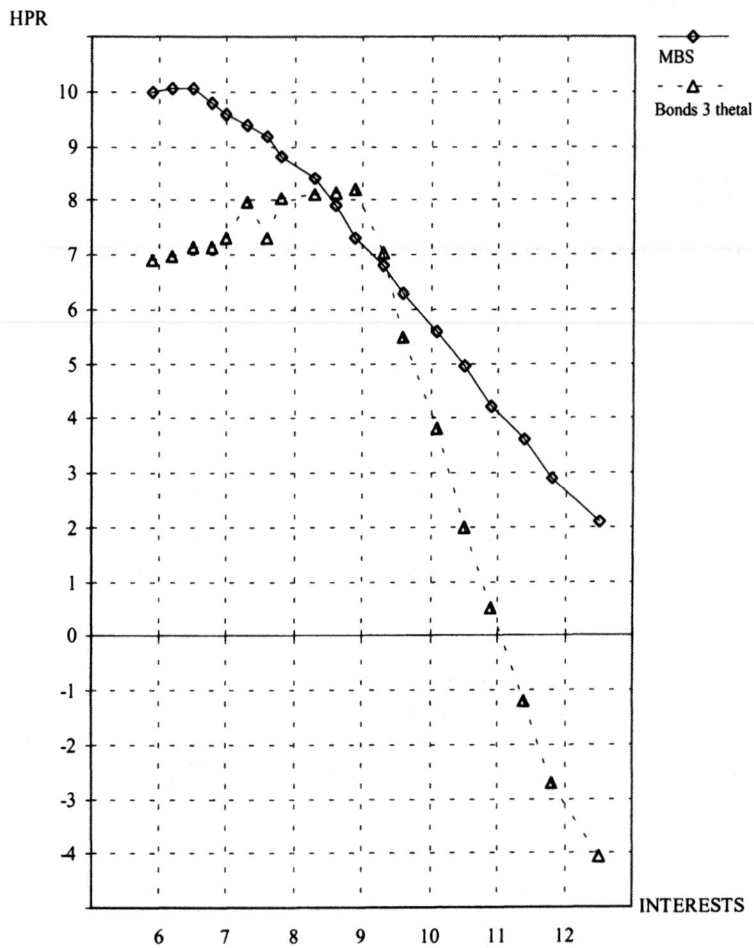

Figure 9: Holding period returns of the overall best bond designed using simulated annealing.

ideas of duration matching do not necessarily reflect the decision makers' best preference under their utility function and for the holding period scenarios.

Finally, we want to bring up an important issue deserving further investigation. There is no reason to expect that a single bond will be used as the liability to fund the assets. Instead, a portfolio of bonds will typically be used. Deciding the composition of the portfolio — assuming that the bond designs are given *a priori* — is in itself a difficult problem. This problem can be solved using models of *stochastic optimization*, such as those developed by Golub et al. [3]. However, it is possible to integrate the bond design framework developed here with stochastic optimization models to design the best combination of bonds to be put in the portfolio. This integration is the subject of a current study by the authors. Preliminary results, applied to the exact same problem addressed in this paper, show that even better *CEROE* can be expected by the optimal design of a portfolio of bonds.

References

[1] F. Black, E. Derman, W. Toy. A one-factor model of interest rates and its application to treasury bond options. Financial Analysts Journal, 1990;33–39.

[2] M.L. Dunetz, J.M. Mahoney. Using duration and convexity in the analysis of callable bonds. Financial Analysts Journal, 1988;53–72.

[3] B. Golub, M. Holmer, R. McKendall, L. Pohlman, S.A. Zenios. Stochastic programming models for money management. European Journal of Operational Research, 1995;85:282–296.

[4] M.R. Holmer. The asset/liability management strategy at Fannie Mae. Interfaces, 1993 (to appear).

[5] M.R. Holmer, S.A. Zenios. The productivity of financial intermediation and the technology of financial product management. Operations Research, 1995;43:970–982.

[6] S. Kirkpatrick, C.D. Gelatt, M.P. Vecchi. Optimization by simulated annealing. Science, 1983;220:671–680.

[7] M. Koenigsberg, J.L. Showers, J. Streit. The term structure of volatility and bond option valuation. The Journal of Fixed Income, Sept. 1991.

[8] R.W. Kopprasch, W.M. Boyce, M. Koenigsberg, A.H. Tatevossian, M.H. Yampol. Effective duration of callable bonds: The Salomon Brothers term structure-based option pricing model. Technical report, Salomon Brothers Inc., Bond Portfolio Analysis Group, April 1987.

[9] R. Litterman, T. Iben. Corporate bond valuation and the term structure of credit spreads. The Journal of Portfolio Management, Spring 1991;52–64.

[10] J.M. Mulvey, S.A. Zenios. Diversifying fixed-income portfolios: modeling dynamic effects. Financial Analysts Journal, Jan-Feb 1994;30–38.

[11] W.H. Press, B.P. Flannery, S.A. Teukolsky, W.T. Vetterling. *Numerical Recipes, The Art of Scientific Computing.* Cambridge University Press, 1989.

[12] A. Consiglio, S.A. Zenios. Designing callable bonds and its solution using Tabu search. Journal of Economic Dynamics and Control, 1997 (in print).

TOWARDS SEQUENTIAL SAMPLING ALGORITHMS FOR DYNAMIC PORTFOLIO MANAGEMENT

Z.Chen, G.Consigli, M.A.H.Dempster, N.Hicks-Pedrón.
Finance Research Group, Judge Institute of Management Studies, University of Cambridge

Abstract. This paper describes in detail the computations required to generate and solve large scale strategic financial portfolio management problems by sequential importance sampling methods. Data and model generation processes are emphasized and expected value of perfect information importance sampling criteria under current development outlined.

Keywords: Dynamic stochastic programming, datapath generation, expected value of perfect information, sequential sampling.

1. Introduction

Dynamic stochastic programming (*DSP*) formulations are particularly suitable for the solution of strategic portfolio management problems requiring the consideration of a large set of state variables. By focussing on a limited set of decision stages, they allow the characterization of portfolios with a rich set of investment and liability classes [8]. At each stage the portfolio manager (of a financial institution, insurance company, industrial conglomerate, etc.) takes a decision – in the form of a portfolio allocation – in the face of uncertainty typically generated by the random behaviour of market prices.

The solution of this dynamic decision problem depends crucially on the stochastic process model adopted to describe the behaviour of the random variables relevant to each stage of the problem. The random behaviour of the rates of return in stock portfolios was first identified by Markowitz [27] as the main source of uncertainty for the definition of an optimal investment decision in static stock portfolio problems. In modern applications stock prices provide only one possible source of risk for the portfolio manager [8, 28]; in many cases the randomness of short and long term interest rates, exchange rates, and other possible factors, needs to be considered for a correct representation of the problem. The definition of a stochastic, possibly multidimensional, financial data process and its inclusion in the generation of a stochastic optimization problem for numerical solution represents an important and controversial aspect of applied stochastic programming techniques for portfolio management [20, 24, 8, 28].

In §2 we consider a set of related issues concerning *scenario generation* in stochastic programming models when arbitrary underlying models of uncertainty are considered. To this end we introduce a distinction between a random vector *data process*, representing a primary source of uncertainty, and a random *coefficient process*, dependent on the former, which is problem-dependent and whose behaviour generates the specific information upon which the portfolio manager bases his strategy.

This distinction clarifies the important interaction between scenario generation and the subsequent *problem solution* when an *importance sampling* criterion based on the *Expected Value of Perfect Information* (*EVPI*) process [14, 10, 16, 9] is introduced, as discussed in §3. The properties of the EVPI process allow the selection of an enhanced set of *relevant representative* data paths in a sequential sampling refinement of an original stochastic optimization problem.

In our applications we consider a constrained stochastic optimization problem in the form of a *dynamic recourse problem* (*DRP*) (*cf.* Dempster [13], Ermoliev and Wets [19]) whose canonical formulation is given by (bold characters denote random elements)

$$\max_{x_1 \in \mathbb{R}^{n_1}} \{f_1(x_1) + \mathbb{E}_{\boldsymbol{\xi}_2}[\max_{\mathbf{x}_2}(\mathbf{f}_2(\mathbf{x}_2) + \ldots + \mathbb{E}_{\boldsymbol{\xi}_T|\boldsymbol{\xi}^{T-1}}[\max_{\mathbf{x}_T} \mathbf{f}_T(\mathbf{x}_T)])]\}$$

$$\begin{array}{llllllll}
\text{s.t.} & & & & & & & \\
A_1 x_1 & & & & & = & b_1 & \\
\mathbf{B}_2 x_1 & + & \mathbf{A}_2\mathbf{x}_2 & & & = & \mathbf{b}_2 & a.s. \\
& & \mathbf{B}_3\mathbf{x}_2 & + & \mathbf{A}_3\mathbf{x}_3 & = & \mathbf{b}_3 & a.s. \\
& & & \cdots & & & \cdots & \\
& & & \mathbf{B}_T\mathbf{x}_{T-1} & + \quad \mathbf{A}_T\mathbf{x}_T & = & \mathbf{b}_T & a.s.
\end{array}$$

$$l_1 \leq x_1 \leq u_1$$

$$\mathbf{l}_t \leq \mathbf{x}_t \leq \mathbf{u}_t \quad a.s., \quad t = 2, \ldots, T \ . \tag{1}$$

In (1) the constraint region is appropriately represented by a set of linear constraints representing financial as well as strategic and regulatory constraints [8]. The process $\boldsymbol{\xi}$ in $(\Omega^\xi, \mathcal{F}^\xi, P^\xi)$ is typically defined as a discrete, possibly autocorrelated, vector process with sample space Ω^ξ. The filtration $\mathcal{F}_t^\xi := \sigma\{\boldsymbol{\xi}^t\}$ generated in Ω^ξ by the history $\xi^t := (\xi_1, \ldots, \xi_t)$ of the random process $\boldsymbol{\xi}$ at time t defines the information set available to the decision maker at the different stages of the problem. In financial planning problems the process $\boldsymbol{\xi}$ is considered to be a function, $\boldsymbol{\xi}_t := \xi(\boldsymbol{\omega}_t)$, of a process $\boldsymbol{\omega}$ defined in a different probability space $(\Omega, \mathcal{F}^\omega, P^\omega)$. We refer to $\boldsymbol{\xi}$ as the *coefficient process* and

ω as the *data process* of the problem.

The objective in (1) is defined through a sequence of *nested* optimization problems corresponding to the different stages. Each decision $x_t \in X_t$ is required to be feasible with respect to a sequence of stage-dependent constraints: $A_1 \in \mathbb{R}^{m_1 \times n_1}$ and $b_1 \in \mathbb{R}^{m_1}$ define deterministic constraints on the first stage decision x_1, while, for $t = 2, \ldots, T$, $\mathbf{A}_t : \Omega^\xi \to \mathbb{R}^{m_t \times n_t}$, $\mathbf{B}_t : \Omega^\xi \to \mathbb{R}^{m_t \times n_{t-1}}$ and $\mathbf{b}_t : \Omega^\xi \to \mathbb{R}^{m_t}$ define stochastic constraint regions for the recourse decisions $\mathbf{x}_2, \mathbf{x}_3, \ldots, \mathbf{x}_T$. $\mathbb{E}_{\boldsymbol{\xi}_T | \boldsymbol{\xi}^{T-1}}$ denotes conditional expectation of the state $\boldsymbol{\xi}_T$ of the coefficient process $\boldsymbol{\xi}$ with respect to the history $\boldsymbol{\xi}^{T-1}$. At each stage previous decisions affect remaining optimization problems through the stochastic matrices $\mathbf{B}_t$, $t = 2, \ldots, T$.

The sequence of random events and decisions is given in Figure 1.

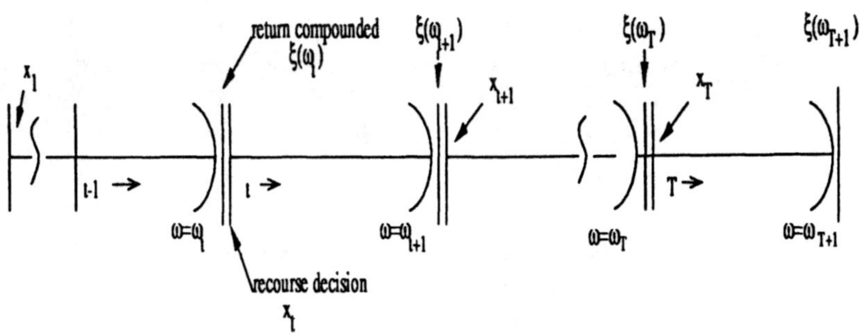

Figure 1. **Sequence of decisions and random events in dynamic stochastic programming**

The decision process $\mathbf{x} := \{\mathbf{x}_t\}_{t=1}^T$ is required to be strictly *adapted* or *nonanticipative*, i.e. $\mathbf{x}_t = \{\mathbf{x}_t | \mathcal{F}_t^\xi\}$ *a.s.*, with respect to the filtration $\mathcal{F}_t^\xi$ generated by the process. This condition can be imposed in the model implicitly [8, 7] or explicitly, leading to a stochastic program in *split-variable* form [14, 3].

Dynamic portfolio problems are easily formulated as a DRP. Applications of this approach can be found in Bradley and Crane [4], Lane and Hutchinson [26], Kusy and Ziemba [25], Dempster and Ireland [15], Mulvey and Vladimirou [29], Zenios [33], Cariño *et al.* [6]. The *CALM* model (Dempster [8]) has been formulated as a linearly constrained mixed integer stochastic programming problem and adopted for the formulation of a 10 year pension fund asset and liability problem – the *Watson* model [8] – with uncertainty generated according to Wilkie's autoregressive model [32] – and a 20 year asset allocation problem – the *FRC* model – with uncertainty generated according to the extended Brennan and Schwartz model [5]. The CALM-FRC model has been developed for a Frank Russell Company sponsored project and is defined with three asset classes: consol bonds, stocks and bank deposits, and an underlying four dimensional Itô process for the short and long interest rates, the dividend yield and the stock price. Due to its simpler associated data generator, it is being used as the reference model for the algorithm development described briefly in the final section.

2. Specification of the data process for dynamic stochastic programmes

The distinction introduced in §1 between the processes $\boldsymbol{\xi}$ and $\boldsymbol{\omega}$ is both conceptual and methodological. Unlike $\boldsymbol{\xi}$ which is constructed as a *discrete time* path-dependent process in accordance with the DRP formulation of the problem, the data process $\boldsymbol{\omega}$ in $(\Omega, \mathcal{F}^\omega, P^\omega)$ may be given different characterizations, all referring to a conceptually underlying *continuous time* process. This is the sense in which we refer to *dynamic* – in contrast to *multistage* – recourse problems. In [5, 17, 20, 28] $\boldsymbol{\omega}$ is an element of the class of real-valued diffusion processes with time set $[1, \tau]$, $\tau < \infty$ and uncertainty is generated by a (multivariate) Wiener process $\mathbf{W}_t$.

In [8], following Wilkie [32], $\boldsymbol{\omega}$ belongs to the class of autoregressive processes of the j-th order with continuous state space and discrete time set $T = \{1, 2, \ldots, \tau\}$, with random behaviour induced by disturbances $\mathbf{e}_t \sim \mathcal{N}(0, \sigma^2(\omega))$ and depending on the financial variable (e.g. the long interest rate) we may have an autoregressive equation up to the third order in the model. In Zenios [33] and Klaassen [24] $\boldsymbol{\omega}$ is a discrete state binomial process.

All these cases may be described in a form suitable for simulation purposes as

$$\omega_{n_t+1} - \omega_{n_t} = \mu(\omega)\Delta n_t + \sigma(\omega)\varepsilon_{n_t} \ , \tag{2}$$

for $t = 1, \ldots, T$ and $n_t = 1, \ldots, N_t$ up to $N_T - 1$, where $\mu(\omega)$ defines the *drift* of the process, $\sigma(\omega)$ its *volatility*, $\Delta n_t := (n_t + 1) - n_t$, and each stage of (1) refers to N_t subperiods.

For N_t sufficiently large and $\varepsilon_{n_t} \sim \mathcal{N}(0, \Delta n_t)$, (2) describes the discrete version of a diffusion process driven by Wiener noise [20, 28]. For smaller N_t and ε_{n_t} with arbitrary probability distribution, we typically have autoregressive models for long term allocation problems or binomial or trinomial models.

The different discretization schemes are all made consistent with a (DRP) characterization of the decision problem by introducing a *compound return* function defined by

$$\Gamma_t := \Pi_{n_t=1}^{(N_t-1)}(1+\omega_{n_t}) - 1 \quad t = 1, \ldots, T , \tag{3}$$

which gives at the end of period t the return of a monetary unit invested at the beginning of the period; where for each t each ending cash position is carried over to the position at the first index of the next period.

The history ω^t of the data process, for $t = 1, 2, \ldots, T$, enters a specific portfolio problem as parameters $\xi_t = \xi(\omega^t)$, required for the recourse decision x_t, as described in Figure 1. The decision maker is assumed to follow the behaviour of the price processes over time, while recourse decisions are allowed only at the end of every period consistent with the nonanticipativity requirement. Inhomogeneous time stages are easily accomodated in this framework and alternative stochastic models as described above can be adopted as individual inputs for the definition of the discrete vector process $\boldsymbol{\omega} := \{\omega_t\}_{t=1}^T$.

The specification of $\boldsymbol{\omega}_t$ is the output of a data generator *datagen*, in terms of a set of random functions with coefficient estimates for its mean and volatility functions, and a random number generator of the type described briefly in §2.2.

Datagen takes as inputs the initial state of the process together with a *nodal partition matrix* identifying the associated tree structure. It is interfaced with the generator of the random coefficients of the problem – the scenario generator *scengen* – needed for the definition of the stochastic program for numerical solution. *Scengen* takes as input the complete data process specification along the scenario tree and generates, as output, the scenario-dependent coefficients required by the mathematical formulation of the problem.

2.1. Datapath generation

We consider in this section an iterative procedure, interfaced with the data simulator, for the correct generation of data paths in the form of a scenario tree.

The definition of a *scenario tree nodal partition matrix* as a two-dimensional array, with number of rows equal to the number of scenarios of the problem and number of columns equal to the number of stages, is at the core of the conditional simulator. The matrix identifies uniquely the tree structure for the associated stochastic program and is used by the data generator in order to derive the states of the data process in conditional mode, and by the model generator *STOCHGEN* [10, 8] for the definition of the corresponding *SMPS* files [1] necessary for the numerical solution of the problem [8].

Figure 2 provides an example of the matrix specification associated with an arbitrary tree structure.

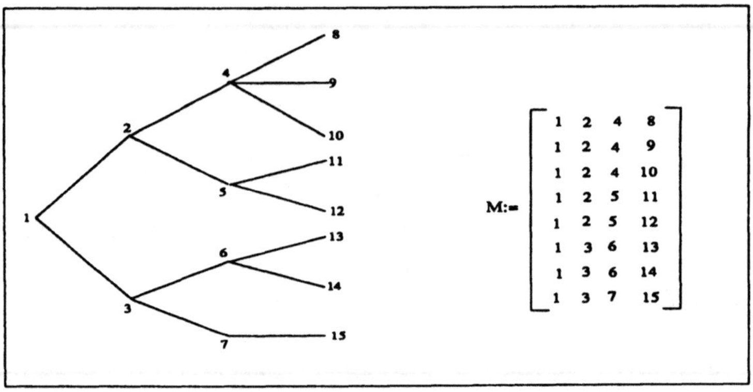

Figure 2. Definition of the nodal partition matrix

Following the matrix order, conditional simulations are run, compound annual rates of return computed and initial conditions passed consistently along the tree. Consistent with the time partition of the planning horizon, the generator runs over the N_t subperiods for $t = 1, \ldots, T$ and the final state ω_{N_t} (see eq. 2) for one simulation is adopted as initial state for the following run. The nodal partition matrix allows both the conditional run of the data generator – with one run for every increment in the matrix entries, columnwise – and the consistent updating of the initial seeds – rowwise. The stage-oriented nodal labeling order is convenient in view of the sequential generation-solution procedure described in §3.

In the FRC problem below a simulator has been constructed for a 4-dimensional diffusion system driven by a Wiener process with state variables representing stock return, short and long interest rate and dividend yield processes and based on estimated instantaneous mean vector and correlation matrix [5]. In this case the Wiener noise was generated by implementing a *congruential* method based on the Park and Miller *minimal standard method* [30] for the generation of normal unit deviates $\nu \sim \mathcal{N}(0,1)$ and applying the transformation $W_t = \nu dt$ leading to $W_t \sim \mathcal{N}(0, dt)$.

The set of data paths generated by *datagen* permits estimation of the joint probability distribution of the return at the horizon of a portfolio which initially invests equal amounts in each asset. Based on 1,024 data paths an estimated joint probability distribution generated at the horizon by the multidimensional conditional generator for the FRC and the Watson problems is displayed in Figure 3. In the case of the Watson problem the generation of the normal random variates from Wilkie's model was based on Marsaglia's *polar method* [32].

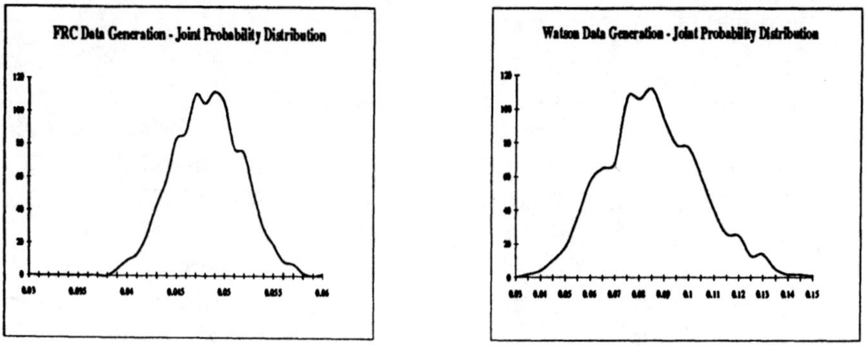

Figure 3. FRC and Watson problems empirical balanced portfolio return probability distributions generated by *datagen*.

The accuracy of the sequential procedure described in §3 relies on the possibility of an unbiased approximation of the form displayed in Figure 3 of the continuous probability density generated by the random process underlying the optimization problem. This density, as shown in Figure 3, is in general not consistent with the usual assumptions of normality or log-normality made in finance theory.

2.2. Coefficient process specification

The distinction between the data process ω in $(\Omega, \mathcal{F}^\omega, P^\omega)$, $\omega := \{\omega_{n_t} : n_t = 1, \ldots, N_t,\ t = 1, \ldots, T\}$ and the corresponding coefficient process ξ in $(\Omega^\xi, \mathcal{F}^\xi, P^\xi)$, $\xi := \{\xi_t : t = 1, 2, \ldots, T\}$ with $\xi_t := \xi(\omega^t)$ is motivated by the following considerations in formalizing financial planning problems:-

- Recent DRP formulations of asset and liability models have adopted a characterization of uncertainty based on complete market arbitrage free models of interest rates and price processes developed and well-established in the financial literature [2, 23, 32]. These, however, explain only in part the risk embedded in financial positions of investors operating worldwide and across different markets [8, 28].

- The specification of an optimal policy in recourse models generally relies on the definition of complex hierarchical forecasting and model generation systems [28, 8] for the definition of a set of (coefficient) scenarios derived in a cascade structure from the specification of a set of underlying *core* random processes possibly defined in different probability spaces.

The filtration $\mathcal{F}_t^{\xi}$ generated for $t = 1, 2, \ldots, T$ by the histories of the coefficient process $\boldsymbol{\xi}$ contains the information necessary for the solution of the corresponding dynamic portfolio problem. In general $\mathcal{F}_t^{\omega} \subset \mathcal{F}_t^{\xi}$.

Important examples of data and coefficient process specification and generation within portfolio management tools are the two instances of the CALM model - Watson and FRC, the Towers-Perrin model of Mulvey [28], the general asset and liability model of Klaassen [24], the MBS model of Zenios [33] and the Yasuda-Kasai model of the Frank Russell Company [7]. All these applications require the derivation from the relevant data generator of a large set of coefficients which are needed for the mathematical specification of the problem.

This step, which generally results in the definition of *ad hoc*, problem dependent, valuation criteria has an impact on the properties of the stochastic program finally generated [9].

In (1) the process $\boldsymbol{\xi}$ is defined by $\boldsymbol{\xi}_t := (\boldsymbol{\zeta}_t, \mathbf{A}_t, \mathbf{B}_t, \mathbf{b}_t)$, with $\boldsymbol{\zeta}_t$ denoting a random parameter in the objective functional given by $f_t(\zeta_t, x_t)$. The specification of the random coefficient matrices $\mathbf{A}_t, \mathbf{B}_t$ and $\mathbf{b}_t$ in $(\Omega^{\xi}, \mathcal{F}^{\xi}, P^{\xi})$ refers to the generation of the complete information structure necessary for the solution of the portfolio problem.

The steps required by conditional scenario generation may be briefly summarized as:-

- Initially the number of scenarios and stages, with associated stage discretization $n_t = 1, \ldots, N_t$, for $t = 1, \ldots, T$, are defined.

- The nodal partition matrix is then specified in order to define the complete tree structure for the problem (note that one simulation here corresponds to one complete data path along the event tree). Then recursively:-

  - the vector of initial conditions is defined and *datagen* is run, travelling the tree forward from the root node to the terminal node;
  - for each such simulation the compounded returns are computed;
  - the simulations are associated with the stages according to the nodal partition matrix and the complete set of conditional data paths specified;
  - for every trajectory of the data process, *scengen* is run, the corresponding set of model coefficients defined and

- a set of scenarios are generated and interfaced with a matrix generator (e.g. MODLER [22]).

Given the model formulation and the generation of the model coefficients, the resulting stochastic programming problem is now defined in standard input format for numerical solution [1]. In our system the SMPS format is generated using STOCHGEN [10].

3. Information flows and the resolution of uncertainty

We consider a stochastic programming system for the solution of financial planning problems based on:-

1. The representation of the decision problem in dynamic recourse form with implicit or explicit characterization of the nonanticipativity condition [14, 9, 18].
2. A data path simulator for an underlying continuous data vector process $\boldsymbol{\omega}$ in $(\Omega, \mathcal{F}^{\omega}, P^{\omega})$ representing the core uncertainty of the portfolio allocation problem.
3. A scenario generator for the specification of the vector stochastic process $\boldsymbol{\xi}$ in $(\Omega^{\xi}, \mathcal{F}^{\xi}, P^{\xi})$ defining the coefficients of the model and interfaced with this simulator.
4. Generation of the SMPS format, for which we use the STOCHGEN library [11] incorporating Greenberg's MODLER [22], required for the numerical solution of the stochastic programming problem.
5. The solution of the problem either by a *primal-dual interior point* (IP) method (CPLEX 4.0, 1996) [12], or by *nested Benders decomposition* (MSLiP-OSL, Version 8.3, 1995) [21, 31].

We intend to show here how the phases 2, 3, 4 and 5 are integrated, based on the valuation of the information generated by the coefficient data process, when the sample space approximation of this process is sequentially refined using estimates of the EVPI process(es) below.

Consider problem (1) in the more compact *dynamic programming* representation which takes advantage of the Markov structure exhibited by the set of constraints. For each $t = 1, \ldots, T$ we have the set of *nodal* problems

$$\begin{aligned} \pi_t(\boldsymbol{\xi}^t) \ := \ & \max_{x_t \in X_t} \mathbb{E}\{f_t(\boldsymbol{\xi}^t, x^{t-1}, \mathbf{x}_t) + v_{t+1}(\boldsymbol{\xi}^t, \mathbf{x}^t) \mid \mathcal{F}_t^{\xi}\} \\ & \text{s.t.} \\ & \mathbf{B}_t x_{t-1} \ + \ \mathbf{A}_t \mathbf{x}_t = \mathbf{b}_t \ \ a.s., \end{aligned} \tag{4}$$

where v_{t+1} expresses the optimal expected value for the remaining optimization problem for the stages from $t+1$ to T. At the horizon $v_{T+1}(\boldsymbol{\xi}^T, \mathbf{x}^T) := 0$. In (4) the dependence of the decision vector $\mathbf{x}_t$ on the filtration $\mathcal{F}_t^{\xi}$ is expressed explicitly.

The *expected value of perfect information* (EVPI) process [14, 10, 16] is defined by

$$\eta_t(\boldsymbol{\xi}^t) := \phi_t(\boldsymbol{\xi}^t) - \pi_t(\boldsymbol{\xi}^t) \quad t = 1, 2, \ldots, T, \tag{5}$$

where $\phi_t(\boldsymbol{\xi}^t)$ corresponds to the set of *distribution* problems associated with the relaxation of the nonanticipativity condition to the case of *perfect foresight*

$$\phi_t(\boldsymbol{\xi}^t) := \mathbb{E}[\max_{x_t \in X_t} \{f_t(\boldsymbol{\xi}^t, x^{t-1}, \mathbf{x}_t) + v_{t+1}(\boldsymbol{\xi}^t, \mathbf{x}^t) \mid \mathcal{F}_T^{\xi}\}]\,. \tag{6}$$

Based on the behaviour of $\boldsymbol{\eta}_t$ we can both assess the level of *stochasticity* of the DRP problem [8, 10] and define a sampling procedure for the selection of a sample set of *relevant representative* data paths in a sequential procedure. From the definition of the EVPI process we have at the horizon, by construction, $\boldsymbol{\eta}_{T+1} := 0$. For the properties of the $\boldsymbol{\eta}$ process which justifies its adoption as an importance sampling criterion for selection of a sample set of *objective-relevant* sample paths we refer to [14, 10, 9, 16]. Of particular importance is the characterization of the process as a nonnegative *supermartingale* [14] which reflects the nonnegative and increasing value associated with early resolution of uncertainty.

This property has two impacts useful in defining a sampling procedure: when the EVPI value is zero at one node in the tree, say ξ^t, it will remain null in all descendant nodes. Furthermore, if $\eta_t(\xi^t) = 0$ for some ξ^t, then there is a decision x_t optimal at t for all subsequent nodes. The future uncertainty is thus *irrelevant* and the local problem can be replaced by a deterministic problem.

The same properties of the EVPI process are shared by the *marginal EVPI*, δ-EVPI, or *shadow price of information* process [14] defined by the dual variables of the stochastic programming problem associated with the nonanticipativity constraints of the model in *split variable* form. Unlike (4) we now consider an *explicit* characterization of the nonanticipativity condition in *conditional expectation form*:

$$x_t(\boldsymbol{\xi}^t) = \sum_{\hat{\boldsymbol{\xi}}^t} p(\hat{\boldsymbol{\xi}}^t) x_t(\hat{\boldsymbol{\xi}}^t) \quad t = 1, 2, \ldots, T, \tag{7}$$

where $\hat{\boldsymbol{\xi}}^t$ denotes, at each stage t, the set of scenarios descending from the current node $\boldsymbol{\xi}^t$. Accordingly $p(\hat{\boldsymbol{\xi}}^t)$ denotes the probability of each such scenario occurring conditional on the fact that the process is in state ξ^t at time t. Definition (7) is referred to as the nonanticipativity condition in *conditional expectation projection* form (*cf.* [14]).

The nonanticipativity condition (7) leads to the specification, for $t = 1, 2, \ldots, T$, of a sequence of stochastic dynamic programs in the form

$$\begin{array}{l} \max_{x_t \in X_t} \mathbb{E}\{f_t(\boldsymbol{\xi}^t, x^{t-1}, \mathbf{x}^t) + v_{t+1}(\boldsymbol{\xi}^t, \mathbf{x}^t) \mid \mathcal{F}_t^{\xi}\} \\ \text{s.t.} \\ \mathbf{B}_t x_{t-1} + \mathbf{A}_t \mathbf{x}_t = \mathbf{b}_t \;\; a.s. \\ (I_t - \mathbf{\Pi}_t)\mathbf{x}_t = 0 \;\; a.s. \,. \end{array} \tag{8}$$

The programme (8) has associated Lagrangean given by

$$\begin{aligned} \mathcal{L}(x_t, y_t', \rho_t') := \;\; & \mathbb{E}\{[f_t(\boldsymbol{\xi}^t, x^{t-1}, \mathbf{x}^t) + v_{t+1}(\boldsymbol{\xi}^t, \mathbf{x}^t)] + \\ & \mathbf{y}_t'(\mathbf{B}_t x_{t-1} + \mathbf{A}_t \mathbf{x}_t - \mathbf{b}_t) + \boldsymbol{\rho}_t'(I_t - \mathbf{\Pi}_t)\mathbf{x}_t \mid \mathcal{F}_t^{\xi}\} \,. \end{aligned} \tag{9}$$

The *marginal EVPI process* $\boldsymbol{\rho} := \{\rho_t\}_{t=1}^{T+1}$ is thus the dual process associated with the nonanticipativity condition in conditional expectation form. At the optimum the δ-EVPI coefficients provide a measure of the value generated by a perturbation of the constraint. Unlike the full EVPI process, the marginal process is defined at every node of the tree up to and *including* the last stage. This property makes the criterion suitable for the solution of two stage problems by δ-EVPI sampling. At present the estimation of the δ-EVPI process requires the generation and solution of the complete deterministic equivalent problem [8, 3] with explicit nonanticipativity constraints.

We are now in position to sketch a sequential procedure based on the solution of the stochastic optimization problem with either the MSLiP-OSL solver [31] or the Cplex IP solver [12]. The two solvers are interfaced respectively with the EVPI sampling algorithm developed by Dempster and Corvera-Poiré [10, 11] and the δ-EVPI sampling algorithm currently under development.

Based on the EVPI information, the sampling procedure allows the sequential refinement of an original tree structure according to the procedure outlined in Table 1.

In both sampling procedures the *permanence* after resampling of the nodal EVPI values in the neighbourhood of 0 leads to a deterministic optimization problem over the remaining periods up to the horizon.

Each iteration with either importance sampling criterion requires: the generation of the data paths for the data process, the derivation of the coefficient scenarios, the definition of the standard input SMPS format and the solution by nested Benders decomposition or the IP

Table I. EVPI-Sampling Algorithm

define	number of iterations in the algorithm: J
define	initial scenario tree structure: T_1
	The Algorithm
	$j = 1$
	while $j \leq J$
1.	**construct** a tree T_j based on EVPI information
2.	**solve** problem T_j and compute its nodal EVPI
3.	**if** EVPI *near* 0, **resample**
4.	**else if** EVPI *near* 0 after resampling
	take one sample scenario
5.	**else if** EVPI > 0, increase branching at the node
6.	$j = j + 1$
	CONTINUE

method including the current estimates of the nodal EVPI values. Sequential refinement of the previous tree structure is based on an analysis of the current EVPI process – full or marginal – and the definition of a new nodal partition matrix that allows *datagen* to run again, as described in the inner loop of Figure 4.

The adoption of the full, as opposed to the marginal, EVPI sampling criterion has been previously reported [10, 8, 16]. Results have been presented in the case of a sampling procedure *independent* of the phase of scenario generation considered in §2.

4. Conclusions and further research

The sequential procedure outlined in Figure 4 calls for a few final remarks.

The system under development relies on the definition of a master program that calls at every iteration of the sampling procedure the subroutines for the data process generation – *datagen*, the coefficient process generation – *scengen*, the model generator – *STOCHGEN* and the solver, analyzes the EVPI estimates and derives the nodal partition matrix for the next iteration. The same framework is adopted for the use of the marginal EVPI importance sampling criterion derived from the solution of the problem with an IP method.

The efficiency of the sequential solution procedure relies heavily on

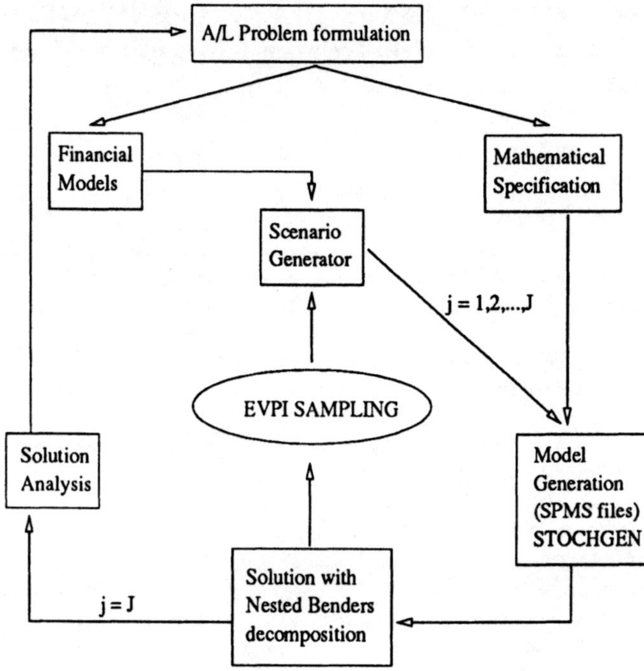

Figure 4. EVPI based sequential solution procedure

the speed and accuracy of the model generation. This step is currently based on MODLER [22] which was not originally designed for sequential matrix generation. We will shortly be in position to integrate the recursive MPS generator AIMS into our system with a very positive impact on the speed and efficiency of the sequential solution procedure.

In previous work [9, 16] we have established the accuracy of the EVPI sampling rule as a criterion for the approximation of large scale stochastic problems with an EVPI-based selection of scenarios sampled from a pregenerated finite population. In this paper the sampling framework has been extended to a dynamic procedure in which the sample of the random process generating the uncertainty in a portfolio allocation problem is associated with an increasingly representative stochastic sample problem.

Acknowledgements

Research partially supported through contract "HPC-Finance" of the INCO '95 (no 951139) project funded by Directorate General III (Industry) of the European Commission, the UK EPSRC and the FECIT Laboratory of Fujitsu Systems (Europe) Limited. Partial support was also provided by the "HPC-Finance" partner institutions: Universities of Bergamo (IT), Cambridge (UK), Calabria (IT), Charles (CZ), Cyprus (CY), Erasmus (ND), Technion (IL) and the "Centro per il Calcolo Parallelo e i Supercalcolatori (IT)".

References

1. J.R. Birge, M.A.H. Dempster, H.I. Gassmann, E.A. Gunn, A.J. King and S. Wallace. A standard input format for multiperiod stochastic linear programs. Mathematical Programming Society, *Committee on Algorithms Newsletter* **17** (1987) 1-20.
2. F. Black, E. Derman and W. Toy. A one factor model of interest rates and its application to Treasury bond options. *Financial Analysts Journal*, Jan-Feb (1990) 33-41.
3. A. Berger, J.M. Mulvey, E. Rothberg and R. Vanderbei. Solving multistage stochastic using tree dissection. Statistics and Operations Research Research Report, Princeton University, Princeton, NJ (1995).
4. S.P. Bradley and D.B. Crane. A dynamic model for bond portfolio management. *Management Sciences* **19.2** (1972) 139-151.
5. M.J. Brennan, E.S. Schwartz and R. Lagnado (1996). Strategic Asset Allocation. *Journal of Economic Dynamics and Control*, forthcoming.
6. D. Cariño, T. Kent, D. Myers, C. Stacy, M. Sylvanus, A.L. Turner, K. Watanabe and W.T. Ziemba. The Russell-Yasuda Kasai Model: an asset/liability model for a Japanese insurance company using multistage stochastic programming. *Interfaces* **24** (1994) 24-49.
7. D. Cariño, D.H. Myers and W.T. Ziemba. Concepts, technical issues, and uses of the Russell-Yasuda Kasai financial planning model. Research Report, Frank Russell Company, Tacoma, Washington, May (1995).
8. G. Consigli and M.A.H. Dempster. Dynamic stochastic programming for asset-liability management. To appear in *Annals of Operations Research*. Proceedings of APMOD95 Conference, Brunel University of West London (1996).
9. G.Consigli and M.A.H. Dempster. Solving dynamic portfolio problems using stochastic programming. To appear in *Zeitschrift fur Angewandte Mathematik und Mechanik*. Proceedings of the GAMM96 Conference, Charles University, Prague, May (1996).
10. X. Corvera-Poiré. Model Generation and Sampling Algorithms for Dynamic Stochastic Programming. PhD Thesis, Dept. of Mathematics, Univ. of Essex, U.K. (1995).
11. X. Corvera-Poiré. STOCHGEN User's Manual. Dept. of Mathematics, Univ. of Essex, U.K. (1995).
12. Cplex Optimization, Inc. *Using the Cplex Callable Library, Version 4.0* Incline Village NE, USA (1996).
13. M.A.H. Dempster. Stochastic programming: An introduction. In M.A.H. Dempster, ed. *Stochastic Programming*. Academic Press, London (1980) 3–59.
14. M.A.H. Dempster. On stochastic programming: II. Dynamic problems under risk. *Stochastics* **25** (1988) 15–42.
15. M.A.H. Dempster and A. Ireland. Object oriented model integration in a financial decision support system. *Decision Support Systems* **7** (1991) 329–340.
16. M.A.H. Dempster and R.T. Thompson. EVPI-Based importance sampling solution procedures for multistage stochastic linear programmes on parallel MIMD

architectures. To appear in *Annals of Operations Research*. Proceedings of the POC96 Conference, Versailles (1996).

17. M.A.H. Dempster. The CALM-FRC Model. Internal Document. Finance Research Group, Judge Institute of Management Studies, University of Cambridge, U.K. (1996).
18. J. Dupacova. Multistage stochastic programs: The state-of-the-art and selected bibliography. *Kybernetica* **31** (1995) 151–174.
19. Yu. Ermoliev and R.J-B. Wets, eds. *Numerical Techniques for Stochastic Optimization*. Springer-Verlag, Berlin (1988).
20. K. Frauendorfer, C. Marohn and M. Schurle. SG-Portfolio Test problems for stochastic multistage linear programming, Institute of OR, Univ. of St. Gallen, Switzerland (1995).
21. H.I. Gassmann. MSLIP: a computer code for the multi-stage stochastic linear programming problem. *Mathematical Programming* **47** (1990) 407-423.
22. H.J. Greenberg. A Primer for MODLER: Modelling by Object-Driven Linear Elemental Relations. Mathematics Department, University of Colorado at Denver (1995).
23. T.S.Y. Ho and S.B. Lee. Term structure movements and pricing interest rate contingent claims. *Journal of Finance* **41** (1986) 1011–1029.
24. P. Klaassen. Stochastic Programming Models for Interest-Rate Risk Management. PhD Thesis, Sloan School of Management, M.I.T., Cambridge, Massachusetts, May (1994). Published as IFSRC Discussion Paper.
25. M.I. Kusy and W.T. Ziemba. A Bank Asset and Liability Management Model. *Operations Research* **34** (1986) 356–376.
26. M. Lane and P. Hutchinson. A model for managing a certificate of deposit portfolio under uncertainty. In M.A.H. Dempster, ed. *Stochastic Programming*. Academic Press, London (1980) 473–493.
27. H.M. Markowitz. Portfolio Selection. *Journal of Finance* **7** (1952) 77–91.
28. J.M. Mulvey. Generating Scenarios for The Towers Perrin Investment System. *Interfaces* **26.2** (1996) 1–15.
29. J.M. Mulvey and H. Vladimirou. Stochastic Network Optimization Models for Investment Planning. *Annals of Operations Research* **20** (1989) 187–217.
30. S.K. Park and K.W. Miller. *Communications of the ACM* **31** (1988) 1192–1201.
31. R.T. Thompson. MSLiP-OSL 8.3 User's Guide. Judge Institute of Management Studies, University of Cambridge, U.K. (1997).
32. A.D. Wilkie. More on a stochastic asset model for actuarial use. Institute of Actuaries, London (1995).
33. S.A.Zenios. Asset-Liability management under uncertainty for fixed-income securities. *Annals of Operations Research* **59** (1995) 77–97.

THE DEFEASANCE IN THE FRAMEWORK OF FINITE CONVERGENCE IN STOCHASTIC PROGRAMMING

Philippe Spieser, Alain Chevalier

Groupe École Supérieure de Commerce de Paris, Finance Department - 79 avenue de la République 75011 Paris.

Abstract: This article deals with the modelisation of defeasance strategies chosen by industrial firms or financial institutions. In the first part, we present the financial concepts and the classical formulation of defeasance based on linear programming, dynamic programming and duality theory. Then we present differential inclusion and develop a practical example dealing with the primal dual differential method and the algorithm of resolution. The third part contains the main novelty of the paper, the method to yield convergence in finite time which leans on the result of Flam and Seeger.

Keywords: Defeasance, linear programming, dynamic programming, stochastic models, duality theory, differential inclusion, convergence.

The defeasance is a process allowing to extract the debt of a firm and to transfer it, at its market value, to a trustee which buys in the same time some bonds or equities. These financial tools serve to the payment of the service of that debt. Minimization of expected net cost in this adequation process should be the goal of any process of that kind in order to make an optimal trade-off between risk, return, and liquidity. The financial riskless assets necessary to reimburse the debt must be bought by the firm. To that purpose, the company must either have cash or create them by contracting an other debt. In the first case, the firm will be able to improve its balance sheet and the debt ratios.

In this paper, our goal, is to present the general linear and dynamic formulations of the problem of defeasance and then to use differential inclusion and the algorithm of resolution on a practical example. In the last part we develop a method to yield convergence in finite time.

1. GENERAL FRAMEWORK

The financial literature has distinguished three categories of discontinuous time models aiming the problem of Assets and Liabilities Management (ALM) (Leberre and Sikorav 1991):

- the first one includes deterministic models using linear programming ;
- the second approach deals with stochastic dynamic programming and analyzes simultaneously new financial instruments ;
- the third one includes stochastic decision trees models which are mathematically and computationnally difficult but operationnal.

1.1. Linear formulation of the problem

The assets must be chosen according to mathematical techniques of linear programming with an objective to minimize the market value under three constraints:
- the sum of the par value of the "principal" must be equal or superior to the total debt ;
- the sum of the interests which are received must be equal or superior to the interest paid on the debt ;
- the time to maturity of the principal and interest payment must be equal or inferior to that of the debt . There is an interest rate risk in the compound or reinvestment process of the available liquidities .

This appears to be the dual problem of asset liability management and the reader is invited to refer to the numerous publications dealing with ALM. Besides, defeasance puts numerous problems in fiscality, banking laws,etc.. in which we will not refer to.

Let us consider a firm issuing a set of bonds. They cause a flow of payments - interest and amortization-until the last payment of the last bond. We suppose that the quantity and the maturities of those flows are perfectly known.

To transfer to a trustee the charge of recovering that set of liabilities, it is necessary to provide it with a portfolio of assets with no defaults : no negative cash can be admitted. In an other approach we could release that constraint. Those assets must have fixed characteristics, that is maturity, redemption date, amount of generated flows.
It will be bonds denominated in the same currency as the debts, without any risk of signature, fixed rated, and without any option attached.

1.1.1. Construction of a bond portfolio

The problem lies in the constitution of a bond portfolio, generating total flows A_t, the bonds being held in proportions a_i. The portfolio is sufficient to cover the CF(t).
We have to determine a structure of assets, which is optimal (that is at the lowest cost) and which replicates the given portfolio of liabilities.
Let us suppose that the quantity of available bonds is fixed and known and let us also suppose that the volume is important enough such as there is no change in the market. The assets which serve to replicate the liabilities may generate a surplus of flows which can be reinvested until the next date of payment. But it is impossible to

know by advance the rate of reinvestment of that possible positive treasury. For the first step let us suppose that the reinvestment rate is zero.

The conditions for matching the liabilities are :

First step : t_1

$$A_1 \geq L_1$$

Second step (t_2), the amount of asset available at the date -ie the cash they get at that date plus the rest of the reimbursement operations complete at date t_1 :

$$A_1 - L_1$$

must cover the liabilities L_2.
So, the following inequality must be satisfied :

$$A_2 + (A_1 - L_1) \geq L_2$$

Final step t =T :

The previous argument is to be generalized at each date t unil the last date T. The disposable amount of assets i.e. the flow at date T plus the rest of the reimbursement operations completed at dates 1,2,...T-1 must cover the liabilities L(T).
We have the following inequality, if we actually suppose that the rest of the operations can be cumulated :

$$A_2 + (A_1 - L_1) \geq L_2 \Leftrightarrow A_2 + A_1 \geq L_1 + L_2$$

$$\Sigma A_k \geq \Sigma L_k \quad \begin{matrix} T. \\ \\ k=1 \end{matrix}$$

The inequalities can be separated into two subsums: on the right side of the inequality the amount of the inflows that we try to calculate, on the other side the amounts of outflows that are known.
Each bond generates a sequence of flows $b_j(t)$ and the price B_j of the bond is linked with the flow of reimbursement by the relation :

$$B_i = \underset{0 \leq t \leq k}{\Sigma} b_j \; (k)$$

If, at time t, the portfolio is made out of N bonds in α_i proportions (nominal value invested in bond "i"), the total available asset is :

$$A(t) = \sum^{N} a_i \left(\sum b_j (k)\right)$$

A priori N will be equal to the number of bonds of the benchmark and no constraint is retained on the α_i, some of them being possibly equal to zero.

The problem is typically the minimization of the acquisition price of the portfolio :

$$\textbf{Min A} = \sum a_i QC_i$$

where QC_i is the price of the bond i with coupon included that is $\sum b_j (k)$
The linear inequalities which have been found before formulating the system of linear constraints are.

$$\sum A_k \geq \sum L_k$$

1.1.2. Calculation of the optimal solution

1.1.2.1. First step (first class of models)

The former system is analogous to many problems of linear optimization and the "simplex algorithm" or "Dantzig method" is very well known: we know that the inequalities in this case represent a convex polyedron with a finite number of summits. The optimal solution is located on one of the summits. This problem does not raise difficulties of any kind if objectives and constraints are supposed to be linear. Dual variables are easy to calculate and have an interesting financial interpretation. (Cohen, Maier, Vander Weide 1981).

1.1.2.2. Second step(second class of models)

Let us suppose that the assets may be reinvested at a reinvestment rate "r". The way to evaluate the new financial product is the same as before, but we allow the dates of inflows and the dates of outflows to be different. The main point is that the actualized difference between the liabilities and the assets added to the asset of the period matches the liabilities of the period under review. We also have to take into account the fact that the dates of payments and receipts can be different. We make the assumption that the reimbursement flows can be reinvested too.
Since then the different inequalities become :

$$A_1 \geq L_1$$

at t=2:

$$A_2 + (A_1 - L_1)(1+r)^{(t2-t1)} \geq L_2 (1+r)^{(t2-t1)}$$

at t=n

$$\Sigma A_k (1+r)^{(t_k - t_{k-1})} \geq \Sigma L_k (1+r)^{(t_n - t_{n-1})}$$

By rearranging the system of inequalities above, we can write, for all possible dates k and for all bonds :

$$\Sigma a_i (\Sigma b_i (t))(1+r)^{(t_n - t_{n-1})} \geq \Sigma L_n (1+r)^{(t_n - t_{n-1})}$$

In the above equations, we suppose:
- that the interest rate and its differential are known in advance ;
- that the different cash-flows cannot be gathered in advance and consequently cannot be invested at "r".

At that step the model remains linear and is not difficult to solve. Dual variables are still easy to calculate and explain.

1.2. DESIGN OF A GENERAL DYNAMIC STOCHASTIC MODEL

We could take into account some simple constraints. Let us try to write a dynamic stochastic model of cost minimization. The defeasance process is an intertemporal decision making process. Asset portfolios are determined after each period of time of payments. The objective function is written in a very general way, that is the optimization of the mathematical expectation of a function. . The decision variable is y(s). It is the investment choice at s. We suppose that there is no bequest function. The function is denoted by utility function which has only one constraint : to be concave.

The two principle methods of solving problems of optimal control are the dynamic programming approach based on the optimality principle (Bellman 1957) and the principle of Pontryagin maximum (1962).

$$J(x,t,T) = Min_{y(s)} E_t \int_t^T U(y(s), x(s)) ds$$

u.c. $dx(s) = \mu(x,y,s)ds + \sigma(x,y,z)\, dZ(s)$

This process refers to the dynamics of the interest rates. $\mu(x,y,s)ds$ is the instantaneous drift leading the interest rates and $\sigma(x,y,z)$ is the instantaneous variance.

The application of Bellmann's optimality principle leads to the different following equalities :

$$J(x,t,T) = \underset{y(s)}{Min}\, E_t \int_t^T U(y(s), x(s)) ds$$

$$= Min\, E_t\left(\int_t^{t+\delta t} U(y,x)ds + Min\, E_{t+\delta t} \int_{t+\delta t}^{T} U(y,x)ds \right)$$

$$= Min\, E_t\left(\int_t^{t+\delta t} U(y,x)ds + J(x,t+\delta t,T) \right)$$

The above equation can be analysed in the following way : the minimal utility obtained in (t,T) results from the choice of control variables y(s) and from the evolution of the state variable. Consequently, the first term refers to the direct effects of the decision taken at "t" and the second term J(x, t+δt,T) prices the indirect effects. The first term may be approximated by U(y(t),x(t)) δt .

2. DIFFERENTIAL INCLUSIONS

We defined in the first part of the paper the general framework in which we want to apply some specific rules of dynamic programming for the problem of defeasance. The general aim of this part is to check if the method of differential inclusions can lead to an algorithm or at least gives us the insurance that an algorithm converges. The content of this part is designed as follows:

- first section shows how useful the differential inclusion method is.

- second section describes the method in the context of control theory and economic theory .

2.1. Differential inclusions: a reappraisal

There is a great variety of motivations that led mathematicians to study dynamical systems having dynamics not only determined by the state system but depending loosenly upon it.
So they were led to replace the classical differential equations

$$x' = f(x)$$

by what they called a differential inclusion

$$x' = F(x)$$

when F is the sert valued map which associates to the state x of the system the set of feasible solutions.
If deterministic models are convenient for describing systems which arise in microeconomics, their use for explainig evolutions of "macrosystems" does not take into account the uncertainty, the absence of controls, and the heterogeneity of possible dynamics. The uncertainty involves the impossibility of a complete description of the dynamics. The absence of controls means also the ignorance of the laws relating to the controls and the states of the system.

We will first study the existence of solutions to classes of differential inclusions and investigate the properties of the set of trajectories. This set of trajectories is rather large and the natural continuation is to devise mechanisms for selecting peculiar trajectories.
A first class of such mecanisms is provided by optimal control theory : it consists in selecting paths to optimize a given criterion as a functional of the space of all such trajectories.
It is implicitly required that:
1)there is a decision maker who controls the system
2) such a decision maker has a perfect knowledge of the future or at least knows the diffusion process , for example, which drives the variables
3) the optimal trajectories are chosen once and for all at the origin of the period of time.

Let us remind that a great impetus to study differential inclusions came from the development of Control Theory, the dynamical systems of the following form:

$$(*) \qquad x'(t) = f(t, x(t), u(t)), \qquad x(0) = x_0$$

"controlled" by parameters $u(t)$ (the "controls"). Indeed, if we introduce the set-valued map

$$F(t,x) = \{f(t,x,u)\}_{u \in U}$$

then solutions to the differential equations(*) are solutions to the "differential inclusions"

$$(**) \qquad x'(t) \in F(t, x(t)), \qquad x(0) = x_0$$

in which the controls do not appear explicitely.
Systems Theory provides dynamical systems of the form

$$x'(t) = A(x(t))\frac{d}{dt}(B(x(t))) + C(x(t)); \qquad x(0) = x_0$$

in which the velocity of the state of system depends not only upon the state $x(t)$ of the system at time t, but also *on variations of observations* $B(x(t))$ of the state.
This is particular case of an implicit differential equation

$$f(t, x(t), x'(t)) = 0$$

which can be regarded as a differential inclusion of the genre (**), where the right-hand side F is defined by

$$F(t,x) = \{v | f(t,x,v) = 0\}$$

During the 60's and 70's, a special class of differential inclusions was thoroughly investigated : those of the form

$$x'(t) \in -A(x(t)), \qquad x(0) = x_0$$

where A is a "maximal monotone" map.

This class of inclusions contains the class of gradient inclusions which generalize the usual gradient equations

$$x'(t) = -\nabla V(x(t)) \quad x(0) = x_0$$

∇ is a differentiable potential .
There are many cases where potential functions are not differentiable, notably if they "equal" to ∞ outside a given closed subset.
First conclusion: the state of the system must belong to the K space :
.When the potential function V is a lower semicontinuous convex function we can replace ∇ V(x(t)) by a generalized gradient also called subdifferential ∂ V(x) which associates to any point x a set of subgradients

The gradients inclusions
$x'(t) \in - \partial V(x(t))$
have an important property : if the state minimizes the potential V then the trajectories x(t) x(t) of the gradient inclusions do converge to such minimizers.

Differential inclusions provide a mathematical tool for studying differential equations

$$x'(t) = f(t,x(t)) \quad x(0) = x_0$$

with discontinuous right hand side by embedding f(t,x) into a set valued map F(x,t) which offers enough regularity to accept trajectories closely related to the trajectories of the original differential equation

2.2. Differential inclusions : control theory and economic theory.

Three remarks are to be made at this step:
1) differential variational inequalities form a special class of differential inclusions; to some conditions related to the space K the equation genre :

$$\sup_{y \in K} < x'(t) - f(x(t())), x(t) - y> = 0$$

can be expressed like a differential equation x'(t) = f(x(t))

2) It is necessary to begin with the problem of the existence (global or loccal) of solutions to a diffrential inclusion. This leads to the investigation of the topological properties of the set S of such solutions and the nature of its dependance upon the initial state x_0 . Some difficulties appear which do not exist in ordinary differential equation.

3) Since we may expect a differential inclusion to have a rather large spectre of trajectories , a second class of problems consists of devising mechanisms for selecting special trajectories. Three methods are available:

a) the equilibria or stationary states which are the constant trajectories $\bar{x}$ solution to $0 \in F(\bar{x})$

b) a selection provided by Optimal Control Theory which starts with the Hamilton Jacobi Bellmann equation: we can use a continuous functional W associating to each trajectory x(.) element of the set S a cost W(x(.)) for singling out optimal theory $\bar{x}$ (.) $\in$S minimizing the functional W over the set of trajectories. It is roughly the path we will use. A more sophisticated approach uses the tools of the game theory.

c) the last way is to use the Viability theory.i.e. to select the trajectories that are viable in the sense that they always satisfy given constraints. We can summarize this by saying that a trajectory x(.) is viable iff when

$$\forall t, \quad x(t) \in K(t)$$

where K(t) is the viability subset at time t, which is closed and compact. From a pure economic or financial point of view , this selection procedure is highly consistent with the behavioral assumption of limited rationality due to Simon (see referencies) where pure optimality is replaced by mere satisfaction

It is far beyond the scope of this paper to describe all the results of the differential inclusion theory, but we can select some results which will be explicitely or implicitely admitted in the following part of the paper.
A major result concerning the relations between the viability problem and the problem of finding equilibria is that under convexity assumptions, the **necessary and sufficient conditions for equilibria and the viability implies the existence of equilibria.**
For the monotone trajectories and by considering the functional W, several informations on the asymptotic behavior of the trajectory when t goes to infinity. may be infered.It is useful to adapt the Lyapunov method for studying the stability of trajectories.

The last point of this introduction will be a consideration concerning the optimal control theory, the viability conditions and the economic theory following the works summarized by Aubin, Cellina and due to Aubin, Cellina, Ekeland, Fillipov, Haddad, Lions , Lyapunov, Wazewski among others. (see referencies)...

If S denotes the subset of trajectories of the differential inclusions $x'(t) \in F(x(t))$ issued from x_0, let us denote by $V(x_0)$ the value function :

$$V(x_0) = \inf_{x(.) \in S(x_0)} \int_0^{\infty} W(x(r), x'(r))dr$$

Monotone trajectories for such a function V are the optimal trajectories of this problem of optimal control and the function $x_0 \rightarrow V(x_0)$ satisfies the Hamilton Jacobi Bellman equation when the usual derivative which may not exist is replaced by the always existing upper contingent derivative.
The application to the economic theory leads to use the viability theory to build a dynamical analog to the static concept of Walras equilibrium. In other words, the the price system is considered as a control which is used by each consumer i=1,...n to govern the evolution of his consumption of a commodity bundle x_i (t) according to a differential equation

$$x'_i(t) = d(x_i(t), p(t)), \qquad x_i(0) = x_i^0$$

The viability constraint in this framework is the requirement that the sum $\sum x_i(t)$ of the consumed commodity bundles lies in the set of available goods. It can be proved that the common financial laws (it is not allowed to spend more than earned) guarantees the existence of price systems p(t) yielding viable trajectories. This dynamical approach retains the good properties of the Walras model of general equilibrium by letting aside the static concept of equilibrium.

2.3. Differential inclusion : a practical example.

This section is directly inspired by Sjur Flam's works (see referencies), who considers more generally planning problems plagued by uncertainty about the (future) outcome w in some event space W.
Its general purpose is to study the convergence conditions of an algorithm designed to select a class of financial vehicles.
Of course, such problems can often be cast in the form of a constrained stochastic program.

(P) : Minimize the expected cost

$$(1.0) \qquad f_o(x) := EF_o(\omega, x_1(\omega), \ldots, x_s(\omega)) := \int F_o(\omega, x_1(\omega), \ldots, x_s(\omega))\mu(d\omega)$$

With respect to the strategy profile $x = x(.) = (x_1(.), \ldots, x_s(.))$ under two types of constraints :

- First, we must cope with "technological" restrictions of the standard genre.

$$(1.1) \qquad F_s(\omega, x_1(\omega),\ldots,x_s(\omega)) \leq 0 \text{ a.e. for } s = 1,\ldots,S$$

Here F_S takes values in R^{m_s}, and (1.1), may reflect variable resource endowments, production possibilities and the like.

- Second, we face informational limitations expressed formally by

(1.2) : x_S (.) should be S_S-measurable for s = 1,...,s.

Two features are incorporated in (1.2) :
First, decisions are implemented **sequentially**. At each stage (decision epoch) s = 1, 2,.. up to the planning horizon S included, an irreversible commitment x_S (w) e R^{n_S} is made.

Second, x_S (w) is committed, - within a time window which opens temporarily at stage s -, under **imperfect information** about the exact state w e W of the world. This stepwise resolution of uncertainty means, in more simple words jargon, that decisions never depend on **future** information. They are all **non-anticipative**, and resemble "sunk investments" once made : historical decisions cannot be modified. All those assumptions are of course coherent with the general framework of defeasance.

By way of example, let the information flow be generated sequentially by a stochastic process E_1,..., E_S on W. Then decision x_S cannot await either E_{S+1} or E_{S+2} ... or E_S. Rather, x_S should only take into account the actual realization of E_1,...,E_S. Thus, S_S is, in this case, the smallest s-algebra rendering all (possibly vector) variates E_1,...,E_S measurable.

It is also worthwhile to emphasize that all strategies x_1(.),..,x_S(.) are laid down (computed) right **here and now.** The feature does not contradict the fact that one must **wait and see** (appropriate information) before these strategies can actually be implemented on line contingent upon how the system unfolds and uncertainty is unveiled.

This completes the heuristic description of the multistage stochastic optimization problem. Technical assumptions are relegated to Part 2.

3. THE ALGORITHM

The purpose of the second part of this paper is to provide an algorithm, described in Section 3, which under broad hypotheses, yields finite convergence to optimal solutions. This algorithm aims to simulate a very large scale, deterministic, differential system.

3.1. The characteristics of the algorithm

In this section we specify the assumptions imposed on problem (P).

The operator E in (1.0) denotes the expectation over W, this set being conceived as a probability space with sigma-algebra S and probability measure m (possibly subjective).

We assume that S_s, s = 1,2,...,S, in (1.2) are complete sub-sigma-algebras of (W,S,m).

Constraints (1.2) will be supplemented by requiring also square integrability, i.e.,

(2.1) $\quad x_s \in L^2 (S, R^{n_s})$ for all $s \geq 1$.

where $L^2 (S, R^{n_s})$ denotes the classical Hilbert space of square integrable, S-measurable random vectors in R^{n_s}.

In short, (1.2) and (2.1) say jointly that no strategy x can be selected outside the set

(2.2) $\quad x := L^2(S_1, R^{n_1}) \times ...... \times L^2(S_S, R^{n_s})$

We reiterate that (2.2) embodies two requirements : strategies must be non-anticipative and square integrable. In accordance with (2.2), we demand that the common place "technological" restrictions (1.1) satisfy, for all $s \geq 1$, the two conditions :

(2.3) $\quad x \in x \Rightarrow F_s (.,x_1(.),..,x_s(.))$ is S_s-measurable

(2.4) $\quad x \in H \Rightarrow F_s(.,x_1(.),..,x_s(.)) \in L^2 (S,R^{m_s})$ and continuous.

Here, for simplicity in notation, H denotes the Hilbert space L^2 (S, R^n) with n : = $n_1 + ... + n_S$.

Motivated by practical examples, and also by the need to remain within the confines of convex analysis,(otherwise the mathematical problems would be too complicated) we assume that:

(2.5) { the cost function F_o (w,.) and all m_s components
{ of the constraint functions F_s (w,.), s = 1,..,S, are convex
{ and finite-valued for all w e W.

Also, to make problem (P) tractable we have incorporated no constraints in the objective function f_o (1.0).

Specifically, we suppose that

(2.6) f_o (x) is finite-valued continuous at all x e H = L^2 (S, R^n).

As customary, violations of (1.1) will be evaluated (or penalized) by means of multiplier vectors y_s e R^{m_s}, s = 1,...,S.

However, these multipliers are random [2]. Specifically, in accord with (2.3) and (2.4), we assume that **all Y_S(.) to be S_S-measurable, square integrable.**

For notational convenience, we shall codify this requirement by saying that any multiplier y = (y_1,...,Y_S) must belong to the Hilbert space

(2.7) $Y : = L^2(S_1, R^{m_1}) \times .... \times L^2(S_S, R^{m_S})$.

Such multipliers y e Y enter into a "functional" **Lagrangian**

(2.8) $L(x,y) : = \int \wedge (\omega, x(\omega), y(\omega)) d\mu(\omega)$.

where the integrand $\wedge$: W $\times R^n \times R^m \rightarrow$ R is a "pointwise" Lagrangian

(2.9) $\wedge(\omega,\xi,n) : = F_o(\omega,\xi) + \sum_{s=1}^{S} n_s . f_s(\omega,\xi_1,...,\xi_s)$

defined for all x = (x_1,...,x_S) e R^n, n = n_1 + ... + n_S, and all h = (h_1,...,h_S.)

A non-standard feature appears in (2.9) : The function $f := (f_s)_{s=1}^{S} := F^+ := (F_s^+)_{s=1}^{S}$ mentioned here is a shorthand for the positive part :

(2.10) $\quad f_s(\omega,.) := \text{Max}\{0, F_s(\omega,.)\}$ a.e.

the maximum operation in (2.10) being taken both poinwise and coordinatewise.

More generally, in (2.9) we can let

$f_s(\omega,.) := \varphi_s(\text{Max}\{0, F_s(\omega,.)\})$ a.e.

with $j_s : R^{m_s} \to R_+^{m_s}$ non-decreasing convex and vanishing only at the origin.

The only essential restriction here is that we want the implication

(2.11) $\quad x \,\varepsilon\, X \Rightarrow f_s(.,x_1(.),\cdots,x_s(.)) \,\varepsilon\, L^2(\Sigma_s, R^{m_s})$

to hold for all s _ 1, as indeed does under (2.3-4) and (2.10).

To insist : the non-conventional property of the Lagrangian L in (2.8-10) is that only strict constraint violations are priced by means of multipliers. No gain is obtained by slackness. In other words, what we invoke is a (one-sided) exterior penalty method employing non-standard multipliers. Moreover, according to (2.7) these multipliers must be non-anticipative and square integrable. As customary, only non-negative multipliers are of interest, i. e., we shall invariably select them from the cone:

(2.12) $\quad Y_{\underline{+}} := \{y \,\varepsilon\, Y : y \geq 0 \text{ a.e.}\}.$

Observe, via (2.3-4) and (2.66-10)), that the integral in (2.8) defines a finite, bivariate, function L over the space H x Y. Furthermore, by the convexity assumption (2.5), this function L is convex-concave on H x Y+.

Not surprisingly, L will be our main object in searching for solutions to problem (P).

3.2. The algorithm resolution : the primal dual differential methcd

We are now prepared to state our algorithm. To solve problem (P) we propose to follow a trajectory (x, y) (t, w) t _ 0, w e W, of the **differential inclusion.**

$$
(DI)\quad
\begin{aligned}
\dot{x}(t) &\,\varepsilon\, -\Pi_x \partial_x L(x(t), y(t)) \\
\dot{y}(t) &\,\varepsilon\, \quad \partial_y L(x(t), y(t))
\end{aligned}
$$

verifying the **viability condition** : $y(t) \geq 0$ a.e. for all t e R+T

Here $\dot{x}(t), \dot{y}(t)$ denote the time derivative, L was defined in (2.8-10), and by a trajectory we mean an absolutely continuous function (x,y) (.) : R+Æ H x Y satisfying (DI) almost everywhere (written now a.e.).

$$F(t,x) = \{v | f(t,x,v) = 0\}$$

During the 60's and 70's, a special class of differential inclusions was thoroughly investigated : those of the form

$$x'(t) \in -A(x(t)), \qquad\qquad x(0) = x_0$$

where A is a "maximal monotone" map.

In the first inclusion DI here, above P_X signifies the orthogonal projection onto the set X (2.2). Also, in (DI), the partial subgradient operators δ_X, δ_y should be understood in the sense of convex analysis [7].

To wit :
$d_X L(x,y) := d\,[L(\cdot,y)]\,(x) = df_0(x) + d\langle y, f(\cdot)\rangle(x),$

$d_y L(x,y) := -\,d\,[-L(x,\cdot)]\,(y) = \{f(x)\}.$

The dynamics (DI) can be interpreted as **a continuous (infinitesimal) steepest feasible direction method in both variables x and y separately.**

It also portrays a process driven by first-order, myopic adjustments.

Observe that the projection operator P_X in (DI) takes care of one viability concern, namely that x (t) e X, for all t e R+. The other concern, that $y(t) \geq 0$ a.e., is not relevant here. It is automatically satisfied as long as the initial guess y(0) is nonnegative a.e.

To make (DI) handy in computations we must evaluate the partial subdifferentials $d_X L(x,y)$and $d_y L(x,y)$. When y e Y+ as defined in (2.12), a general rule for computing subgradients of convex integral functionals (cf. (6) and (9),p. 442), yields.

$$(3.1) \qquad \partial_Y L(x,y) = \{u \varepsilon H : u(\omega) \ \varepsilon \ \partial_\xi \wedge (\omega, x(\omega), y(\omega)) \quad \text{a.s.}\}$$

In (3.1) the partial subdifferential $d_X \wedge$ of $\wedge$ with respect to $x = (x_1, \ldots, x_S)$ can be evaluated directly from (2.9-10).

Similarly, one gets.

$$(3.2) \qquad \partial_y L(x,y) = \begin{cases} y \ \varepsilon Y : y(\omega) \ \omega \ \partial_\eta \wedge (\omega, x(\omega), y(\omega)) & \text{a.e.}) \\ y \ \varepsilon Y : Y_S(\omega) = f_S(\omega, x_1(\omega), .. x_s(\omega)) & \text{a.e. for all } s = 1,..,S \end{cases}$$

We shall see shortly that the representations of (3.1-2) of the partial subdifferentials take us a long way to make (DI) tractable.

However, we need first to spell out the projection operator Π_X in (DI). We know that all sigma-algebras Σ_S (s=1),...,s) are complete. Then, evidently, X is a closed linear subspace of H. But, orthogonal projection of H onto $H_S := L^2 (\Sigma_S, R^{n_S})$ amounts to conditional expectation with respect to Σ_S, i.e.,

$$\pi_X [X] = (\pi_{H_1} [X], \ldots\ldots, \pi_{H_S} [X])$$

with

$$\pi_{H_s}[x] = E_s[x_s] : = E(x_s | \Sigma_s) \quad \text{for all } x = (x_1, .., x_s) \ \varepsilon H.$$

Thus, by this last observation and formulas (3.1-2), the "functional" differential inclusion (DI) splits finally into a system of "pointwise" inclusions :

$$(DI)_\omega \qquad \begin{aligned} &\dot{x}_s(t)(\omega) \ \varepsilon \ -E_s[\partial_{E_s} \wedge (\omega, x(t)(\omega), y(t)(\omega))] \\ &\dot{y}_s(t)(\omega) = f_s(\omega, x_1(t)(\omega), .., x_s(t)(\omega)). \end{aligned}$$

In computations this latter system $(DI)_W$ is the one which should be solved (integrated numerically) for all stages s = 1,...,S and for almost every w e W. Any (x(0), y(0)) in the set X x Y+ (see (2.2) and (2.12)) can be used as initial point. Clearly, system $(DI)_W$, w e W, may be very large. Therefore, in practice one may have to contend with discrete probability measures approximating the original m.

Computations are however, beyond the scope of this paper. Rather, in the next part we only explore the convergence properties of (DI).

4. CONVERGENCE

This part contains the main novelties of this paper. It is concerned with the theoretical efficiency of (DI) as a computational method. Specifically, we shall show, under broad hypothesis, that (DI) can be expected to yield convergence in finite time which is practically essential for the practician. Our analysis leans on the following Flam and Seeger's result.

4.1. Theorem .1 (Convergence)

Suppose that the set S of saddle points of L with respect to X x Y+ is bounded and converges monotonically in norm to S. This trajectory (x,y) (.) of (DI) emanating from X x Y+ is bounded and convergences monotonically in norm to S. This trajectory stays within X x Y+, and y(t) converges weakly and monotonically upwards to some $\bar{y} \varepsilon Y_+$.

Brief outline of proof: Let L(x,y) = -∞ whenever y ∉ Y+, and +∞ whenever x ∉ X, y e Y+. Write z = (x,y). Then the correspondence M(z) : = $(d_xL,-d_yL)$ (z) is maximal monotone .

Consequently the gradient system

$$z(t)\varepsilon - M(z(t)), z(0)\varepsilon X \times Y_+$$

admits a unique infinitely extendable bounded trajectory which lives in X Y+ forever.

This trajectory also solves (DI). Since y(.) is bounded and monotone it converges weakly and monotonically upwards to some $\bar{y}$ e Y+. Next, all weak accumulations points of x(t) , t e R + must be feasable because otherwise $\|y(t)\| \uparrow \infty$

Finally consider the Lyapunov function

$$l(t) := \text{dist}\,(x(t), S_p)^2 / 2 + \|y(t) - \bar{y}\|^2 / 2$$

where dist(·,S_p) denotes the distance to the set S_p of optimal solutions to problem (P). The right hand derivative of l(t) is majorized by inf(P) - $f_0(x(t))$. Therefore, l(t) ↓ −∞ unless all weak accumulation points of x(t), t e R+, belongs to S_p.

We desire here not only convergence as ensured by Thm.4.1, but also that this occurs in finite time. For that purpose we need an assumption (On sharpness of constraints)

We say that the constraints of problem (P) are sharp with modulus $a > 0$ if

(4.1) $E1\cdot f(x) \geq \alpha \operatorname{dist}(x,C)$ **for all** $x \in X$.

Here X is as defined in (2.2) ; C denotes the feasible set for problem (P) ;

$$\operatorname{dist}(x,C) := \inf_{c \in C} \|x - c\|$$

is the L^2-distance between $x \in X$ and C ; the constant vector $1 = (1,...,1)$ belongs to R^m, $m = m_1 + ... + m_S$; and finally, $f = (f_s)_{s=1}^{S}$.

THEOREM 4.2 (Feasibility in finite time)

Suppose all coordinates of the initial multipliers $Y_s(0)$, $s = 1,..., S$ are minorized a. e. by a positive number g. Also suppose constraints are sharp with modulus $a > 0$, and that

$$\sup_{x \in B} \sup_{g_0 \in \partial f_0(x)} \|g_0\| - \alpha\gamma =: -\Delta < 0,$$

where B is a ball centered at the optimal solution $\bar{x}(0)$ nearest to $x(0)$ with radius majorizing $\|(x,y)(0) - (\bar{x},\bar{y})(0)\|$. Then $x(t)$ is feasible for all $t \geq$ $\operatorname{dist}(x(0), C)/\Delta$.

PROOF in Flam & Seeger (see references) it is shown that

$$\|x(t) - \bar{x}(0)\| \leq \|(x,y)(t) - (\bar{x},\bar{y})(0)\| \leq \|(x,y)(0) - (\bar{x},\bar{y})(0)\|.$$

Therefore $x(\cdot)$ stays within the ball B mentioned here above. Consider the distance $d(t) := \operatorname{dist}(x(t),C)$ between the current point $x(t)$ and the feasible set C. As long as $x(t) \notin C$ we have

$$\delta(t)\,\dot{\delta}(t) = d(\delta(t)^2/2)/dt$$

$$= \langle x(t) - x^{*}(t),\ \dot{x}(t) \rangle$$

(where the derivative is taken from the right, and where $x^*(t)$ denotes the feasible point which is closest to $x(t)$)

$$= E \sum_{s=0}^{S} < x(t) - x^*(t), -g_s(t) >$$

for $y_0 \equiv 1$ and appropriate subgradients

$$g_s(t) \in \partial_x[Y_s(t).f_s(x(t))], \quad s = 0,..,S)$$

$$\leq \delta(t) \|g_o(t)\| - E\sum_{s=1}^{S} y_s(t).f_s(x(t))$$

$$\leq \delta(t) \|g_o(t)\| - \gamma E1.f(x(t))$$

$$\leq \delta(t) [|g_o(t)| - \alpha\gamma] \leq - \delta(t)\Delta.$$

It follows that :

$$\dot\delta(t) \leq - \Delta, \text{ when } x(t) \notin C,$$

and now the conclusion is immediate.

REMARK Thus, to obtain feasibility in a finite lapse of time one should choose all initial values $y_1(0), \ldots, y_S(0)$ large a.e. Conceptually one might contemplate to set $y_s(0) = +\infty$ for all s d 1. In practice, this is impossible however, and large values $y_s(0)$ may yield a fairly stiff system.

In the light of Theorem. 4.2 it is natural to inquire when constraints are indeed sharp. The next result, inspired by (4), gives a sufficient condition in this direction. For its statement some notation is needed. We introduce the cone

$$y_{s+} := \{y_s \in L^2(\Sigma_s, R^{m_s}) : y_s \geq 0 \text{ a.e.}\}$$

of non-negative, Σ_S-measurable, square integrable random vectors in R^{m_s}. Let the correspondence $G = (G_S)_{s=1}^{S}$ from X (2.2) to Y (2.7) be defined by

$$G_S(x) := F_S(x_1(\cdot), \ldots, x_S(\cdot)) + Y_{S+}.$$

Note that feasibility of x in problem (P) amounts to the statement that 0 e G(x). Thus, the feasible set C equals $G^{-1}(0)$. Recall that any L^2 space of (square integrable) random vectors may be regarded as a subset of the corresponding space L^1 of absolutely summable random vectors. Thus, on L^2 we also have a relative topology induced by the L^1-norm.

PROPOSITION 4.1 (Shap constraints)

Suppose the range of the correspondance G contains the origin as interior point and is closed in the L^1-completion of $L^2(d, R^m)$. Then the constraints of problem (P) are sharp on any bounded set.

PROOF On the L^1-completion of the range space of G, which is Banach, we temporarily use the L^1-norm, and denote it by $\|\cdot\|_1$. Observe, using this norm, that

$$\text{dist}\,(G(x), 0) = \|F^*(x)\|_1 = \|f(x)\|_1 = E1'f(x)$$

for every $x \in X$. For any $x_0 \in G^{-1}(0)$ there exists $g > 0$ such that

$$\text{dist}(x, C) = \text{dist}(x, G^{-1}(0)) \leq \text{dist}(G(x), 0)(1 + \|x - x_o\|) / \gamma$$

$$= E1'f(x)(1 + \|x - x_o\|) / \gamma \leq E1'f(x)(1 + \|x\| + \|x_o\|) / \gamma$$

for every $x \in X$. Proof is given by Robinson-Ursescu theorem, as Flam recalls. The conclusion is then immediate provided that all vectors x in question are uniformly bounded in norm.

REMARKS

(i) Suppose Σ is finite (so that μ has finite support). Then constraints are sharp under the Slater condition requiring that (P) has to be strictly feasible, i.e., there should exist $x \in X$ such that (1.1) holds with strict inequality in every coordinate. In this case the hypothesis of Prop. 4.1 is satisfied.

(ii) The conditions imposed in Prop. 4.1 are very strong. Essentially, they imply that Σ is finite, so that the L^1- and the L^2-topologies coincide. Otherwise, when Σcontains a sequence of events A_k, $k = 1,2,..$ such that $\mu(A_k)$ is strictly decreasing to zero, one may easily show that L^2 is not closed in L^1.

(iii) The most important practical instances of (P) are linearly constrained. Then (1.1) reads.

$$A_s(\omega)[x_1(\omega), .., x_s(\omega)]^T \leq b(\omega) \text{ a.e. for } s = 1, .., s.$$

The possibly random technology matrix $A(w) = [A_1(w),..,A_S\ (w)]$ defines here a linear mapping.

$$(Ax)_S(\omega) : = A_S(\omega)\ [x_1(\omega),\dots,x_s(\omega)]^T$$

from x (2.2) into Y (2.7). Then using the so-called Hoffmann inequality one may then show, again provided that d is finite, that the constraints are sharp.

Once we have obtained feasibility, it is time to rais the question of optimality. To this end consider the derivative

$$f'_o(x;d) : = \lim_{h\downarrow 0} \frac{f_o(x+hd)-f_o(x)}{h}$$

in any direction

$$(4.2) \qquad d \in D : = \Pi_X\left[- E\sum_{s=0}^{s} \partial_x[y_s f_s(x)]\right]$$

as prescribed by (DI). To reduce f_0 swiftly it is safe to select a direction

$$x \in \operatorname{argmin}_{d\in D} f_o\ (x;d).$$

Such a choice yields a **directional derivative**

$$f_o(x;x) = \min_{d\in D} f'_o(x;d)$$

$$= \min_{d\in D} \max_{g_o\in\partial f_o(x)} < g_o, d >$$

In particular, when x(t) is feasible, we may select the direction d(t) such that the contribution from every term $d_X[Y_S(t)\cdot f_S(x(t))]$, s ≥ 1, in (4.2) is nil. It follows then that

$$f_o(x,x) \leq -\|g_o\|^2 \text{ for all } g_o \in \partial f_o(x).$$

To reflect this we say that f_0 **descends at least linearly on** C if x e C implies

$$f_o(x,x) \leq -\|g_o\|^2 \text{ for some } g_o \in \partial f_o(x).$$

THEOREM 4.3 (Finite convergence)

Suppose $x(\cdot)$ generated by (DI) is feasible for all $t \geq d$ some $t \geq 0$. Also, suppose that problem (P) is essentially constrained in the sense that

$$\mu := \inf_{x \varepsilon C, g_o \varepsilon \partial f_o(x)} \|g_0\|^2 > 0.$$

Then, if f_0 descends at least linearly on C, $x(t)$ is optimal no later than time

$t = [f_0(x(t))\text{-}\inf(f_0/C)]/m + t.$

PROOF When $x(t)$ e C we have

$$\frac{df_o(x(t))}{dt} = f'_o(x(t)\,;\,\dot{x}(t))$$

$$\leq -\|g_o(t)\|^2 \text{ (for some } g_o(t) \ \varepsilon\ \partial f_o(x(t))) \leq -\mu.$$

Hence, before optimality has occured it holds that

$$f_o(x(t)) - f_o(x(\tau)) \leq -\mu(t-\tau)$$

for all $t \geq t$, hence the conclusion is immediate.

5. CONCLUDING REMARKS

We tried to develop the entire theoretical and practical approach of defeasance. The usual models are not difficult to solve. but as soon as stochastic process is added regarding the evolution of interest rates, the models become more difficult. When the constraints are developed remaining roughly linear, the convergence is not guaranteed. The classical Bellmann equation may not be sufficient to solve properly all models.

Stochastic programming is quite challenging : Neither modeling nor computation is straight-forward. Regarding the latter issue most effort has naturally been directed towards decomposition in one form or another (10). Here we have gone very far in that direction : Problem (P) is explored by means of a very large scale differential system. That system updates all decisions (primal variables) and multipliers simultaneously. If data are smooth, the system dynamics (DI) involve "kinks" which are "few" and easy to handle. Moreover, it is only the asymptotic behavior of (DI) which presents interest. Is is a matter of satisfaction that (DI) present a good stability properties provided constraints are sharp.

BIBLIOGRAPHY :

J.P.AUBIN A. CELLINA , *Differentials inclusions,* Springer Verlag 1984.

S. P. BRADLEY and D. B. CRANE, *A dynamic model for bond portfolio management,* Management Science, Vol. 19, n° 2, Oct. 1972.

K. J. COHEN, S. F. MAIER and J. H. VANDER WEIDE, *Recent developement in management science in banking*, Management Science, n° 27, Oct. 1981.

J. EKELAND, R.TEMAM, *Eléments d'économie mathématique*, Hermann, Paris, 1979.

A.F. FILIPOV, *A minimax inequality and applications, in : Inequalities III,* O Sisha Ed., Academic Press, 103-113, 1972.

S. FLAM, *Finite convergence in stochastic programming* . Bergen University Preprint 1994.

S. FLAM, R. SCHULTZ, *A new approach to stochastic linear programming* . Bergen University. Preprint 1993.

W. FLEMING and R. RISHEL, *Deterministic and stochastic optimal control,* Springer-Verlag, New York, 1986.

G. HADDAD, J.M. LASRY, *Periodic solutions of functional differential inclusionsand fixed of Selectionable correspondences*, J. Math. Anal. Appl., 1983.

M. I. KUZY and T. ZIEMBA, *A bank asset and liability management model,* Operation Research, 1986.

J. Y. LEBERRE et J. SIKORAV, *Gestion actif-passif : une approche dynamique,* Les entretiens de la finance, AFFI, Paris, Dec. 1991.

P.L. LIONS, B. MERCIER, *Splitting algorithms for the sum of two nonlinear operators*, SIAM J. Num. Anal. 16, 964-979, 1979

A. LYAPUNOV, *Problème général de la stabilité du mouvement*, Annales de la Faculté des Sciences de l'Université de Toulouse, 9, 27-474, 1910

H.A. SIMON "*Rationality and Process as a Product of thought*" American Economic Review 1978 vol. 69

H.A. SIMON *Rational Decision Making in Business Organization* " American Economic Review 1979 vol 69

C. S. TAPIERO, *Applied stochastic models and control in management*, North-Holland, 1988.

T. WAZEWSKI, *On an optimal control problem*, Proc. Conference " Differential equations and their applications, Prague 1964, 229-242, 1964.

MATHEMATICAL PROGRAMMING AND RISK MANAGEMENT OF DERIVATIVE SECURITIES

Les Clewlow, Stewart Hodges, Ana Pascoa
Financial Options Research Centre*
The University of Warwick, Coverty, UK

Abstract: In this paper we discuss the use of mathematical programming techniques linear, dynamic, and goal programming to the problem of the risk management of derivative securities (also known as contingent claims or options). We focus on the problem of the risk management of complex or exotic options in the presence of real market imperfections such as transaction costs. The advantages and disadvantages of the various approaches which have appeared in the literature are discussed including a new approach which we are developing.

Keywords: Mathematical Programming, Optimisation, Options, Contingent Claims, Derivatives, Risk Management.

1 INTRODUCTION

In this chapter we discuss the application of mathematical programming techniques such as linear programming, dynamic programming, and goal programming to the problem of the risk management of derivative securities (otherwise known as contingent *claims or options*). Derivative securities are those who value depends on the value of fundamental securities or assets such as stocks or bonds. In this chapter we will be concerned in particular with complex or exotic options. For example path dependent options whose value depends on the path the underlying asset price tool over the life of the option rather than just its final value as is the case with standard European options.

In a perfect market the Black-Scholes model (Black and Scholes (1973)) provides the recipe for the risk management of standard European options. A perfect market is one in which there are no transaction costs and no taxes, the market operates continuously, and the price of the underlying asset is continuous (that is there are no jumps in the asset price). In addition Black and Scholes assumed that the asset price follows a geometric Brownian motion (GBM) stochastic process with constant volatility and the risk free interest rate is constant[1]. The behaviour of the asset price under GBM can be characterised by its stochastic differential equation

$$dS(t) = \mu S(t)dt + \sigma S(t)dz(t) \tag{1.1}$$

---

* This work was partially supported by sponsors of Financial Options Research Centre; HSBS Markets, Tokyo Mitsubishi International, Deutsche Morgan Grenfell, SBC Warburgs, Tradition (UK) and Programa PRAXIS XXXI.

[1]This was generalised by Merton (1973) to allow the volatility to be a deterministic function of time and the interest rate to be stochastic.

where μ is the expected return on the asset and σ is the volatility of returns on the asset. Black and Scholes showed that options could be priced by constructing a perfectly riskless portfolio with an option, the underlying asset and cash. Itos lemma allows us to write down the stochastic differential equation govening the price of an option $c(S(t),t)$ which only depends on the asset $S(t)$ and time t

$$
\begin{aligned}
dc(S(t),t) = & \frac{\partial c(S(t),t)}{\partial t} dt \\
& + \frac{\partial c(S(t),t)}{\partial S(t)} \left(\mu S(t) dt + \sigma S(t) dz(t) \right) \\
& + \frac{1}{2} \frac{\partial^2 c(S(t),t)}{\partial S(t)^2} \sigma^2 S(t)^2 \, dt
\end{aligned}
\tag{1.2}
$$

If we form a portfolio P in which we are short the option and long an amount $\frac{\partial c(S(t),t)}{\partial S(t)}$ of the asset, the equation governing the price of the portfolio is

$$
dP(S(t),t) = -dc(S(t),t) + \frac{\partial c(S(t),t)}{\partial S(t)} dS(t) \tag{1.3}
$$

Substituting into equation (1.3) using equations (1.1) and (1.2) gives

$$
dP(S(t),t) = \frac{\partial c(S(t),t)}{\partial t} dt + \frac{1}{2} \frac{\partial^2 c(S(t),t)}{\partial S(t)^2} \sigma^2 S(t)^2 \, dt \tag{1.4}
$$

The portfolio P is riskless, that is it has no random component, and must therefore earn the riskless rate of interest

$$
\frac{dP(S(t),t)}{P(S(t),t)} = r dt \tag{1.5}
$$

Substituting into equation (1.5) for dP using equation (1.3) and for P leads to the Black-Scholes partial differential equation

$$
\frac{\partial c(S(t),t)}{\partial t} + rS(t) \frac{\partial c(S(t),t)}{\partial S(t)} + \frac{1}{2} \sigma^2 S(t)^2 \frac{\partial^2 c(S(t),t)}{\partial S(t)^2} = rc(S(t),t) \tag{1.6}
$$

The solution to this partial differential equation subject to the boundary condition of a standard European call option $c(S(T),T) = \max(0, S(T) - K)$ is the Black-Scholes equation

$$c(S(t),t) = SN(d_1) - Ke^{-r(T-t)}N(d_2) \tag{1.7}$$

$$d_1 = \frac{\ln\left(\frac{S}{K}\right) + \left(r - \frac{1}{2}\sigma^2\right)(T-t)}{\sigma\sqrt{T-t}}, \quad d_2 = d_1 - \sigma\sqrt{T-t}$$

where K is the strike price and T is the maturity date of the option. The important point to note about equation (1.6) is that it does not depend on any parameters, in particular the expected return of the asset μ, which depend on investors risk preferences. This is because the option can be perfectly hedged, that is the risk due to the underlying asset completely eliminated, by a continuously rebalanced position in the underlying asset. However, this result relies critically on the assumptions of continuous trading and no transaction costs.

The quantity of the underlying asset which must be held in the hedge portfolio of a short option position, called the *delta* of the option, is the partial differential of the option price with respect to the underlying asset. This generalises to the case of multiple risky state variables, the quantity of the risky state variable which must be held is the partial differential of the option price with respect to the state variable. The partial derivatives of the Black-Scholes formula with respect to all the parameters have standard names which are defined in Table 1.

Table 1: Black-Scholes Risk Measures

Delta	$\dfrac{\partial c(S(t),t)}{\partial S(t)}$
Gamma	$\dfrac{\partial^2 c(S(t),t)}{\partial S(t)^2}$
Vega or Lambda	$\dfrac{\partial c(S(t),t)}{\partial \sigma}$
Theta	$\dfrac{\partial c(S(t),t)}{\partial t}$
Rho	$\dfrac{\partial c(S(t),t)}{\partial r}$

Note that for the case of Vega and Rho these are derivatives with respect to constant parameters in the model. Practitioners routinely do this because they find themselves using models which assume parameters are constant which they know are risky.

The Black-Scholes riskless hedge idea extends to more complex or exotic options as long as we remain in the perfect market world. For example, consider the case of a down and out call option, which is a type of barrier option. This is a standard European option except that if the asset price falls below a pre-determined level, H, called the barrier then the option disappears. The price of this option is also governed by the Black-Scholes partial differential equation (equation (6)) but with the additional boundary condition $c(H,t) = 0$.

Consider hedging this down-and-out call option. When the asset price is far from the barrier the option behaves like a standard call option, both in price and sensitivities. However, as the barrier is approached the option value begins to fall more and more rapidly. Consequently the delta rises near to the barrier and can be greater than one. If then the barrier is hit, the option instantly disappears and the delta changes discontinuously from a large value to zero. With no transaction costs this is of course not a problem, we simply rebalance our holding in the underlying asset to zero. With transaction costs, having to sell a large holding in the underlying asset is a significant problem. However the real problem is more subtle, it is that the delta changes rapidly with changes in the underlying asset price near the barrier (this is called gamma risk, see Table 1). Therefore, in the presence of transaction costs the delta hedging strategy will incur large transaction costs even if the barrier is not hit which can lead to the hedging strategy costing far more than the Black-Scholes price of the option. The reason for the failure of the Black-Scholes hedging strategy is that it should be taking into account the expected future hedging costs inclusive of transaction costs.

Thus, when we introduce market imperfections, in particular non-continuous trading and transaction costs the whole nature of the problem and solution change. The Black-Scholes delta hedging approach relies critically on being able to continuously rebalance the delta hedge without incurring transaction costs. In a real market we cannot trade continuously so we cannot make the hedge portfolio riskless and every trade incurs transaction costs so that if we try and trade close to continuously we will lose a large amount of money. For a typical market even daily rebalancing of the hedge can lead to costs which exceed the Black-Scholes value the option. It is possible to measure the performance of this hedging strategy, for example expected profit against variance of profit. But this approach is extremely sub-optimal (Clewlow and Hodges (1996)).

The problem of hedging under transaction costs was first tackled by Leland (1985) for standard European options. His solution was an adjustment to the volatility in the Black-Scholes formula which accounted for the presence of proportional transaction costs The intuition behind his approach is helpful in understanding the nature of the problem. Imagine we have written a standard European call on non-dividend paying stock and are delta hedging this liability. We will therefore have a long position in the underlying stock. Imagine now that the stock price goes up. The delta of the option will increase and we will therefore buy more stock. Now in the presence of transaction costs it will cost us more than the value of the stock to obtain the required amount of stock. That is, it will be as if the stock price had increased slightly more than it did. Now imagine that the stock prices goes down. The delta of the option will decrease and we will sell some stock. Again, the transaction costs will mean that we get slightly less than the actual value of the stock. That is, it will be as if the stock price decreased slightly more than it actually did. Overall the effect of transaction costs on our delta hedge will be similar to the stock having a slightly higher volatility than it actually has. Leland's adjustment of the Black-Scholes model has since been extended by Whalley and Wilmott (1993); and Hoggard, Whalley and Wilmott (1994). Boyle and Vorst (1992) proposed a model similar to Leland (1985)'s, although making different assumptions about the distribution of the changes in the underlying assets. They use a binomial tree to represent the stochastic behaviour of the underlying asset. Note that none of the above models are in any sense optimal. Approaches based on optimisation will be reviewed in the following sections.

2 ONE PERIOD HEDGES - THE LINEAR PROGRAMMING APPROACH

Since the problem with the Black-Scholes delta hedging strategy in the presence of transaction costs stems from the continuous trading, one solution is to move to single period or static hedging strategies. This approach leads to a linear programming framework. There are essentially two ways in which the problem can be formulated in this framework. The first is to keep the idea of the Black-Scholes risk measures but to only consider the next rebalancing period. We imagine we have a position x_T (usually -1) in a target option c_T, and we wish to solve for the set of holdings $\{x_i ; i=1,\ldots,n\}$ in a set of more basic securities $\{c_i ; i=1,\ldots,n\}$ (which may also be options) in order to neutralise a set of risk measures $\{R_j(c); j=1,\ldots,m\}$. We therefore obtain the following constraints:

$$x_1 R_j(c_1) + x_2 R_j(c_2) + \ldots + x_n R_j(c_n) = x_T R_j(c_T) \; ; \quad j=1,\ldots,m \tag{2.1}$$

$$x_i \geq 0 \; ; \; i=1,\ldots,n^2 \tag{2.2}$$

We would then, for example, minimise the cost inclusive of transaction costs $(x_1 c_1 + x_2 c_2 + \ldots + x_n c_n)$, subject these constraints:

$$\underset{\{x_i ; i=1,\ldots,n\}}{\text{Min}} \left(x_1 c_1 + x_2 c_2 + \ldots + x_n c_n\right) \quad \text{subject to (2.1) and (2.2)} \tag{2.3}$$

[2] It is also possible to allow negative positions with a lower bound.

The form of the optimisation problem is an expression of the risk preferences of the hedger. This is an important point to bear in mind when formulating these problems. The solution may be very different for different choices of the objective function. With this formulation we obtain a solution which is only optimal over the next period so if it is applied repetitively over multiple periods it may be seriously sub-optimal.

Alternatively, we may only be concerned with the liability at a single future date (the maturity date of the target assuming it generates no intermediate liabilities). In this case we imagine we have a set of scenarios $\{s_j ; j = 1,\ldots,m\}$ for the underlying state variables at the future date and we wish to solve for the set of holdings $\{x_i ; i = 1,\ldots,n\}$ in the set of basic securities $\{c_i ; i = 1,\ldots,n\}$ in order to meet the target in all scenarios. We therefore obtain the following constraints:

$$x_1 c_1(s_j) + x_2 c_2(s_j) + \ldots + x_n c_n(s_j) \geq x_T c_T(s_j) \; ; \quad j = 1,\ldots,m \tag{2.4}$$

$$x_i \geq 0 \; ; \; i = 1,\ldots,n \tag{2.5}$$

Note that equation (2.4) is now an inequality so that the hedge portfolio super-replicates the target. A strict equality could be used but this usually leads to less robust and more expensive hedges. We would then, for example, minimise the initial cost inclusive of transaction costs $(x_1 c_1(0) + x_2 c_2(0) + \ldots + x_n c_n(0))$, subject these constraints (see for example Aparicio and Hodges (1996)):

$$\underset{\{x_i ; i = 1,\ldots,n\}}{\text{Min}} \left(x_1 c_1(0) + x_2 c_2(0) + \ldots + x_n c_n(0)\right) \quad \text{subject to (2.4) and (2.5)} \tag{2.6}$$

Dembo (1991) introduces stochasticity into these types of models by using a scenario optimisation approach in two stages. First he computes solutions to the deterministic problems under all scenarios and then he solve a co-ordinating or tracking model to find a single, feasible solution. The tracking model satisfies all the constraints and minimises the overall difference from the optimal solutions to the deterministic problems.

Another approach is the minimax hedging strategy of Howe, Rustem and Selby (1996). They aim to minimise the maximum potential hedging error between time periods. That is, at the rebalancing date, they finds the worst-case scenario of the underlying asset price giving the worst hedging error and then solve for the holding in the asset which minimises this. This is therefore most relevant where the underlying asset is highly volatile and crosses the exercise price frequently. If the worst possible scenario does not occur, it usually would have been better to use Black-Scholes delta hedging. Merton's 'ideal portfolio' is the benchmark to the hedging error in the objective function. They show how the rebalancing strategy can be used at the end of a time interval, at the beginning and end of the time interval (two-period minimax) or a variable minimax where the hedger monitors the hedging error and rebalances it whenever finds it unacceptable.

3 MULTI PERIOD HEDGES - THE DYNAMIC PROGRAMMING APPROACH

It is possible to solve the problem of delta hedging options in the presence of transaction costs using a dynamic programming approach. Since it is not possible to form a riskless delta hedge when transaction costs are incurred in trading the underlying asset the solution now depends on the risk preferences we assume for the hedger. As we saw in section 2, these risk preferences manifest themselves in the form of the objective function over which we optimise. For example we could choose to maximise expected utility[3] of wealth at a future date or minimise the initial cost of the hedge subject to super-replication (i.e. guaranteeing a pay-off at least as large as the liability at a future date). This approach makes the very important assumption that we can specify very accurately the probability distribution of future states of the world. Given this distribution, the optimal solution we obtain is valid whatever future state of the world occurs. But, there is a significant computational effort involved in obtaining this kind of solution and if our probabilities are not correct then the solution may be severely sub-optimal (more so as we look over longer time horizons). The approaches in section 2 and 4 tend to be more robust to imperfectly specified probability distributions. Furthermore, for more sophisticated models and hedging strategies it may be computationally impractical to solve the problem directly in a dynamic programming framework.

Hodges and Neuberger (1989) (see also Davis and Panas (1991), Davis *et al* (1993), and Clewlow and Hodges (1996)) were the first to formulate this problem with proportional transaction costs[4] as one of stochastic optimal control and show how to solve it using dynamic programming. By careful choice of the utility function Hodges and Neuberger were able to obtain a formulation in which the only state variables were the asset price $S(t)$, the holding in the asset $x(S(t),t)$, and time t. The solution method is based on constructing a binomial tree approximation for the asset price. At each node in the tree a vector of possible holdings in the asset is held together with an associated vector of values of the portfolio. The solution method consists of working backwards from the option maturity date boundary

---

[3] A utility function of wealth $U(w(t))$ expresses individuals preferences for levels of wealth $w(t)$ in dimensionless units.

[4] By proportional transaction costs we mean the costs are proportional to the value of the asset traded.

condition, computing the portfolio value and applying the optimal control boundary conditions. The optimal control strategy consists of upper and lower limits on $x(S(t),t)$ within which $x(S(t),t)$ must be maintained. Figure 1 illustrates the typical control limits which are obtained. The optimal delta hedging strategy consists of doing nothing while $x(S(t),t)$ remains between the control limits. But as soon as $x(S(t),t)$ reaches either the upper or lower limit the asset is traded continuously to control $x(S(t),t)$ never to be outside the limits. Also shown in Figure 1 is the Black-Scholes delta, the control limits lie roughly either side of the Black-Scholes delta, although not always.

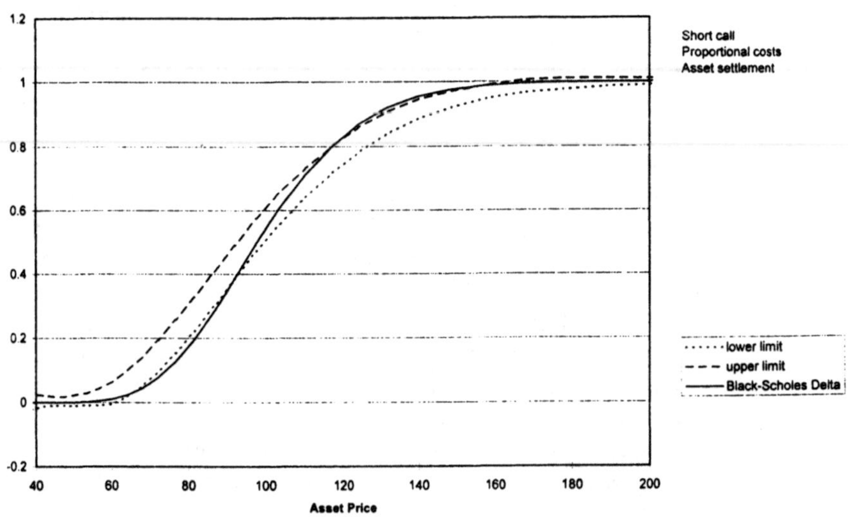

Figure 1: Delta Hedging under Proportional Transaction Costs Control Limits

Since the solution is obtained by a numerical procedure this model can be used for hedging mixed portfolios of long and short positions, mixed maturity dates and with general transaction cost structures (see Clewlow and Hodges (1996))

Edirisinghe *et al* (1993) took the alternative approach of minimising the initial cost obtaining a payoff at least as large as the liability at a future date (super-replication). The authors claim that this approach is independent of investor preferences but, although investor preferences are not explicitly modelled, the chosen formulation implicitly defines investor preferences. The important difference between this two approaches is that typical utility functions allow profits in good future states of the world to be traded off against losses in bad future states. The approach of minimising the initial cost of super-replication is applicable to situations where the cost of not meeting the liability are unacceptable under any circumstances. This result can be approximated by using a utility function which assigns negative wealth a relatively very low utility.

Dempster (1995) uses stochastic, multiperiod models to maximise expected utility of terminal wealth. He represents the stochasticity of the financial world through huge trees with thousands of paths. Dempster has developed techniques for shortening these tree. He shows that in most problems where we have huge trees of possible scenarios, one can find a sub-tree that represents the bulk of stochasticity. It is then feasible to solve these problem using advanced computational systems.

4 HEDGING WITH MULTIPLE OBJECTIVES - THE GOAL PROGRAMMING APPROACH

Consider an investment bank which is writing complex options and thus has a large book of derivative instruments which it needs to hedge. This involves many conflicting objectives. The primary objective is to minimise the risk of the book and maximise profits. However, transaction costs, non-continuous trading, discrete lot sizes, etc. means that the risk can not be reduced to zero and the greater the risk reduction the greater the cost. Furthermore real assets do not follow GBM, at the very least they have jumps and their volatility is also stochastic. There are also sometimes restrictions on short selling and the interest rate at which borrowing can be obtained.

Jumps and stochastic volatility are sources of risk, in addition to that from the Weiner process driving the underlying asset, which cannot be hedged with a position in the underlying asset. This situation is referred to as an incomplete market and options must be introduced into the market to complete it and allow these additional sources of risk to be hedged. However, options markets have higher transaction costs and are less continuous than the underlying markets (for example the maturities and strike prices of exchange traded options, see Clewlow and Hodges (1994)) and so managing the transaction costs is very important.

In principle we could extend the optimal delta hedging approach to gamma/vega hedging. However, we would need to solve for the optimal holdings in all the available options. Solving this in a dynamic programming framework is very difficult if not impossible in practice.

The market imperfections lead to conflicting goals. These can be stated generally as:

1) Risk minimisation.
2) Transaction cost minimisation.
3) Minimisation of the opportunity costs of capital tied up in hedging.
4) Cash flow minimisation (hedge management cost minimisation).

2, 3, and 4 corresponding to profit maximisation.

These conflicting goals constitute a multi-objective problem, motivating the use of goal programming which allows the formulation and solution of multi-objective problems. Recently Clewlow and Pascoa (1996) used this approach to hedge barrier options in incomplete markets under transaction costs. They used the market prices of standard European options to obtain an implied discrete time, discrete state evolution of the underlying asset. This discrete time and state structure approximates the jumps and stochastic volatility of the real market and allowed them to obtain approximate prices for the standard options and barrier option in the future. They also use this implied evolution to generate a set of scenarios at the boundaries of the standard and barrier options by Monte Carlo simulation. Using these scenarios they solve the following LP problem; The goals are to minimise the hedge error and to minimise transaction costs. These goals are represented by the following constraints:

$$\sum_{i=1}^{n} \left(x_l(i) - x_s(i)\right) * V_i(s) = V_T(s) + e_p(s) - e_m(s) \tag{4.1}$$

$$\left(\sum_{i=1}^{n} x_l(i) + x_s(i)\right) * \text{tcost} = \text{total\_tcost} \tag{4.2}$$

$$\begin{aligned} 0 \le x_l(i) &\le \text{large\_number} * dummy(i) \\ 0 \le x_s(i) &\le \text{large\_number} * (1 - dummy(i)) \end{aligned} \tag{4.3}$$

where

$V_T(s)$ is the price of target in scenario $s = 1, \ldots, ns$;
$V_i(s)$ is the price of standard security $i = 1, \ldots, n$ in scenario s.
$V0_i$ is the current price of standard security $i = 1, \ldots, n$.
tcost is the cost per unit of option bought or sold.
$x_l(i)$ is the quantity of option i to buy (≥ 0).
$x_s(i)$ is the quantity of option i to sell (≥ 0).
$dummy(i)$ = integer variable $\{0,1\}$.
$e_p(s), e_m(s)$ is the positive and negative hedge error respectively.
total_tcost is the total transaction costs.

Equation (4.1) is the hedge error minimisation constraint, (4.2) is the transaction costs minimisation constraint and (4.3) prevents the simultaneous buying and selling of the same security.

The objective function is

$$\sum_{s=1}^{ns} e_p(s) + e_m(s) + \text{total\_tcost} \tag{4.4}$$

Clewlow and Pascoa show that this approach can provide accurate and robust hedges for complex options under realistic conditions including jumps, random volatility and transaction costs. Figure 2 gives an example of the hedge obtained for an up-and-out call option. The advantage of this approach is that it is very flexible many realistic constraints can be added such as bounds on the short and long positions, no short fall in the hedge, upper or lower bounds on the cost of the hedge whilst maintaining a realistic model of the market.

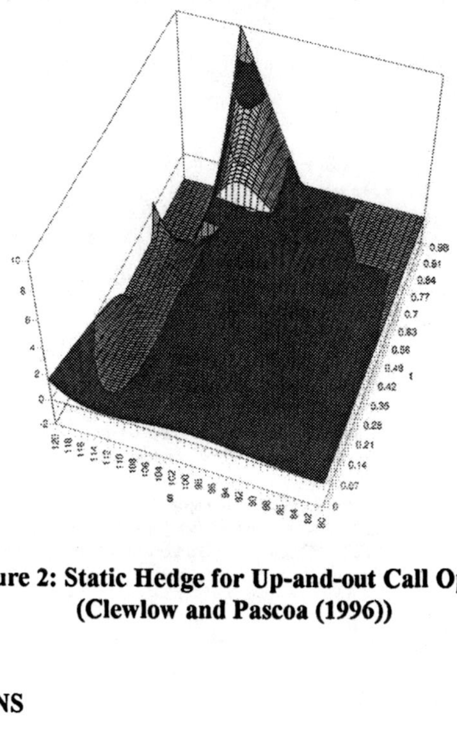

Figure 2: Static Hedge for Up-and-out Call Option (Clewlow and Pascoa (1996))

5 CONCLUSIONS

In this chapter we have reviewed the use of mathematical programming and optimisation in the risk management of derivatives securities. In section 1 we introduced the traditional Black-Scholes approach and discussed its flaws. The Black-Scholes approach assumes perfect markets whereas in reality there are transaction costs, non-continuous trading, jumps in asset prices and other risk sources. Failure to deal with these factors in a rigorous way can lead to large losses. In section 2 we introduced simple static replication alternatives to the Black-Scholes dynamic replication approach. The dynamic programming approach was considered in section 3, which gives us insights into the nature of the problem of risk management in an imperfect market. However, it was pointed out that this approach becomes impractical for realistic problems where the number of state variables becomes large very quickly. We then went on, in section 4, to describe a new approach we are developing which uses goal programming as the framework. We use the latest ideas of implying the future stochastic structure of the world from the market prices of standard European options together with a multi-objective formulation. This approach has the potential to allow realistic modelling of the risks in financial markets and allowing realistic constraints on the risk management process to be incorporated.

REFERENCES

Aparicio, S.D., and S.D. Hodges, (1996), "Estimating Implied Distributions and Issues in Static Hedging", Financial Options Research Centre Ninth Annual Conference 5-6 September, University of Warwick, Coventry, UK.

Black, F. and M. Scholes, (1973), "The Pricing of Options and Corporate Liabilities", *Journal of Political Economy* Vol. 81, pp 637-654.

Boyle, P. and T. Vorst, (1992), "Option Replication in Discrete Time with Transactions Costs", *Journal of Finance*, Vol. 47, pp 271-293.

Clewlow, L. and S.D. Hodges, (1994), "Gamma Hedging in Incomplete Markets under Transaction Costs", *Financial Options Research Centre Preprint 94/52*, University of Warwick, Coventry, England.

Clewlow, L. and S.D. Hodges, (1996), "Optimal Delta-Hedging Under Transaction Costs", *Financial Options Research Centre Preprint 96/68*, University of Warwick, Coventry, England.

Clewlow, L. and A. M. Pascoa, (1996), "Static Hedging of Barrier Options in Realistic Markets", *Financial Options Research Centre Working Paper*, University of Warwick, Coventry, England.

Davis, M. H. A. and V. G. Panas, (1991), "European Option Pricing with Transactions Costs", *Proc 30th I E E E Conference on Decision and Control*, pp 1299-1304.

Davis, M. H. A., V. G. Panas and T. Zariphopoulou, (1993), "European Option Pricing with Transactions Costs", *SIAM Journal of Control and Optimisation*, Vol. 31, pp 470-493.

Dembo, R., (1991), "Scenario Optimization", *Annals of Operations Research*, Vol. 30, pp 63-80.

Dempster, M.A.H., and X.C. Poiré, (1995), "Stochastic Programming: A New Approach to Asset/Liability Management", Financial Options Research Centre Eighth Annual Conference 22-23 June, University of Warwick, Coventry, UK.

Edirisinghe, C., V. Naik and R. Uppal, (1993), "Optimal Replication of Options with Transactions Costs and Trading Restrictions", *Journal of Financial and Quantitative Analysis*, Vol. 28, pp 117-138.

Hodges, S. D. and A. Neuberger, (1989), "Optimal Replication of Contingent claims Under Transactions Costs, *The Review of Futures Markets*, Vol. 8, pp 222-239.

Hoggard, T., A. E. Whalley, and P. Wilmott, (1994), "Hedging Option Portfolios in the Presence of Transaction Costs", *Advances in Futures and Options Research*, Vol. 7, pp 21-35.

Howe, M. A., B. Rustem, and M. J. P. Selby, (1996), "Multi-Period Minimax Hedging Strategies", *European Journal of Operational Research*, Vol. 93, pp 185-204.

Leland, H. E., (1985), "Option Pricing and Replication with Transactions Costs", *Journal of Finance*, Vol. 40, pp 1283-1301.

Merton, R. C., (1973), "Theory of Rational Option Pricing", *Bell Journal of Economics and Management Science*, Vol. 4, pp 141-183.

Whalley, A. E., and P. Wilmott, (1993), "Counting the Cost", *Risk*, Vol. 6 (10), pp 59-66.

IV. FUZZY SETS AND ARTIFICIAL INTELLIGENCE TECHNIQUES IN FINANCIAL DECISIONS

FINANCIAL RISK IN INVESTMENT

Jaime Gil-Aluja
Departament d'Economia i Organització d'Empreses
Facultat de Ciències Econòmiques i Empresarials
Universitat de Barcelona
Avgda. Diagonal, 690
08034 Barcelona
SPAIN

Abstract: In finance, the concept of risk is used to refer to phenomena of a diverse nature giving to it very different and sometimes diverse, meanings. Many definitions have been established at these effect, definitions, which with more or less fortune succeed in linking the term with its meaning. The work we present starts from the definition of the term "risk in investment" with a particular meaning: "the possibility of not being able to face the financial requirements of the investment process". Concepts such as "financial capacity of the investment", "financial pre-diagnosis", "financial diagnosis", "financial pathologies in the investment" amongst others are raised. in our view, such concepts enable the development of a proper model with the aim of expressing in numerical terms the degree of risk assumed if the investment activity is carried out.

Keywords: Diagnosis, Financial, Investment, Risk, Uncertainty.

FINANCIAL ASPECTS OF THE INVESTMENT PROCESS

The activities needed in order to carry out an investment process begin with the decision of investing. These activities imply the necessity of having financial capitals available, that is to say, monetary stocks, in particular moments of time. In the strict sense, the financial aspect of the process has the aim of obtaining these resources, so that they can be at the disposal of those responsible for making the collections.

It is common to find in the specialised books the concept of risk of the investment, which is defined according to the capacity of the purchased item to generate earnings. In some way, one of the measures or assessments of the possible economic profit is taken as a basis for the financial notion of risk. In this paper of research, we have no wish to indulge in a polemic about the suitability of the word "financial risk" for representing such a phenomenon. Instead, we want to put forward a different notion which we think it has been less studied and deserves, however, a special attention.

Let us begin by saying that the normally unusual characteristics of the investment process makes the own sources of finance not enough to deal with the required disbursements. That is why the necessity of external sources of finance arises.

We have pointed out several times that one of the important aspects of the investment programmes results from the possibility of forecasting in time the sums of money required to meet the obligations taken at the beginning of the process. It is obvious that modern programming patterns are a strong support for financial management. However, they cannot be considered as a barrier which always prevents the possibilities of default, because of the several aspects which can give rise to friction in the system. From that it follows the existence of a financial risk which results from the investment activities.

We define the financial risk of an investment as " the possibility of not being able to meet the payments needed for the end of the process to take place in the moment of time previously fixed".

In the study of the financial risk of an investment two major aspects arise. The first one refers to the disposability at the proper time of the monetary stocks needed for each one of the activities which make up the programme. The second one refers to the aggregation of the financial possibilities of all the activities, which will make up the degree of risk assumed at the beginning of the process. These two aspects reveal the dependence of the concept of risk on the time in which the activities are fulfilled. It should be taken into account that the lack of money supply in a particular moment of time does not compulsory stop its incorporation to the enterprise afterwards, and thus, the project can be finished late. That is why it is necessary to include the notion of "insolvency". Insolvency will imply the definitive stoppage of the activities and, thus, the rupture of the investment process.

Everything we have mentioned above shows that the most elementary rules of wisdom recommend an "economic and financial diagnosis" of the enterprise before beginning the activities which make up the project. It should be verified this way if the enterprise is reasonably ready to carry out the project.

As we approach this aspect of the problem, it is extremely useful the choice of the proper technique to make the programme because, starting from it, it is possible to know the quantity of means of payment required to face the disbursements derived from the activities of the project. Thus, there will be immediate needs in the short, medium and long run. How to face each one, and to what extent it is possible to do so will enable to determine the degree of risk assumed if the project is carried out. The "health" of the enterprise in a particular moment of time can result in more or less serious "illnesses" depending on assuming the higher or lower risk involved in each investment.

There are several ways to be followed in order to know the economic and financial situation of an enterprise. In this piece of research, we are going to put forward a method which we have already developed[1] with good results.

In order to follow our path, it seems wise to decide which are the fundamental elements that determine the solvency of an enterprise in relation to the financial

---

1 Gil-Aluja, J. "Ensayo sobre un modelo de diagnóstico económico-financiero". Actas de las V Jornadas Hispano-Lusas de Gestión Científica. Vigo, Septiembre 1990, pág. 26-29.

needs of the investment. The duality health-illness in the living beings has points of similarity in the field of the enterprise. As an example, it can be pointed out:

Health	→	Illness
A. Proper liquidity	→	Lack of immediate liquidity
B. Comfortable discount lines	→	Filled discount lines
C. Solvency with regard to suppliers	→	No credit from suppliers
D. Open-ended short credit	→	Limited short credit
E. Long-term accessibility	→	Exhausted long-term credits
F. Option to mortgaging credits	→	Goods fully mortgaged
G. Resort to guarantees	→	Impossibility of new guarantees
H. Easy enlargement of the common equity	→	non-enlarged common equity

Once these financial pathologies are established (we do not seek to include in an exhaustive way of all the weaknesses of the enterprise), we would like to enumerate just like in a medical diagnosis those "symptoms" by means of which we are able to make a good diagnosis. Sometimes, these symptoms arise. because some magnitude does not reach a particular level. In other cases, the illness manifests itself because the magnitude exceeds the limits considered to be dangerous or it falls beyond a certain interval. These magnitudes have diverse nature. In some cases, they are absolute magnitudes, in other cases, they are relative. They can be estimated by means of measurements or valuations, but the assessment must be made through numerical assignment as far as possible.

These considerations lead us to the question: where to search for the symptoms? If we take into account that the diagnosis is made in an initial moment of time and that it refers to one or several stages of the future (period of time in which the investment process takes place), it is not crazy to take the symptoms out of the balance sheet and out of the informations directly or indirectly derived from it. As an example, we are going to list in detail the ones which we are going to use later on:

a) Cash and bank
b) Realisable assets
c) Conditional realisable assets
d) Generated cash flow
e) Quotient between the circulating assets and the short-term liabilities
f) Short-term indebtedness ratio
g) Medium and long-term indebtedness ratio

Each of these elements can be valuated in different units. In our example, a, b and c will be expressed in monetary units, whereas e, f and g will be in ratios.

We think it is unnecessary to insist in the fact that the symptoms we have enumerated are not immutable as far as their concept and quantity are concerned. Other symptoms can be considered, those which are considered to be important can be added to these ones, or some of them can be annulled if they are thought to be superfluous.

Moreover, bearing in mind that in order to obtain the risk of the investment, the enterprise or institution must be checked in relation to several "illnesses". Some symptoms may be significant for some of them and not very important for others. It is normal to think that the level each symptom must reach to detect the economic and financial health or illness will be different in one or other "health" manifestation.

Let us move on to study the set of activities which make up the project and let us approach the economic and financial diagnosis. In order to do this, the graph we reproduce afterwards must be used. This graph refers to a standard investment programme in which the "area" of strictly financial activities has been indicated.

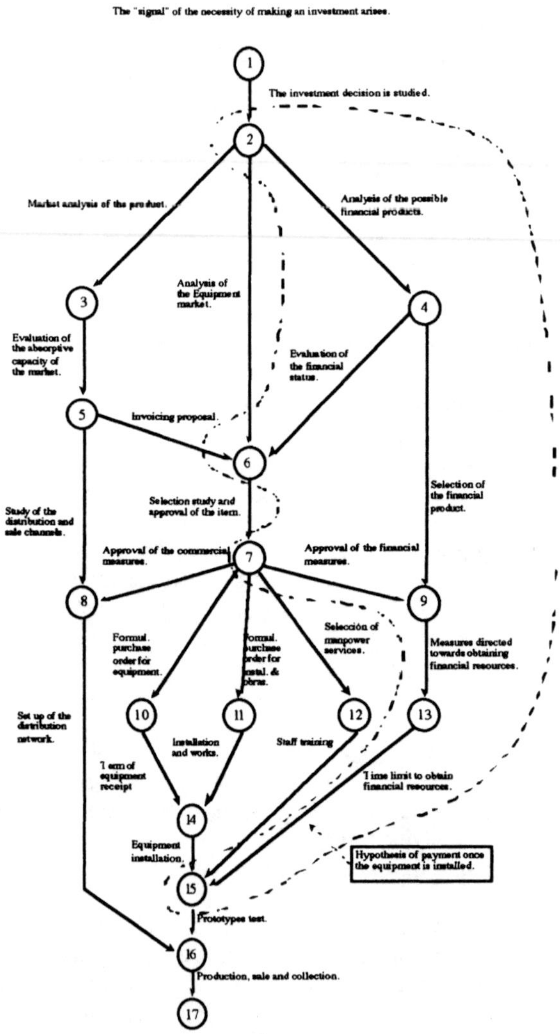

In this graph, we have started from the hypothesis that management directed towards obtaining external resources has the aim of financing in a total or partial way the purchase of the investment item (resources come as a result of the vertex (13, 15), once the equipment is installed). This statement implies that the rest of activities in the project are financed with common equity . However, there is no problem in applying this scheme to the assumption that the need of means of payment force us to agree with the financial institutions on consecutive partial deliveries when the own resources are not enough or it is not wise to use them with this function. As a result, a modification of the graph architecture in the arcs involving the financial tasks of the process would appear.

After this specification, it should be pointed out that the resort to "extraordinary" financing (either internal or external) in order to face the investment process requires some tasks previous to the fund-raising. Those tasks are the following activities: (2, 4) “Analysis of the possible financial products” and (4, 6) “Evaluation of the financial status”. The first one implies a knowledge of the financial and monetary markets. Starting from these markets, it is possible to get to know the possible products they offer that can be used by the enterprise. Each of these products has its own payment terms, price, warrants, etc. Only when this information is available it is possible to carry out the determination of the financial capacity taking a diagnosis as a basis. This task is the fundamental aspect of the activity (4,6) in the graph.

DETERMINATION OF THE FINANCIAL CAPACITY OF THE INVESTMENT

In order to know the financial capacity required to face the payments derived from the activities, we will start from two sets: the set R, which includes those elements representatives of the health required by the enterprise to begin the tasks of the project. In our case:

$$R = \{ A, B, \dots , H \}.$$

S will be the second set, which includes the symptoms that lead to determine the degree of health. Then:

$$S = \{ a, b, \dots, g \}.$$

After that, we are going to determine for each element of the set R, which show the economic and financial health required to carry out the investment project, those levels each symptom must exceed, or must not reach (the position between two of them is also possible) in order to consider that there is no financial risk. Let us see these elements one by one:

A. Immediate

a	b	c	d	e	f	g
≥200	≥350	≥400	≥500	1.3≤	≤0.6	≤0.8

B. Comfortable discount lines

a	b	c	d	e	f	g
≥50	≥250	≥500	≥550	1.5≤	≤0.5	≤0.6

C. Solvency with regard to suppliers

a	b	c	d	e	f	g
≥100	≥300	≥600	≥600	1.5≤	≤0.4	≤0.7

D. Open-ended short credits

a	b	c	d	e	f	g
≥80	≥400	≥600	≥600	1.6≤	≤0.4	≤0.6

E. Long-term accessibility

a	b	c	d	e	f	g
≥40	≥200	≥500	≥700	1.4≤	≤0.5	≤0.3

F. Option to mortgaging credits

a	b	c	d	e	f	g
≥40	≥200	≥500	≥650	1.3≤	≤0.5	≤0.4

G. Resort to guarantees

a	b	c	d	e	f	g
≥150	≥300	≥600	≥550	1.5≤	≤0.5	≤0.7

H. Easy enlargement of the own resources

a	b	c	d	e	f	g
≥50	≥150	≥500	≥800	1.2≤	≤0.6	≤0.5

Let us go now to express this data in a unitary way though a matrix. We will put the "symptoms" in its lines, and the elements representative of the economic and financial "health" of the enterprise required to carry out the investment in its columns. In order to do so, we are going to put the lines represented above as columns one next to the other.

[M] =

	A	B	C	D	E	F	G	H
a	≥200	≥50	≥100	≥80	≥40	≥40	≥150	≥50
b	≥350	≥250	≥300	≥400	≥200	≥200	≥300	≥150
c	≥400	≥500	≥600	≥600	≥500	≥500	≥600	≥500
d	≥500	≥550	≥600	≥600	≥700	≥650	≥550	≥800
e	1.3≤	1.5≤	1.5≤	1.6≤	1.4≤	1.3≤	1.5≤	1.2≤
f	≤0.6	≤0.5	≤0.6	≤0.4	≤0.5	≤0.5	≤0.5	≤0.6
g	≤0.8	≤0.6	≤0.7	≤0.6	≤0.3	≤0.4	≤0.7	≤0.5

It should be taken into account that the values of the balance accounts which stand for the liquidity of the enterprise cannot be used only for the payment of the investment tasks. These values should also be used to face the disbursements derived from the usual activity.

The table represented above as a matrix shows what we could call "standards of financial comfortability for an investment". These standards are valid for a wide range of enterprises located within a geo-economic space in a particular period of time. The matrix will become the basis for the comparisons with the actual situation of the enterprise that wishes to invest. Then, a comparison between what it should be and what it is should be established. The financial analyst will take the data and the valuation of the potentially investing enterprise as a "check". Let us suppose that the check has already been done and that the information for each symptom has resulted in the following vector:

[P] =

	Enterprise Z
a	[160, 180]
b	[320, 340]
c	[530, 580]
d	[660, 720]
e	[1.3, 1.4]
f	[0.2, 0.3]
g	[0.8, 0.9]

In order to make our pattern more general, we have had recourse in establishing the valuations of the enterprises' data by means of confidence intervals. The assumption of accurate numbers will be a particular case.

At this point, the comparison between the health standards and the situation of the enterprise takes place. There are some paths which can be followed. We have chosen amongst them the one which leads to the search for the intersection between health requirements and the foreseen situation of reality for each symptom regarding each illness. This fact implies that we have to consider each column of the matrix [M] and compare it to the vector [P]. The use of the operator $(\cap)$ will lead to the following result:

$[M] \cap [P]$ =

	A	B	C	D	E	F	G	H
a	∅	[160, 180]	[160, 180]	[160, 180]	[160, 180]	[160, 180]	[160, 180]	[160, 180]
b	∅	[320, 340]	[320, 340]	∅	[320, 340]	[320, 340]	[320, 340]	[320, 340]
c	[530, 580]	[530, 580]	∅	∅	[530, 580]	[530, 580]	∅	[530, 580]
d	[660, 720]	[660, 720]	[660, 720]	[660, 720]	[700, 720]	[660, 720]	[660, 720]	∅
e	[1.3, 1.4]	∅	∅	∅	1.4	[1.3, 1.4]	∅	[1.3, 1.4]
f	[.2, .3]	[.2, .3]	[.2, .3]	[.2, .3]	[.2, .3]	[.2, .3]	[.2, .3]	[.2, .3]
g	0.8	∅	∅	∅	∅	∅	∅	∅

Everything we have developed so far must enable the transformation of the matrix $[M] \cap [P]$ in a fuzzy relationship $[\underset{\sim}{B}]$ as a result of the following rules:

1. The elements of the matrix $[\underset{\sim}{B}]$ will be expressed through the hendecadarian system.
2. The valuation segment [0, 1] will be divided into three sub-segments taking the values α and β as a basis and will become thresholds.
   a) The threshold α will set the passage from evident illness to a state of health more and more better.

b) The threshold β will become the passage from partial health to a fully healthy state.

3. Those relationships (x, y) with strict inclusion will be assigned valuations in [β, 1] according to a specific criterion.
4. For those relationships (x, y) with non strict inclusion a valuation in [α, β] will be taking according to a specific criterion as well.
5. Those relationships placed in a limit position will be assigned a valuation which equals α.
6. Valuations in [0, α] will substitute for the values ∅ of the matrix [M] ∩ [P] according to a criterion which should be established.

Let us see from a general point of view more specific aspects and criteria that can be adopted in this transformation algorithm.

Firstly, we are going to establish the semantic hendecadarian scale which will relate the segment [0, 1] to words more widely used to refer to the gradation from the most exulting health to the most galloping illness in the enterprise or institution.

0: galloping illness
.1: seriously ill
.2: very ill
.3: relatively ill
.4: more ill than healthy
.5: neither ill nor healthy
.6: more healthy than ill
.7: relatively healthy
.8: very healthy
.9: extremely healthy
1: exulting health

As we have pointed out, the correspondence we have put forward does not want to become the one and only. The praxis and customs of each country, or even of each person, can recommend another different one. We have assigned different words to the extreme cases of section 2, a) and b).

Secondly, let us divide the segment [0, 1] into three sub-sets [0, α], [α, β] and [β, 1] by means of the next figure:

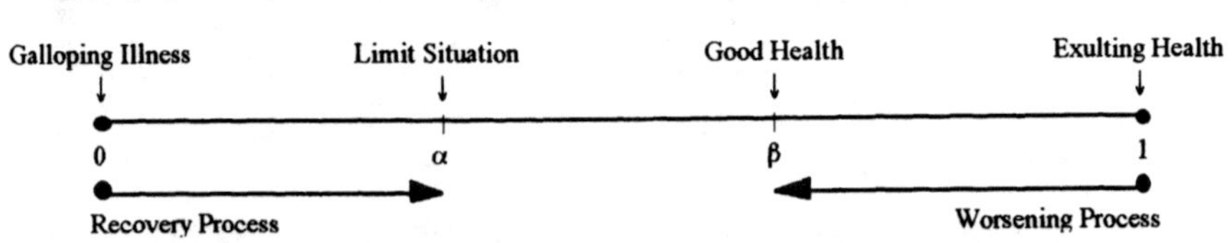

In each particular assumption, the proper values within the segment [0, 1] should be assigned to α and β, according to the sense given to the words in the accepted semantic scale. We are aware of the hypothetical difficulty in knowing (in people and also in enterprises or institutions) when the passage from health to illness takes place. Thus, we are also aware of the difficulty in determining the value of the threshold regarding the limit situation α. However, this determination, full of

unavoidable subjectivity, has in this case a lower impact than if done in the Boolean field.
Thirdly, once the values of α and β are established, we can assign valuations to the segment $[\beta, 1]$. In order to do so, the criterion to be followed should be decided. Before approaching this aspect, it is important to point out that two different cases can arise depending on whether the normality symptoms are expressed by intervals or by a figure (accurate number) which in some cases must be exceeded and, in other cases, not reached. This is the case of the assumptions in which health is conditional on the results of the check falling within the interval $[m_1, m_2]$. It is also the case of the healthy state leading to results higher (or lower) than m, the figure designed in the matrix of accepted normality. We will establish the respective criterion of numerical assignment for each of this two cases:

a) The relationship symptom-health in the matrix is given by a confidence interval. In this assumption, it can be accepted that the most perfect state of health is placed in the middle point of this interval $[m_1, m_2]$. The check is also expressed by another confidence interval $[p_1, p_2]$[2].

As a previous element, we are going to obtain an index expressed in $[0, 1]$ which will become later a number in $[\beta, 1]$. In order to do so, we will lower the entropy by finding the middle point of the interval $\underset{\sim}{m} = [m_1, m_2]$, that is to say $\overline{\underset{\sim}{m}}$ which will give way to the perfect situation. The closer the result of this check gets to this point, the better the state of health. If the actual situation of the enterprise or institution is given by $\underset{\sim}{p} = [p_1, p_2]$, the middle point of this interval $\overline{\underset{\sim}{p}}$ can also be taken as the comparison point , because its representation is better in the precise sphere. Starting from these considerations, we put forward as index the complement to the unit of the double of the difference between $\overline{\underset{\sim}{m}}$ and $\overline{\underset{\sim}{p}}$ with reference according to the width of the $\underset{\sim}{m}$ interval, in absolute values. Let us see it:
given:

$$2 \cdot \left(\frac{m_1 + m_2}{2} - \frac{p_1 + p_2}{2} \right) = (m_1 - p_1) + (m_2 - p_2)$$

once the reference according to the width of the interval is made, it will be :

$$\frac{(m_1 - p_1) + (m_2 - p_2)}{m_2 - m_1}$$

However, bearing in mind that right deviation is as important as left deviation, this value must be taken in absolute terms. The need of giving higher values to the closeness of $\underset{\sim}{m}$ will require the complement of the unit.

2 If it is possible to express the result of the check through an accurate number, and no problem at all will be posed, since this will be a particular case (an accurate number is also a confidence interval).

Finally, we will get:

$$J_{(x,y)} = 1 - \left| \frac{(m_1 - p_1) + (m_2 - p_2)}{m_2 - m_1} \right|$$

Due to its own construction, $J_{(x,y)}$ is a value in [0, 1]. In order to pass to a valuation $K_{(x,y)}$ in [β, 1], it will be enough to use the lineal formula[3]:

$$K_{(X,Y)} = \beta + (1-\beta).J_{(X,Y)}$$

b) The relationship symptom-health of the normality matrix is given by a figure which must be exceeded (interval [m, ∞]) or <u>not</u> exceeded (interval [-∞, m] when the field is $\mathcal{R}$)[4]. Under these circumstances, it is useful to use the notion of distance, which, moreover, gives a remoteness index in relation to a figure. As a result, the higher the distance, the closer to the unit the value we get is and the better the "health" state. The distances obtained, obviously in [0, 1], must be turned afterwards into valuations regarding the segment [β, 1]. In the first place, let us see the assumption of the interval [m, ∞] (m must be exceeded). The absolute distance is:

$$d_{(x,y)}([p_1, p_2], m) = \frac{(p_1 - m) + (p_2 - m)}{2}$$

and with the aim of obtaining a distance value which could be compared to other distances, a relative distance will be obtained, taking as a basis the subtraction between the higher value, between the figures and the higher extreme of $\underset{\sim}{p}$, and the lower value, between the figures and the higher extreme of $\underset{\sim}{p}$. That is to say:

$$\delta_{(x,y)}(\underset{\sim}{p}, m) = \frac{d}{Max\{p_2, m^{(j)}\} - Min\{p_1, m^{(j)}\}} = \frac{(p_1 - m) + (p_2 - m)}{2 . (Max\{p_2, m^{(j)}\} - Min\{p_1, m^{(j)}\})}$$

$$j \in \{A, B, C, ...,\}$$

Let us now go to the assumption of the interval [-∞, m] (the figure m cannot be exceeded). The absolute distance will be:

$$d_{(x,y)}([p_1, p_2], m) = \frac{(m - p_1) + (m - p_2)}{2}$$

---

[3] It is not compulsory, although it is useful in practical terms, the use of the lineal formula. If necessary, any other sigmoid form could also be used.

[4] In the assumption that the validity ambit would be $\mathcal{R}^+$, the interval will be [0, m] and we would be then in the previous assumption.

and the relative distance:

$$\delta_{(x,y)}(\underset{\sim}{p},m) = \frac{d}{Max\{p_2, m^{(j)}\} - Min\{p_1, m^{(j)}\}} = \frac{(m - p_1) + (m - p_2)}{2 \,.\, (Max\{p_2, m^{(j)}\} - Min\{p_1, m^{(j)}\})}$$

$$j \in \{A, B, C, ...,\}$$

It goes without saying that both assumptions can be summarised in a single formula, taking into account the differences $p_1 - m$ and $p_2 - m$ in absolute terms. Finally, we get:

$$\delta_{(x,y)}(\underset{\sim}{p},m) = \frac{|p_1 - m| + |p_2 - m|}{2 \,.\, (Max\{p_2, m^{(j)}\} - Min\{p_1, m^{(j)}\})},$$

Anyway, due to its construction, it will be:

$$0 \leq \delta_{(x,Y)}(\underset{\sim}{p}, m) \leq 1$$

This is the reason why it will be necessary to turn the values into other values placed in the segment [β, 1]. In order to do so, we use again the following lineal formula:

$$L_{(x,Y)} = \beta + (1-\beta) \,.\, \delta_{(x,Y)}$$

Forthly, we are going to establish the criterion in order to obtain the valuations in [α, β].

In this assumption, it seems suitable, to calculate the higher or lower possibility of the symptom implying the existence of an illness. In order to do so, we can find some procedures. We put forward a procedure which combines simplicity and an easy practical use. An index $H_{(x,y)}$ must be created, so that, the closer it gets to the unit, the possibility of the symptom showing a state more and more healthy is considered. The remoteness and the subsequent nearness to zero leads to think in the precariousness and closeness to a situation of illness. In general, this index could be expressed if, as usual:

$[p_1, p_2]$ is the interval resulting from the check
$[m_1, m_2]$ is the interval of accepted normality

and we find the intersection of both intervals:

$$[p_1,p_2] \,(\cap)\, [m_1, m_2] = [n_1, n_2]$$

through this expression:

$$H_{(x,y)} = \frac{n_2 - n_1}{p_2 - p_1}$$

The replacement of m_1 or m_2 by $-\infty$ or $+\infty$, in the case of values not exceeding m_1 or exceeding m_2 poses no problem at all, since the intersection of the intervals $\underset{\sim}{p}$ and $\underset{\sim}{m}$ leaves automatically outside the finite limits of one figure or the other.

It only remains to make the translation of the segment [0, 1] to the segment $[\alpha, \beta]$. We are going to use once more the lineal expression:

$$I_{(x, Y)} = \alpha + (\beta - \alpha) \cdot H_{(x,Y)}$$

Fifthly, we are going to assign the valuations α previously set to those relationships (x, Y) in the situation we have called limit.

It is easy to observe that this phase is a particular case of the previous one because if $m_1 = p_2$ or $m_2 = p_1$, the intersection results in $\varnothing$ and, thus, it can be considered that $H_{(x, y)}$ equals zero and $I_{(x, y)}$ equals α.

Finally, we see how the $\varnothing$ of the matrix $[M] \cap [P]$ can be placed in what we could call illness interval $[0, \alpha]$. In this case a total control of the interval $[m_1, m_2]$ or the accurate figure m over the confidence interval $[p_1, p_2]$ takes place (it is also possible the other way round). The concept of distance can become a good criterion to obtain the values which enable the subsequent valuation. This fact can happen in the assumption of intervals of normality as well as accurate values to exceed or not reach. Let us see the relative distance formulas we have previously found. In the case of intervals:

$$\delta_{(x,y)}(\underset{\sim}{p}, m) = \frac{|p_1 - m| + |p_2 - m|}{2 \cdot \left(\operatorname{Max}\{p_2, m_2^{(j)}\} - \operatorname{Min}\{p_1, m_1^{(j)}\}\right)},$$

Nevertheless, given that the longest distance in relation to $[m_1, m_2]$ implies a worse state of health, the complement to the distance unit, i.e. a "closeness index", should be used in this case. Thus, it should be considered:

$$\bar{\delta} = 1 - \delta_{(x, Y)}$$

In the case of an accurate figure:

$$\delta_{(x,y)}(\underset{\sim}{p}, m) = \frac{|p_1 - m| + |p_2 - m|}{2 \cdot \left(\operatorname{Max}\{p_2, m^{(j)}\} - \operatorname{Min}\{p_1, m^{(j)}\}\right)},$$

Now, for the same reason, it should be taken into account:

$$\bar{\delta} = 1 - \delta_{(x, Y)}$$

in which:

$$\bar{\delta}_{(x, Y)} \in [0, 1]$$
$$j \in \{A, B, C, \ldots, \}$$

In order to move from [0, 1] to [0, α], a lineal transformation will be enough:

$$F_{(x, Y)} = \alpha \, . \, \bar{\delta}_{(x, Y)}$$

We have put forward some criteria which will enable the fuzzy relationship [B]. We insist once more in the fact that the adoption of this criteria it is not absolutely essential. Obviously, there are some other criteria which can turn out to be really useful in some cases. We have used this criteria on several occasions[5]. There is good reason to state that their use is very simple. As an example, we are going to go through the matrix [M] ∩ [P] in our example. This matrix shows four types of results regarding its elements.

If α= 0.3 and β= 0.7 are set as thresholds, the following pre-diagnosis matrix [B] is obtained:

[B] =

	A	B	C	D	E	F	G	H
a	.075	.957	.850	.893	.979	.979	.743	.957
b	.225	.826	.747	.037	.905	.905	.747	.984
c	.958	.791	.107	.107	.791	.791	.107	.791
d	.959	.891	.823	.823	.433	.754	.891	.064
e	.775	.150	.150	.050	.300	.775	.150	.925
f	.962	.887	.812	.812	.887	.887	.887	.962
g	.300	.175	.225	.175	.025	.075	.225	.125

We assume that this fuzzy relationship [B] is the representation of the financial diagnosis of an enterprise or institution. In effect, if we bear in mind that each column represents a fuzzy sub-set of the symptom's referential regarding the subsequent illness, it will be clear to what extent, from 0 to 1, the symptom indicates the healthy situation of the enterprise or institution. Thus, in the assumption we are developing, the pathology A involving liquidity will result in the subsequent fuzzy sub-set (first column of the matrix):

A =

a	b	c	d	e	f	g
.075	.225	.958	.959	.775	.962	.300

It can be observed that the state of liquidity is practically optimal as for the symptoms c, d, f. The state of liquidity is good regarding e and f, and in the limit regarding g. The symptom a and, above all, b indicate serious health problems.

This example shows a fact which happens usually in reality: not all the symptoms lead unequivocally to a clear conclusion about the degree of health in

5 Gil-Aluja, J. "Ensayo sobre un modelo de diagnóstico económico-financiero". Actas de las V Jornadas Hispano-Lusas de Gestión Científica. Vigo. Septiembre 1990, pág. 26-29.

relation to a pathology. It is a general case which involves the assumption of clarity in the pre-diagnosis as a particular assumption.

FROM PRE-DIAGNOSIS TO DIAGNOSIS

The hesitation resulting from these situations forces us to go on with this study. We are going to put forward a scheme which we think will enable to move from pre-diagnosis to diagnosis. In order to do so, we point out that in each illness, in animals or in enterprises or institutions , the set of symptoms capable of detecting it do not have the same importance but there are some symptoms more significant than others. This fact must be formulated in order to include it in the model we are developing. There are some paths which can lead to this goal. The majority of them points out the necessity of assigning a weight to each criterion regarding an illness.

Nevertheless, the problem does not stop here, since the goal we search, to determine the risk of not being able to face the payments derived from the investment process, cannot be forgotten. This is the reason why there will be particular pathologies linked with each activity, depending on whether they imply immediate short, medium or long-term payments. It is common to think that some financial illnesses could have an effect on several payments that can be made within some periods of time (long-term payments) and they affect little the immediate payments. Each expiration date is thus related to the different financial pathologies.

Let us approach in the first place the assignment of weights to the different symptoms in each pathology. We will go on with our example in which there are seven symptoms. We are going to focus our attention in one pathology, As for example, regarding the existence of proper liquidity or not . We are going to establish next the importance of each symptom in comparison with the others, showing the cases in which a symptom has a value higher than others. Let us suppose it is accepted , starting from a, that a equals 3 times b, 5 times c, 8 times d, 2 times e, 2 times f, 10 times g. We go on with b: b equals two times c, 3 times d, 1/2 times e, 1/2 times f, 4 times g. Now, c: c equals 2 times d, 1/3 times e, 1/3 times f, 2 times g. Let us go on with d: D equals 1/4 times e, 1/4 times f and equals g. Finally, f equals 6 times g.

We have assigned some valuations to these correspondences which are not strictly coherent, since reality shows the difficulty of this assumption to be fulfilled. Even in this case, totally coherent relationships will not only confirm the scheme but become a particular case. Starting from the previous valuations, we are going to set a reciprocal matrix, as the following one:

[T] =

	a	b	c	d	e	f	g
a	1	3	5	8	2	2	10
b	1/3	1	2	3	1/2	1/2	4
c	1/5	1/2	1	2	1/3	1/3	2
d	1/8	1/3	1/2	1	1/4	1/4	1
e	1/2	2	3	4	1	1	5
f	1/2	2	3	4	1	1	6
g	1/10	1/4	2	1	1/5	1/6	1

This is a reciprocal matrix due to its construction. However, it is not totally coherent[6]. We are going to calculate now the dominant eigenvalue and subsequent eigenvector. In order to do so, we make the multiplication [T]. [1] to get:

$$\begin{pmatrix} 1 & 3 & 5 & 8 & 2 & 2 & 10 \\ 0.333 & 1 & 2 & 3 & 0.500 & 0.500 & 4 \\ 0.200 & 0.500 & 1 & 2 & 0.333 & 0.333 & 2 \\ 0.125 & 0.333 & 0.500 & 1 & 0.250 & 0.250 & 1 \\ 0.500 & 2 & 3 & 4 & & & \\ 0.500 & 2 & 3 & 4 & 1 & 1 & 6 \\ 0.100 & 0.250 & 2 & 1 & 0.200 & .0166 & 1 \end{pmatrix} \bullet \begin{pmatrix} 1 \\ 1 \\ 1 \\ 1 \\ 1 \\ 1 \\ 1 \end{pmatrix} = \begin{pmatrix} 31.000 \\ 11.330 \\ 6.360 \\ 3.455 \\ 16.500 \\ 17.500 \\ 4.716 \end{pmatrix}$$

We are going to normalise this result in the sense given to this term in the fuzzy sub-set theory (to divide each element of the vector by the highest value of themselves). We get:

$$\begin{pmatrix} 31.000 \\ 11.330 \\ 6.360 \\ 3.455 \\ 16.500 \\ 17.500 \\ 4.716 \end{pmatrix} = 31 \bullet \begin{pmatrix} 1 \\ 0.365 \\ 0.205 \\ 0.111 \\ 0.532 \\ 0.564 \\ 0.152 \end{pmatrix} = 31 \bullet [U_1]$$

The next step is the product of [T] and the normalised vector $[U_1]$. That is to say [T] . $[U_1]$. Then, we go on with the products [T] . $[U_2]$ and [T]. $[U_3]$. We get respectively:

$$[T] \bullet [U_1] = 7.72 \bullet \begin{pmatrix} 1 \\ 0.336 \\ 0.191 \\ 0.115 \\ 0.537 \\ 0.537 \\ 0.138 \end{pmatrix} = 7.72 \bullet [U_2]$$

6 A theoretical justification can be found in Saaty, T.L. : "Exploring the interface between hierarchies, multiple objectives and fuzzy sets". Fuzzy Sets and Systems. Vol 1. n°1, 1978, pág. 57-68.

$$[T] \bullet [U_2] = 7.451 \bullet \begin{pmatrix} 1 \\ 0.334 \\ 0.191 \\ 0.115 \\ 0.535 \\ 0.554 \\ 0.137 \end{pmatrix} = 7.451 \bullet [U_3]$$

$$[T] \bullet [U_3] = 7.425 \bullet \begin{pmatrix} 1 \\ 0.334 \\ 0.191 \\ 0.115 \\ 0.535 \\ 0.554 \\ 0.137 \end{pmatrix} = 7.425 \bullet [U_4]$$

At this point, we stop the process, since $[U_3]= [U_4]$. Having in mind that the matrix order is n=7, the relationship is $\frac{\lambda - n}{n} = \frac{7.425 - 7}{7} = 0.06$. We can assume that 6% is lower enough to accept that the dominant eigenvalue is $\lambda= 7.425$ and the subsequent eigenvector, expressed as a normal fuzzy sub-set:

$\underset{\sim}{K}_A$ =

a	b	c	d	e	f	g
1	.334	.191	.115	.535	.554	.137

If each of these valuations is divided by the addition of all of them (i.e. 2.866) the corresponding weights of a convex ponderation (probabilistic normalisation) will be obtained:

$\underset{\sim}{Q}_A$ =

a	b	c	d	e	f	g
.349	.116	.067	.040	.187	.193	.048

This weights, presented as a fuzzy sub-set of the symptom's referential, show the relative importance of each of the symptoms. In our opinion, the joint observation of $\underset{\sim}{A}$ and $\underset{\sim}{Q}_A$ becomes a good support for the diagnosis. However, if one wants to join together in one figure the representation of the degree of health of an enterprise or institution in relation to a particular pathology, A in this case, it will be enough to calculate the composition addition ordinary-product of the vectors $[\underset{\sim}{A}] \,.\, [\underset{\sim}{Q}_A]$. We get:

$$[\underset{\sim}{A}] \bullet \underset{\sim}{K}_A = \begin{array}{ccccccc} a & b & c & d & e & f & g \\ \hline \multicolumn{1}{|c|}{.075} & \multicolumn{1}{c|}{.225} & \multicolumn{1}{c|}{.958} & \multicolumn{1}{c|}{.959} & \multicolumn{1}{c|}{.775} & \multicolumn{1}{c|}{.962} & \multicolumn{1}{c|}{.300} \\ \hline \end{array} \bullet \begin{array}{c|c|} \cline{2-2} a & .349 \\ \cline{2-2} b & .116 \\ \cline{2-2} c & .067 \\ \cline{2-2} d & .040 \\ \cline{2-2} e & .187 \\ \cline{2-2} f & .193 \\ \cline{2-2} g & .048 \\ \cline{2-2} \end{array} = 0.5$$

The result we have obtained indicates that the analysed enterprise or institution is in a delicate position regarding "proper liquidity", although in the matrix $[M] \cap [P]$ only two illness symptoms, $\underset{\sim}{a}$ and $\underset{\sim}{b}$, have been detected. This fact is due to the great importance of the symptom $\underset{\sim}{a}$, "cash and banks", and to a lower extent b, "realisable assets".

The figure which summarises the state of health in relation to the pathology A, 0.5 in this case, can be expressed in words using the semantic scale, by saying: "The liquidity of the enterprise or institution is neither ill nor healthy" (greater hesitation).

We think it is important to underline, as it can be clearly seen in the example, the essential difference between the results obtained in $[M] \cap [P]$, $[\underset{\sim}{B}]$ and $[\underset{\sim}{A}].[\underset{\sim}{Q}_A]$, the three stages of our goal: the economic and financial diagnosis of enterprises or institutions.

It goes without saying that, starting from the same elements, $[\underset{\sim}{B}] \,.\, [\underset{\sim}{Q}_B]$, $[\underset{\sim}{C}] \,.\, [\underset{\sim}{Q}_C]$, ..., $[H] \,.\, [\underset{\sim}{Q}_H]$ will be successively obtained. Once this data is obtained, the economic and financial analysis would be finished and we would be in position to approach a last aspect, the essential part of this study . In this case, the calculation of the risk assumed by the enterprise when approaching the investment process.

NUMERIC DETERMINATION OF THE FINANCIAL RISK OF THE INVESTMENT

Before approaching this aspect of the problem, let us say that the diagnosis, as it has been put forward, has an interest which goes further than the ambit in which it has been placed. We think its usefulness embraces a wide range of situations in which a deep knowledge of the enterprise is essential for taking financial decisions. After this brief reflection, let us determine the risk according to the previous definition. Two aspects must be taken into account: the moment of time when the payment must be made and its amount. The first aspect is linked with the set of pathologies. The second one is related to the degree of health of the pathologies.

Throughout this work, it has been possible to observe that there are particular pathologies which affect the financial needs in the short run and others which have an effect in the medium and long run. Then, depending on the investing activity studied, some pathologies or others must be considered. Thus, it is accepted the existence of the relationship "time in which an activity finishes-pathology to be considered".

In the sake of simplicity, we are going to develop this idea through an example. Let us suppose we want to analyse the activity (6,7) in the graph presented in the first section. The graph represents the "study of the choice and approval of the investment item". Its beginning is estimated within the time interval [22,30] and its end, within the interval [37,48]. Its cost and subsequent payment has been valuated in [180, 270], for instance.

The fist question to be raised is: taking into account that the estimated payment must be made in the [37,48] time units following the beginning of the project, which are the pathologies affecting this payment? If the time unit is the concept of day, it is clear that it is a disbursement in the short run. A, B, C and D are the pathologies more suitable to know the possibilities of facing the payment. With the aim of a greater simplicity, we can practically leave out the rest of them.

As far as the second question is concerned, it is necessary to consider the level of payment of the activity in order to determine the degree of risk assumed when beginning the activity. It must be taken into account that the payment can be indistinctly made through one of the four health indicators (liquidity, discount, supplier credit, short-term credit). As a result, the possibility of making the payment will be under the influence of the health levels in relation to these concepts. If the influence of the health level over the possibility of payment is accepted, the problem will find its solution by using logic inferences.

In effect, the relationship between the health level regarding a pathology and the possibility of facing the payment of an activity can be determined as a result of a logic reasoning, bearing in mind the propositions P and S and the inference P→S. Thus, for the pathology A:

P_A: Proper liquidity
S_A: Possibility of paying the activity (6,7)
($P_A \rightarrow S_A$): If there is proper liquidity, then there is the possibility of paying the activity (6,7) through this channel.

As an experiment, we are going to use Lukaciewicz inference:

$$\upsilon(P \rightarrow S) = 1 \wedge (\upsilon(\bar{P}) + \upsilon(S))$$

being respectively $\upsilon(P \rightarrow S)$, $\upsilon(P)$, $\upsilon(S)$ the valuations of the inference and that of the propositions P and S.

We see how we have obtained a valuation $\upsilon(A) = 0.5$ for the pathology A. Let us suppose that the corresponding calculations have been made and the results are $\upsilon(B) = 0.8$, $\upsilon(C) = 0.7$ and $\upsilon(D) = 0.8$. That is to say, the valuations corresponding to $[\underset{\sim}{B}] \cdot [Q_B]$, $[\underset{\sim}{C}] \cdot [Q_C]$, $[\underset{\sim}{D}] \cdot [Q_D]$.

In order to get the valuations corresponding to the inferences (P→S) we turn to the opinion of the experts, who express it by the following values[7]:

$$\upsilon(P_A \rightarrow S_A) = 1$$
$$\upsilon(P_B \rightarrow S_B) = 0.9$$
$$\upsilon(P_C \rightarrow S_C) = 0.7$$
$$\upsilon(P_D \rightarrow S_D) = 0.8$$

---

[7] It is obviously a pedagogical example and, then, this data is arbitrary.

This four expressions imply the following reasoning: "If there is proper liquidity, then the possibility of payment is total", "if there are comfortable discount lines, then the possibility of payment is 0.9",...

With this data, we are in a position to use the inference chosen. Let us see it. To get $\upsilon(S_A)$:

$$1 = 1 \wedge (0.5 + \upsilon(S_A))$$

in which $\upsilon(S_A)$ is the valuation of the proposition S resulting from the proposition A. The result is:

$$\upsilon(S_A) = [0.5, 1]$$

For $\upsilon(S_B)$:

$$0.9 = 1 \wedge (0.2 + \upsilon(S_B))$$
$$\upsilon(S_B) = 0.7$$

For $\upsilon(S_C)$:

$$0.7 = 1 \wedge (0.3 + \upsilon(S_C))$$
$$\upsilon(S_C) = 0.4$$

For $\upsilon(S_D)$:

$$0.8 = 1 \wedge (0.2 + \upsilon(S_D))$$
$$\upsilon(S_D) = 0.6$$

Nevertheless, as we have previously pointed out, the payment of the activity (6,7) can be made through A and/or B and/or C and/or D. Then, the operator to use will be the operator of maximisation of the valuations. Thus:

$$\begin{aligned}\upsilon(S) &= \upsilon(S_A)\ (\vee)\ \upsilon(S_B)\ (\vee)\ \upsilon(S_C)\ (\vee)\ \upsilon(S_D)\\ &= [0.5, 1]\ (\vee)\ [0.7, 0.7]\ (\vee)\ [0.4, 0.4]\ (\vee)\ [0.6, 0.6]\\ &= [0.7, 1]\end{aligned}$$

If the risk is valuated as the complement to the unit of the possibility of payment, the financial risk of the activity (6,7), which we are going to call $R_F^{(6,7)}$, will definitively be:

$$R_F^{(6,7)} = 1 - [0.7, 1] = [0, 0.3]$$

In this particular case, it is an uncertain result but limited by means of a confidence interval. This fact is due to the use of Lukaciewicz inference.

The practically unlimited possibilities of the multivalent logics give this scheme great flexibility and adaptability to the requirements of each particular case. It will be possible to choose the inference which adapts better to the conditions that reality creates.

FINAL CONSIDERATIONS

As a summary and brief conclusion, we can say that the development of the subject has been made through three stages. The aim of the first stage has been to get a "pre-diagnosis matrix", which already enables to have an idea about the situation of the enterprise or institution analysed. However, this information is not always enough to make a reliable report, since the degree of importance of each symptom over the different pathologies is not taken into account.

In order to achieve this second aim, we have moved to a second stage. We have showed the way of finding "weights" which, at the same time, enable to have a global result summarised in a figure (in some cases, a certain figure is possible, in other cases it will be an uncertain number) which represents the degree of health in a particular pathology. We have underlined that the operation of composition addition ordinary-product is a way of lowering the entropy.

At this point we have introduced an important hypothesis, as far as the model is concerned. This hypothesis is to accept the existence of an inference relationship between the concept of degree of health in a pathology and the concept of "possibility of facing the payment" through the element this pathology represents. Once this hypothesis is accepted, we are in a position to move to the third and last stage.

In this stage, we have found a valuation of the degree of possibility of facing the payment derived from an activity by an element under an eventual pathology with a particular level of health. In order to do so, we have decided to represent the reasoning which links two propositions by means of an inference resulting from the multivalent logics. This aspect has special characteristics, since there are some inference operators in the field of uncertainty which do not lead to the same result and this fact is perfectly normal. In each particular case, one operator or the other one must be choosen according to the circumstances of the problem we are dealing with. Once the possibility for each element under the pathology is obtained and, taking into account that the payment can be made through one channel or any other, the global possibility is obtained by using the maximisation operator. The next step is automatically made by considering the degree of risk as the complement to the unit of the possibility of payment.

One of the aspects which deserves special consideration and which makes this model an instrument of conceptual and technical interest is the fact that the financial illness seldom arises clearly, as a result of the symptoms pointing out the illness in a clear way. In short, absolute truths and falseness only appear in those minds which, arisen in irreality, behave as automatons. The economic and financial context of society in the end of this century makes human beings, as well as organised groups, capable of reasoning starting from quasi-infinite shades. This is possible thanks to the amazing tool of imagination.

REFERENCES

1. Dinh Xuân Bá
A method for estimating the membership function of a fuzzy set. Revue Busefal nº 19 L.S.I Univ. Paul Sabatier. Toulouse, 1984.

2. Gil-Aluja, J.
Ensayo sobre un modelo de diagnóstico económico-financiero. Actas de las V Jornadas Hispano-Lusas de Gestión Científica. Vigo, Septiembre 1990.

3. GIl-Aluja, J.
Programming investment activities. Proceedings of the International Conference on Intelligent Technologies in Human-Related Sciences. León, 5-7 july 1996. p. XVII-XXXIII.

4. Kaufmann, A. and Gil-Aluja, J. *Técnicas operativas de gestión para el tratamiento de la incertidumbre*. Barcelona: Ed. Hispano-Europea, 1987.

5. Kaufmann, A. and GIl-Aluja, J. *Técnicas de gestión de empresa. Previsiones, decisiones y estrategias*. Madrid: Ed. Pirámide, 1992.

6. Saaty, T.L. "Exploring the interface between hierarchies, multiple objevtives and fuzzy sets." In *Fuzzy Sets and Systems*, Vol. I, nº1, 1978.

THE SELECTION OF A PORTFOLIO THROUGH A FUZZY GENETIC ALGORITHM: THE POFUGENA MODEL

Enrique López-González, Cristina Mendaña-Cuervo, Miguel A. Rodríguez.-Fernández
University of León
Economy and Business Management Department
Campus de Vegazana, s/n, E-24071 León
Spain

Abstract: The selection of a portfolio encounters several extremely complex situations. From among them, it has to be highlighted, due to its difficulty and transcendence, the Financial Assets selection when interrelations (positive and/or negative) occur among the expected profitabilities of each one of them. The tools traditionally used have tried to approach it by simplifying reality and, therefore, the obtained results are not fully satisfactory. This situation has encouraged the authors to questioning whether better solutions can be reached by applying the so called *Intelligent Technologies*. Thus, one of the available tools is the one constituted by *Genetic Algorithms*, due to its utility when offering solutions to complex optimization problems. Furthermore, by using the *Fuzzy Sets Theory,* we intend to obtain a closer representation for the uncertainty that characterises Financial Market. This way, it is intended to outline an approach to solve Financial Assets selection problems for a portfolio in a non-linear and uncertainty environment, by applying a *Fuzzy Genetic Algorithm* to optimize the investment profitability.

Keywords: financial assets, decision making, portfolio analysis, fuzzy numbers, relation, fuzzy genetic algorithms, intelligent technologies applications.

1. INTRODUCTION

In this essay we introduce a new tool to improve the selection of a portfolio. First of all, we present the traditional approach to portfolio management. Second, we explain a new approach to the problem that considers the use of fuzzy logic. After that, we introduce the application of Genetic Algorithms to optimize the expected return from the portfolio with fuzzy information. At the end of the paper we include an example of the problems that could be solved with. Finally, in the last section of the paper we suggest some conclusions and future developments.

2. TRADITIONAL APPROACH TO PORTFOLIO MANAGEMENT

The concept of Portfolio Analysis spreads from the financial area. A financial investment does not usually have a constant profitability, except for a few exceptions, but changes according to certain variables. This variability determines the investment risk, so the bigger risk investment have the higher profitability opportunities, in most of cases.

Traditionally, the measures used to valuate a portfolio profitability and risk are the arithmetic mean and the standard deviation of the different financial assets. The former shows the average annual profitability, being for the portfolio the weighted average of each asset profitability times the asset investment risk. The latter indicates how the financial investment has varied regarding the average of the past data analysis. Thus, when a portfolio has to be managed, the decision maker will have to choose those investments maximizing their profitability with the minimum possible risk.

In this respect, empirical studies [MARKOWITZ, 1952; LEVY, 1970] have demonstrated that, when several investments are brought together into the same portfolio, the assumed risk does not correspond, except for a few given cases, with the weighted average of each one of the investments risk. The idea of correlation between the different financial assets profitabilities emerges here and, therefore, it raises the portfolio risk reduction through diversification.

According to this approach, the mathematical formulas to determine a portfolio expected profitability and risk are:

$$\textit{Profitability} = \sum_{i=1}^{n} w_i \cdot r_i \quad \text{where: } \sum_{i=1}^{n} w_i = 1$$

$w_i =$ **i** investment percentage

$r_i =$ **i** average of return on investment

$$\textit{Variance} = \sum_{i=1}^{n}\sum_{j=1}^{n} w_i \cdot w_j \cdot \sigma_{ij} \quad \text{where: } \sigma_{ij} = \textbf{i} \text{ variance when } \textbf{i=j}$$

and σ_{ij} = correlation coefficient **ij** when **i≠j**

The risk that can be eliminated through diversification is called specific risk. Its reason d'être is based in the fact that many of the threats surrounding a given firm stem out from the own firm and, perhaps, from its immediate competitors.

There is also a risk, called market risk, that cannot be avoided, no matter how much it is diversified. This market risk arises from the fact that there are other perils in the Economy as a whole threatening all the business. This is the reason why investors are exposed to the uncertainties of the market, no matter how many shares they hold.

Once exposed the reasons to have a diversified investment portfolio in which maximum synergies among assets are obtained, the arising question is how to achieve that portfolio. The first contribution in this field was made by Markowitz (1952), who proposed a investment portfolio evaluation method based on risk and profitability analysis. According to him, every investor facing two portfolios with the same risk will choose the one with the bigger profitability, and facing two portfolios with the same profitability will choose the one minimising the risk. It is intended to obtain the optimum decision in the selection of the financial assets.

The different methods consist basically, of three stages. The first one deals with determining all the available financial assets and, subsequently, generating every possible portfolio from these elements. In the second stage the efficient portfolios, that is, those not dominated by others, are selected using some rules, such as the Mean-Variance [MARKOWITZ, 1952], or the Stochastic Dominance [MAHAJAN and WIND, 1984]. Then, in the third one, the optimum portfolio is chosen among them. This final decision is proposed in this method as purely intuitive, and therefore it will depend on the greater or smaller risk aversion of the investor.

3. A NEW APPROACH TO PORTFOLIO MANAGEMENT

This paper endeavours to find a representation of the available information in the financial market , as reliable as possible, since using both mean and standard deviation as indicators for investment profitability and risk, may bring out inefficient decisions. With this, it's not intended to suggest their non-validity but, in addition to the information supplied by those indicators, the investor can complete his knowledge of the financial market with other sources. In particular, it all deals with including estimates from market behaviour experts, economic variables forecasts that can determine the financial assets behaviour through the sensibility they show towards them, government policies, business strategies, etc., or even it may be considered subjective aspects such as the broker's accuracy when focusing a certain portfolio on determined financial assets. With this it is aimed to include both objective and subjective criteria, provided by the financial market itself.

On the other hand, as an additional element to this set of issues, it has to be highlighted a relation or interconnection among the available financial assets. This fact is due to synergies existing among them, that cause changes in profitability because of certain situations, such as modifications in share prices of the enterprises depending on the same economic variables, joint ventures, shares of enterprises whose profitability is interfered with a certain currency price, fixed interest stocks depending on the interest rate set by each country's Central Bank, etc.

In order to fit all the available information together it is proposed to use the Fuzzy Sets Theory [ZADEH, 1965; KAUFMANN and GIL-ALUJA, 1986], since as a mathematics branch dealing with objective and subjective matters, it attempts to take a phenomenon as presented in reality, and handle it to make it certain or accurate.

The reason to use logic and fuzzy technology is based in author's perception that, since portfolio selectors realise that their environment and, therefore, the information they handle, is uncertain and diffuse, it seems obvious they prefer realistic representations rather than just models assumed to be exact.

In this paper, we have decided to represent uncertain values of the different variables taken into account by the decision help system supported by Trapezoidal Fuzzy Numbers (TFN). The membership functions that define then, $\mu(x)$, are linear, as shown in Figure 1.

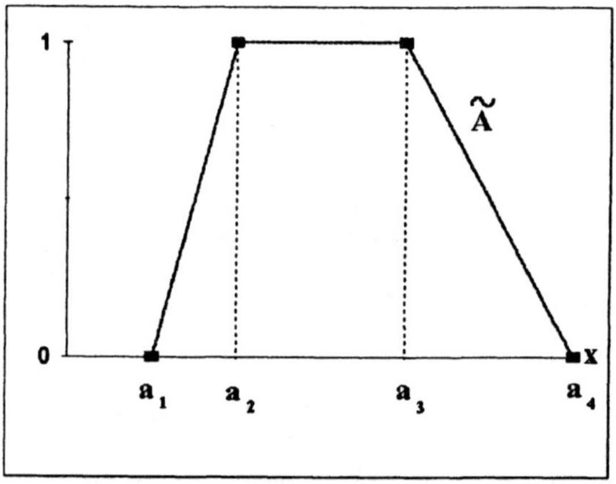

Figure 1

A TFN has four points that represent or define it; in Figure 1 they are: a_1, a_2, a_3, a_4. So, a TFN can be represented in a quaternary way:

$\tilde{A} = (a_1, a_2, a_3, a_4)$ with $a_1, a_2, a_3, a_4 \in \Re$ and $a_1 \leq a_2 \leq a_3 \leq a_4$

These four numbers involve that:

$\forall x \leq a_1 \qquad \mu(x) = 0$

$\forall x \geq a_4 \qquad \mu(x) = 0$

$\forall\ a_2 \leq x \leq a_3 \qquad \mu(x) = 1$

and that function $\mu(x)$ for the remaining values is the line that joins point (a_1, 0) with point (a_2, 1), and the line joining point (a_3, 1) with (a_4, 0).

Consequently, a TFN membership function, can be noted as follow:

$$\mu(x) = \begin{cases} 0 & x < a_1 \\ \dfrac{x - a_i}{a_2 - a_1} & a_1 < x < a_2 \\ 1 & a_2 < x \leq a_3 \\ \dfrac{a_4 - x}{a_4 - a_3} & a_3 < x < a_4 \\ 0 & x > a_4 \end{cases}$$

The very conceptualisation of TFN allows a good suitability to different real situations, particularly to economic and entrepreneurial estimates. Thus, for instance, the expected return for a financial asset in a given period of time can be established as:

$$\widetilde{R} = (4\%;\ 5{,}5\%;\ 6\%;\ 8\%)$$

which means that, the profitability of this asset will be at least four percent, that it is likely to be between five point five and six percent and that it will not be higher than eight percent. This way, not only the expected profitability can be represented, but also the risk run when investing in a financial asset, for in that TFN it is represented the whole set of possibilities where its profitability is bound to be found.

Apart from a TFNs great adaptability to the human mind structure, it is also important to consider how easy to use it is, due to the simplicity of its membership function, which is defined by linear functions.

Once the representation has been decided, a study of the qualities of the financial market must be done. That analysis aims to gather the existence of variables affecting the profitability of the portfolio as a whole.

Therefore, each available financial asset will bear investment's materialization financial expenses, which can be distinguished between those with a flat amount and those whose value varies according to the invested quantity as it happens, for instance, in commissions. It has to be taken into account that the higher investing capital, the lower influence of those expenses on the decision. Their representation can be carried out through matrixes containing the different values they take.

For **n** assets can be considered the following fixed expenses:

$$FEi = \{FE_1, FE_2, FE_3, \ldots\ldots\ldots FEn\}$$

and the following variable expenses:

$$VEi = \{VE_1, VE_2, VE_3, \ldots\ldots\ldots VEn\}$$

Through a specialist's opinion or through the knowledge the current asset manager has about the capital market, estimates can be accomplished about the expected profitability for each asset. These values will be represented by TFNs, that allow to fulfil properly the economic variables estimates. The fuzzy profitability matrix for each of the **n** financial assets would be:

$$\widetilde{R}_i = \{\widetilde{R}_1, \widetilde{R}_2, \ldots\ldots\ldots \widetilde{R}_n\}$$

In addition, different profitabilities can be considered regarding the invested amount and, consequently, there will be as many such like matrixes as different investment levels shown by the assets.

Deepening into the analysis, it is possible to consider the modifications taking place in the portfolio's expected profitability when financial assets related to it are incorporated. This is aimed at contemplating interconnections among enterprises shares, shares with currencies, etc., found out at analysing thoroughly the financial market. Those modifications can be represented as a matrix containing variations of the four values that determine each assets expected profitability.

$$\begin{aligned} RV_{ij} = \{&—, RV_{12}, RV_{13}, \ldots\ldots\ldots RV_{1m} \\ &RV_{21}, —, RV_{23}, \ldots\ldots\ldots RV_{2m} \\ &\ldots\ldots\ldots\ldots\ldots\ldots\ldots\ldots\ldots\ldots \\ &RV_{m1}, RV_{m2}, RV_{m3} \ldots\ldots —\}, \end{aligned}$$

which indicates how the expected profitability of asset **i** varies when we invest also in asset **j**.

At the same time, it might work out to consider those situations in which financial assets have some kind of relationship and then, savings can result in the initial expenses incurred to start the investment. This is why it has to be taken into account the fixed expenses and initial commissions when, in the portfolio, assets from the same enterprise, group, financial entity , etc., have been included. This information can be represented through two matrixes, one indicating the fixed assets variation:

$$\begin{aligned} FEV_{ij} = \{&—, FEV_{12}, FEV_{13}, \ldots\ldots\ldots FEV_{1m} \\ &FEV_{21}, —, FEV_{23}, \ldots\ldots\ldots FEV_{2m} \\ &\ldots\ldots\ldots\ldots\ldots\ldots\ldots\ldots\ldots\ldots \\ &FEV_{m1}, FEV_{m2}, FEV_{m3} \ldots\ldots —\}, \end{aligned}$$

and the other one containing commissions or variable expenses variations:

$$\begin{aligned} VEV_{ij} = \{&—, VEV_{12}, VEV_{13}, \ldots\ldots\ldots VEV_{1m} \\ &VEV_{21}, —, VEV_{23}, \ldots\ldots\ldots VEV_{2m} \\ &\ldots\ldots\ldots\ldots\ldots\ldots\ldots\ldots\ldots\ldots \\ &VEV_{m1}, VEV_{m2}, VEV_{m3} \ldots\ldots —\}. \end{aligned}$$

With this information it is possible to establish the optimum portfolio selection problem, where it is intended to find out the financial assets combination maximizing the total expected profitability of the portfolio as a whole.

Traditional methods do not analyse such a complex problem, but set it up in a *linear* way, escaping from reality.

On the other hand, in this paper it is established the alternative, trying to find any method that brings up an acceptable solution to the so settled problem. So, due to the success obtained by applying Genetic Algorithm in the search of good solutions for optimization problems [GOLDBERG, 1989; DAVIS, 1991; LÓPEZ-GONZÁLEZ et al, 1995a, 1995b, 1995c, 1996], the POFUGENA model is developed using that technology and trying to help in the decision making of assets selection to be hold in a portfolio when the environment is uncertain and non-linear.

4. APPLICATION OF FUZZY GENETIC ALGORITHMS TO THE SELECTION OF FINANCIAL PORTFOLIOS

4.1. Genetic Algorithm as Method of Approaching Optima

The Genetic Algorithms constitute optimization tools based on natural selection and on the genetics mechanisms. In natural selection, the evolution processes happen when the following conditions are satisfied:

- An entity or individual is able to reproduce.
- There is a population of such entities or individuals able to reproduce.
- Some differences in the capacity to survive in the environment are associated to that diversity.

Such diversity is shown in changes in the *chromosomes* of the individuals of a population, and transfers into the variation of the structure and behaviour of the individuals in their environment, which is reflected in the degree of survival, adaptation and in the level of reproduction. The individuals that adapt better to their environment are those who survive longer and reproduce more.

In a period of time and after many generations, the population gets more individuals, whose *chromosomes* define structures and behaviours adapted to their environment, surviving and reproducing in a higher level , so that in the course of time, the structure of individuals in the population changes due to the natural selection.

According to the explanations above and though there are many possible variations of Genetic Algorithms, their fundamental mechanisms are: to operate over a population of individuals, usually generated in a random way, and changing the individuals in every iteration, according to the following steps:

a) Evaluation of the individuals of a population.
b) Selection of a new set of individuals.
c) Reproduction on the base of their relative adaptation or fitness.
d) Re-combination to create a new population from the *crossover* and mutation operators.

The set of individuals resulting of these operations conform the next population, iterating this process until the model cannot produce any fitness improved situation.

Generally, each individual is represented by a binary or decimal string of fixed length, *chromosome*, that codifies the values of the variables that take part in the problem, so that the representation of the data and the operations can be manipulated to generate new strings fitting better to the problem to solve.

Acting this way over a population of individuals, an essential component of every Genetic Algorithm is introduced, the *Fitness Function*, that constitutes the link between such Algorithm and the problem to solve. A *Fitness Function* takes a *chromosome* as input and returns a number that shows the appropriateness of the solution represented by the *chromosome* to the analyzed problem.

The *Fitness Function* plays the same role as the environment in the Natural Selection, due to the fact that the interaction of an individual with its environment gives a measure of its fitting and it determines that the best adapted individual has a higher probability to survive.

Right after the selection, a process of *crossover* is performed, trying to imitate the reproduction of individuals according to the laws of Nature, exchanging the genetic information of the *parents* (selected individuals), in order to obtain the *chromosomes* of the offspring, possibly producing better or more adapted individuals.

Besides the exchange of *chromosomes*, Nature often produce sporadic changes in the genetic information, denominated by biologist *mutations*. That is the reason why, in the execution of the Algorithm, this process is introduced and performs small random modifications in the *chromosomes* of the individuals resulting from the crossover.

When the above described operative is performed correctly within this evolutive process, an initial population will be improved by its successors and therefore, the best fitting individual of the last population can be a very appropriate solution for the problem.

On the other hand, in this paper, and considering the inaccuracy of the information used by Genetic Algorithms, we will use fuzzy numbers to represent that information and so, the different operators of the designed Algorithm have to be adapted to that point which involves a Fuzzy Genetic model [HERRERA, 1996].

4.2. POFUGENA Model (Portfolio-Fuzzy-Genetic-Algorithm)

4.2.1. Problem data

Some data concerning the financial assets to select are necessary to establish the characteristics of the problem.

First, the number of financial assets in which the capital can be invested must be entered. The total capital to invest must be established as well. Afterwards, and according to the financial assets investment conditions, the minimum investment can be defined. This datum will fix the smallest investment level for each asset. All this information can be entered in the first screen of the software designed, as Figure 2 shows.

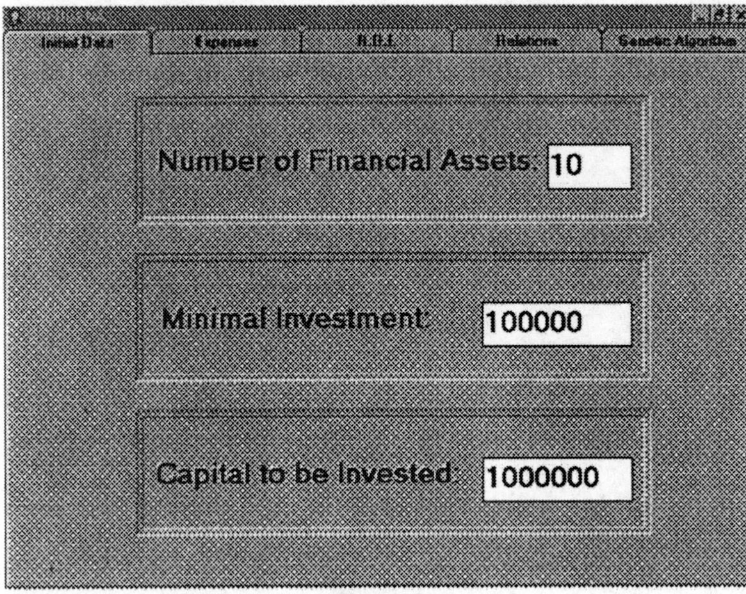

Figure 2

Data concerning the initial expenses are also necessary. Two kinds of expenses can be distinguished: those that are independent from the invested amount, the so called fixed expenses; and those that vary with such quantity, variable expenses. In our model the parameters concerning both types of expenses are entered in the second screen, as shown in Figure 3.

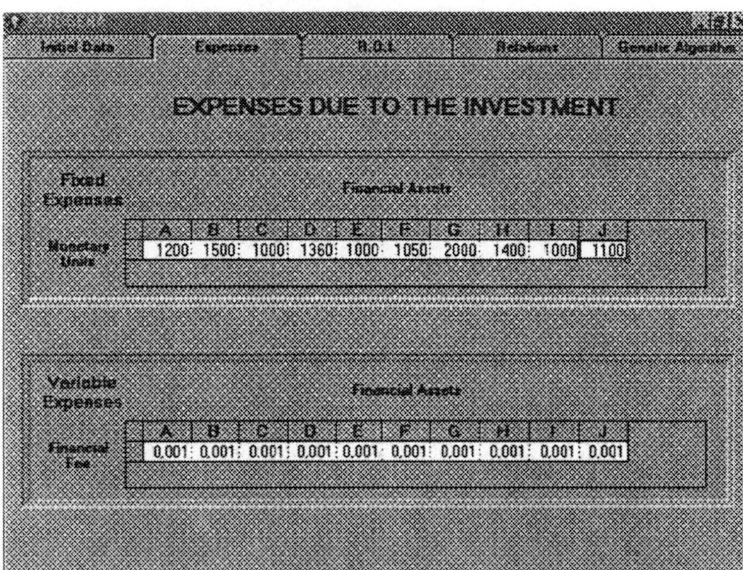

Figure 3

On the other hand, the expected returns on investment of every financial asset are necessary to obtain an optimum portfolio through the Genetic Algorithm. POFUGENA model uses this data as TFNs. An example of it is shown in Figure 4.

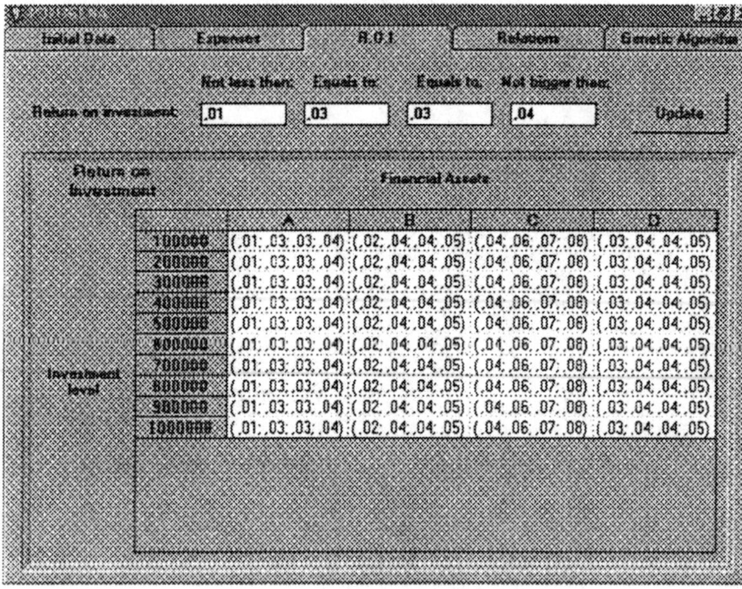

Figure 4

Finally, the data concerning how the assets relate to each other are entered in the software. First of all, we must take into account the deviation on the expected return of each asset when a portfolio contains any other related asset. Some fixed and variable expenses deviations can appear as well. The deviation of the expected return is established by changes on the four TFN values of each financial asset that involve both risk and profitability synergies. For example, if we invest in dollars and in a transportation enterprise we are reducing the risk shown by the first and last numbers of the fuzzy return. Obviously, the changes on the return of each asset must result in a TFN. In the model, the changes of fixed and variable expenses are established as *crisp* or certain numbers. A screen has been designed to enter all this data. An example of it is shown on Figure 5.

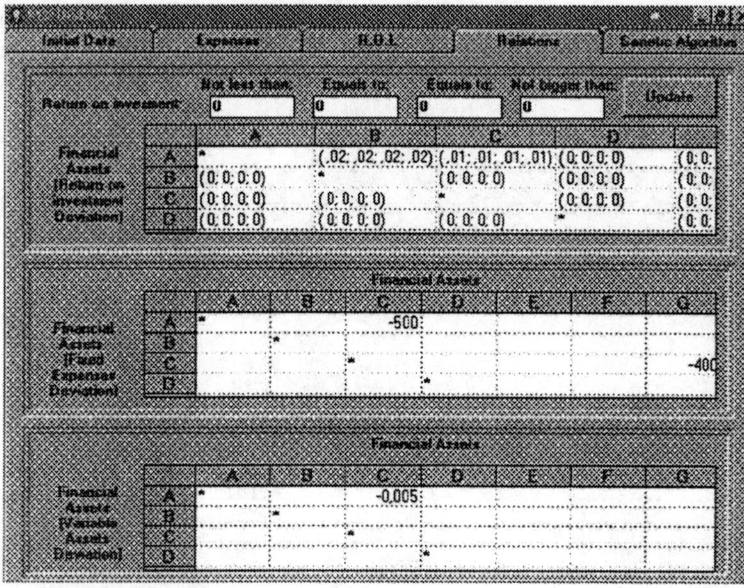

Figure 5

4.2.2. Operative development

Once the data collection is completed, the features of the Algorithm to be used have to be established. This Algorithm differs from traditional ones, where the binary codification is commonly used, in the choice of a decimal representation of the *chromosomes*, for, according to the authors' opinion, such representation suits better to the intended solutions.

Thus, in the case of five assets to choose with five levels of investment a possible solution could be, for example:

$$S_1 = \{2, 4, 2, 3, 5\}$$

This solution shows that the minimum investment is invested twice in the financial asset **2**, while only once in the assets **4**, **3**, and **5**.

Once the codification of solutions is decided, a series of them is generated through a randomized process.

Afterwards, the fitness of each solution to the problem is calculated. In order to do this, it is enough to know the net profitability of the assignation. This is obtained calculating the return of the capitals invest on each assets taking into account the variations produced by the inclusion of a related asset in this portfolio. This amount, less fixed and variable expenses of each asset represents the profitability of each solution in monetary units.

Once the utility of each solution is calculated, a hierarchy between all the generated solutions can be established. In order to make such hierarchy and since we are dealing with TFN always bigger than zero, the models uses as a measure of the solution fitness, the addition of the fuzzy distances (left and right) from each profitability to the origin, TFN (0, 0, 0, 0),.

Those solutions presenting a higher return in the best situation will have a bigger distance to the right, however, in those solutions with a higher risk, the left distance will be smaller. When adding both distances up, we get better solutions corresponding to greater added distances, which will favour higher possibilities to pass the selection procedure. Also, the model can use only one of the distances an so emulate one decision taker that have aversion or propension to risk.

The following step will be to choose through a *Selection Ranking* the most fitted individuals, who will become *parents* of the next generation, as Figure 6 shows.

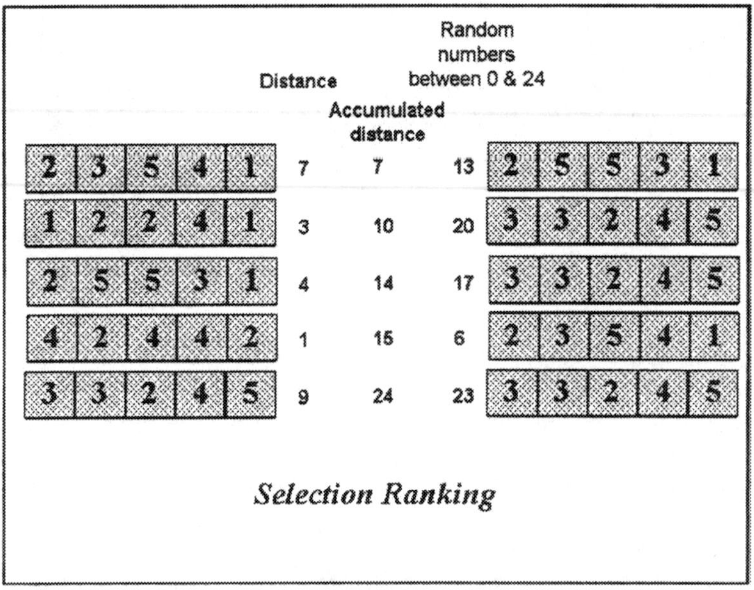

Figure 6

The figure shows to the left five string representing five portfolio' selection. In the following column, the fuzzy distance of the expected profitability of each portfolio has been calculated. After that, the accumulated distance is obtained adding the distance of the previous solutions to the distance of each one. Finally, the model generates random numbers between zero and the sum of all the distances and chooses the solution that has that number within the accumulated distance of the previous solution and its own accumulated distance. So, the solutions with bigger distances, better solutions, have a higher probability to be chosen.

For the parents crossover, and after several tests with different methods, the uniform crossover has been chosen. Its performance is described below. Once two parents have been selected, it is randomly fixed which one of them is going to determine each of the assignments of each offspring. At the end, we obtain two descendants whose assignments can be all the possible combinations of those of the *parents*. The reason to choose this crossover as the best, in authors' opinion, lies in the fact that the most profitable or fitted assets of each solution can appear in any positions. A graphic example of the *Uniform Crossover* is shown in Figure 7.

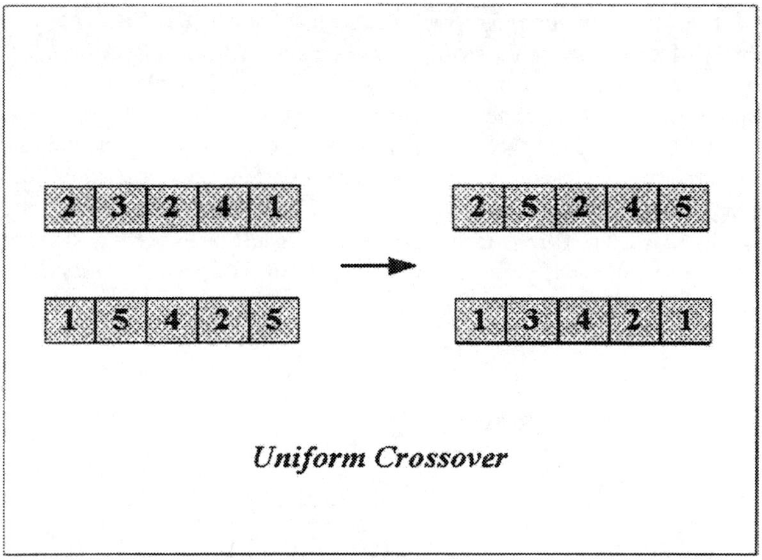

Figure 7

Once the crossover has finished, the process goes ahead with the mutation. To perform it, the model chooses randomly a position of the string, and then changes the asset in that position for any other possible one.

This procedure allows the parents' strings to change a part of their structure in order to achieve any of their descendants improve its parents' returns. An example of this process is shown in Figure 8.

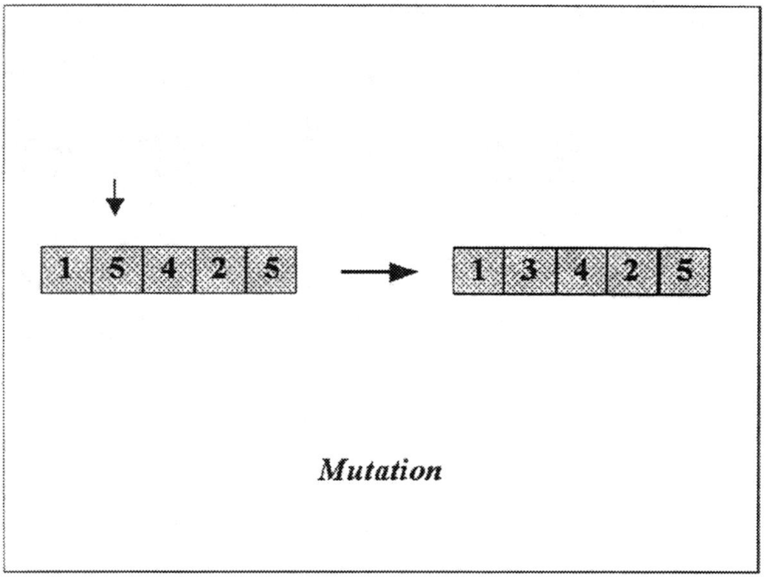

Figure 8

Once the mutation is performed, the *fitness*, that is, the expected profitability of each population string, is calculated, so finishing an iteration. The whole process is repeated as many times as it is found advisable in order to get a solution that either is the optimum one or it is very close to it.

As a last feature of the Algorithm, we have decided to include the so called *Elitism* characteristic, which deals with keeping the best individual of each population throughout the following ones until the model gets a new one that improves its scoring in fitness to the problem. This *elitism* procedure prevents the lost of the best solutions of previous populations until they are not surpassed by any other one beating its fitness, as it is shown in Figure 9.

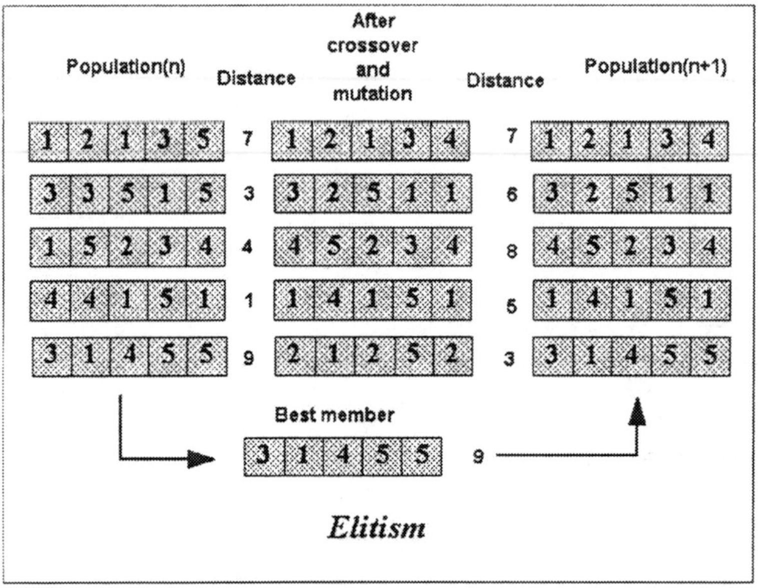

Figure 9

In this paper, the application of the defended model allows a process of selection of a portfolio when the available financial assets are related, considering the variation of the returns, fixed and variable expenses that such relations may produce.

As a summary, Figure 10 shows all the steps of the above explained process.

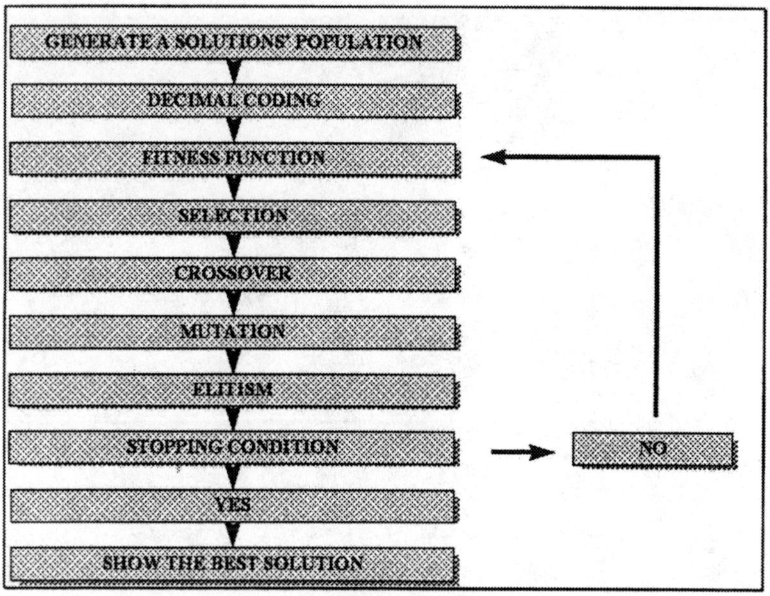

Figure 10

Within the sample program, there is a screen where the parameters of the Genetic Algorithm (crossover probability, mutation probability, number of generations and number of individuals) are entered, and where the model is able to put into operation.

A double parameter of mutation probability, initial and final, has been added. The initial one determines the mutation of the early generations. This mutation probability will change towards the final one to be used within the latest generations. The model also includes an input box with the selection speed of the solution that establishes the exponent of the fuzzy distance of the profitability of each solutions. The screen designed to contain all the mentioned data is shown in Figure 11.

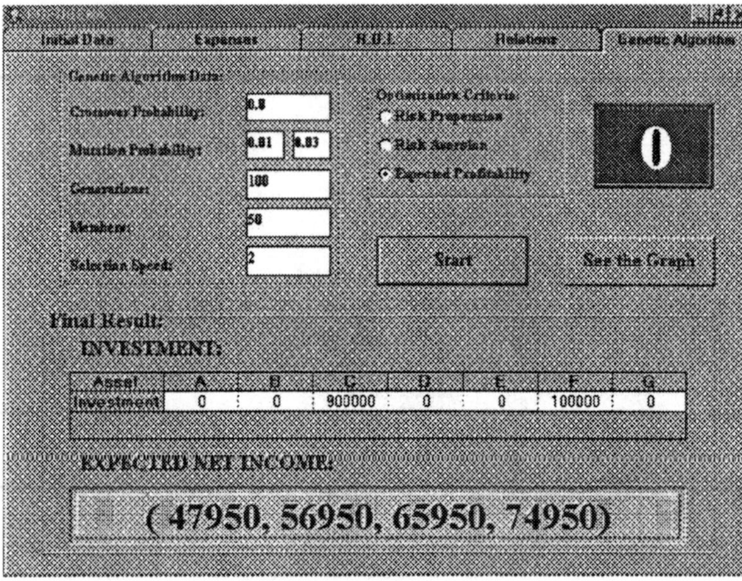

Figure 11

Finally, in the POFUGENA model, a graph has been included displaying the evolution shown by the fuzzy distance of the best individuals of each population. An example of such graph is provided in Figure 12.

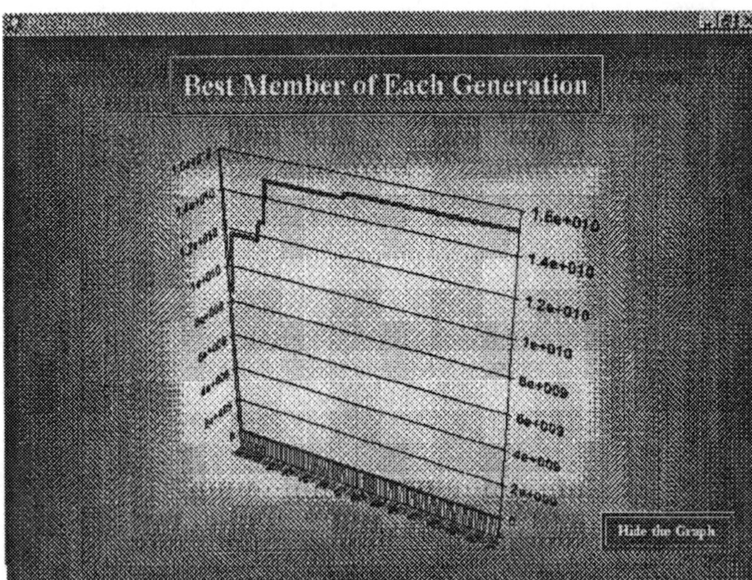

Figure 12

5. STUDY CASE

The software of the Fuzzy Genetic Algorithm has been tested with several problems increasing the difficulty degree. Here following we include an example illustrating the utility of the POFUGENA model.

One investor has 1,000,000 monetary units. There are ten different financial assets available in the market. The minimum investment is fixed in 100,000 monetary units. Every financial asset has the same fixed and variable expenses. The expected returns of the assets are shown below:

$\tilde{A}$ = (0.01; 0.03; 0.03; 0.04) $\tilde{B}$ = (0.02; 0.04; 0.04; 0.05)

$\tilde{C}$ = (0.04; 0.06; 0.07; 0.08) $\tilde{D}$ = (0.03; 0.04; 0.04; 0.05)

$\tilde{E}$ = (0.04; 0.05; 0.05; 0.07) $\tilde{F}$ = (0.05; 0.05; 0.05; 0.05)

$\tilde{G}$ = (0.01; 0.03; 0.03; 0.04) $\tilde{H}$ = (0.02; 0.04; 0.04; 0.04)

$\tilde{I}$ = (0.02; 0.02; 0.02; 0.02) $\tilde{J}$ = (0.02; 0.03; 0.04; 0.05)

Furthermore, assets C and F are related. Due to this relation the deviations on the TFNs of the assets when included in the same portfolio are the following:

RVcf = (0.01; 0; 0; 0) RVfc = (0.01; 0.01; 0.01; 0.01)

The solution obtained using the POFUGENA model was to invest 900,000 monetary units in the financial asset C and 100.000 monetary units in the financial asset F. The profitability of this solution will be (48,000; 65,000; 74,000; 83,000).

The values entered for the parameters of the Algorithm are:

· Crossover profitability: 80%
· Initial mutation profitability: 1%
· Final mutation profitability: 3%
· Number of generations: 100
· Number of individuals in each generation: 50
· Selection speed: 2
· Optimization criteria: expected profitability

We also observed that after fifty generations the best member was always the same.

This is only a simple problem but obviously real problems are more complex. The model can also analyse that problems and obtain good solutions.

6. CONCLUSIONS AND FUTURE DEVELOPMENTS

The solutions obtained with this Fuzzy Genetic Model of portfolio selection are more correct, in our opinion, because it tries to considerate the reality of the financial market without transforming it or reducing its high complexity.

The fuzzy treatment of the information allows the representation of the expected profitability of the assets including the risk that their selection bears as well as the higher yield that can be obtained within the best situation.

Also, the use of a Fuzzy Genetic Algorithm allows us to include the relations between the assets and the resulting modifications that such relations originate in profitabilities and expenses and, therefore, complete and extend the field of application of the portfolio selection to the everyday reality of the problem introducing a useful tool for cash management.

7. BIBLIOGRAPHY

BEETIS , R.A. and MAHAJAN, V. (1985): "Risk/Return Performance of Diversed Firms", *Management Science*, July, pp. 136-167.

DAVIS, L. (1991): "*Handbook of Genetic Algorithms*", Van Nostrand Reinhold, New York.

GOLDBERG, D.E. (1989): "*Genetic Algorithms in Search, Optimation & Machine Learning*", Addison-Wesley, Massachusetts.

HERRERA, F. and VERDEGAY, J.L. (1996): "*Genetic Algorithms and Soft Computing*", Physica-Verlag.

KAUFMANN, A. and GIL ALUJA, J. (1986): "*Introducción a la Teoría de los Subconjuntos Borrosos a la Gestión de las Empresas*", Milladoiro, Santiago de Compostela.

KAUFMANN , A and GIL ALUJA, J. (1987): "*Técnicas Operativas de Gestión para el Tratamiento de la Incertidumbre*", Hispano Europea, Barcelona.

KAUFMANN, A.; GIL ALUJA, J. and TERCEÑO, A. (1994): "*Matemática para la Economía y la Gestión de Empresas*", Ediciones Foro Científico, Barcelona.

KOZA, J.R. (1994): "*Genetic Programming*", Brad Ford, Cambridge, Massachusetts.

LEVY, H. and MARSHALL, S. (1990): "*Capital Investment and Financial Decisions*", Prentice Hall, London.

LÓPEZ-GONZÁLEZ, E. and RODRÍGUEZ-FDEZ., M.A. (1995a): "GENia: A Genetic Algorithms for Inventory Analysis. A Spreadsheet Approach", *AMSE'95*, vol. IV, pp. 200-223.

LÓPEZ-GONZÁLEZ, E.; MENDAÑA, C. and RODRÍGUEZ-FDEZ., M.A. (1995b): "GENIAVIS: Modelo de Algoritmo Genético para el Análisis de Inventarios con Programación Visual", *ESTYLF'95*, pp. 101-102.

LÓPEZ-GONZÁLEZ, E.; MENDAÑA, C. and RODRÍGUEZ-FDEZ., M.A. (1995c): "CAJAGEN: Un Algoritmo Genético para la Gestión Económica de los Cajeros Automáticos en Programación Visual", *SIGEF'95*, vol. II, pp. 233-242.

LÓPEZ-GONZÁLEZ, E.; MENDAÑA, C. and RODRÍGUEZ-FDEZ., M.A. (1996): "Aplicación de los Algoritmos Genéticos en el Control de Gestión del Personal: El Modelo TARAG", *SIGEF'96*, vol. III.

MARKOWITZ, H.M. (1959): "*Portfolio selection: Efficient Diversification of Investments*", Jonh Wiley, New York.

VAN HORNE, J.C. (1986): "*Financial Management and Policy*", Prentice-Hall, Englewood Cliffs.

ZADEH, L.A. (1965): "Fuzzy Sets", *Information and Control*, vol. 8, pp. 338-353.

PREDICTING INTEREST RATES USING ARTIFICIAL NEURAL NETWORKS

ThemistoclesPolitof and **Dan Ulmer**
Department of Decision Sciences and MIS
Concordia University
1455 De Maisonneuve Blvd. W.
Montreal, Que., Canada H3G 1M8

Abstract: The aim of this paper is to asses the effectiveness and easiness of use of Artificial Neural Network (ANN) models in predicting interest rates by comparing the performance of ANN models with that of simple multivariate linear regression (SMLR) models. The task undertaken is to predict the yield of the Canadian 90-day Treasury Bills (TBs) one month ahead using information from eighteen indices of the current economic data, as for example the level of economic activity (GDP), inflation, liquidity etc. Following various approaches, models of both types, SMLR and ANN, are constructed that make use of the same input information. Their performance is compared from the mean absolute percentage error of twelve monthly forecasts. In all cases the ANN models outperform the SMLR models by a wide margin. In addition the absence of need to check the validity of data with respect to assumptions as linearity and normality makes handling the data for ANN models easier and their applicability wider.

Key-words: Neural Networks, Interest Rates, Forecasting.

1. INTRODUCTION

During the last decade there has been an impressive growth in the studies using Artificial Neural Network (ANN) models in business, industry and science [13] and in particular in finance and economics [12]. There are also many reports of ANN models used with great success by corporations in the area of financial analysis and forecasting [2, 9, 10] but unfortunately because of proprietary reasons not much is revealed about these models. ANN's are able to capture functional relationships between the input and output data, learn and generalize the essential characteristics. They perform comparatively better in tasks involving ambiguity. In finance they are best applied in situations involving unstructured problems with incomplete data [3]. They exhibit adaptability by changing automatically their parameters to fit the pattern of the data presented. They can handle problems with thousands of variables for which other nonlinear techniques could be impractical. They have though the drawback of acting as black boxes by not providing much information about the specific relationships between the dependent and independent variables. In finance the applications of ANN models include assessing corporate bankruptcy risk, credit approval, bond rating, predicting currency exchange rates, stock selection and forecasting [12]. Not much has been reported regarding the prediction of interest rates using ANN's. A study of whether it is possible to predict future spot rates based on current forward interest rates is carried in [11] and includes the use of ANN models with promising results.

In this paper we study the performance of ANN models in forecasting short term interest rates based on information derived from the current economic data. The task undertaken is to predict the yield of the Canadian 90-day Treasury Bills (TBs) one month ahead using eighteen economic indices, as for example the level of economic activity (GDP), inflation, liquidity etc. We asses the effectiveness and easiness of use of Artificial Neural Network (ANN) models in predicting interest rates by comparing the performance of ANN models with that of simple multivariate linear regression (SMLR) models. Linear regression analysis is the most widely used quantitative tool in business and finance [5]. A general comparison of the performance of ANN models and linear regression models in estimating simple functional forms is done in a simulated experiment in [7]. The ANN models performed very well and were comparatively better for some but not all types of functions. The authors conclude that the ANN models have considerable potential as an alternative to regression models. In this paper we construct models of both types, SMLR and ANN, following various approaches that make use of the same input information. Their performance is compared from the mean absolute percentage error of twelve monthly forecasts. We find that in all cases the ANN models outperform the SMLR models by a wide margin. In addition the applicability of the ANN models is wider because the data does not have to satisfy certain assumptions as linearity and normality.

2. THE ANN MODEL

We give a brief description regarding the artificial neural network models and some key concepts related to them. More details can be found in [8, 14].

2.1 Components of an ANN

An ANN is composed of artificial neurons, represented by nodes, which are the processing units of the system. The neurons are arranged in three types of layers, one input layer, one output layer and one or more intermediate layers, called hidden layers. The neurons of different layers may be connected. The strength of the connection is represented by a weight. The number of layers, the number of neurons in each of them and the connections determine the architecture of the network. Each neuron receives inputs, processes them (through the activation and transfer functions) and delivers a single output. The neurons of the input layer receive data from the outside world. The other neurons receive as inputs the outputs of the neurons with whom they are connected weighted by the strength of the connection. The output of the neurons of the output layer is the output we receive. The inputs and outputs of the neurons in the hidden layers are not seen by us. The weighted inputs are processed inside a neuron first through what is called an activation function that yields an activation value based on which the neuron may or may not produce an output. A common activation function is given by:

$$S_i(t)=\Sigma_j w_{ji}\cdot x_j(t) + b_i \qquad (1)$$

where $S_i(t)$ is the activation value for neuron i at time t, w_{ji} is the strength of the connection from neuron j to neuron i, $x_j(t)$ is the output value of neuron j at time t and b_i is the bias value of neuron i. Note that positive weights increase the activation level (excitation) while negative weights lower it (inhibition). Subsequently the output of a neuron is calculated from the activation value through what is called a transfer function. Among the popular transfer functions are the step, signum, hyperbolic tangent, linear and threshold-linear functions. A frequently used nonlinear transfer function is the sigmoid given by the formula:

$$Y_i(t)=1/(1+\exp(-S_i(t)/T)) \qquad (2)$$

where $Y_i(t)$ is the output of neuron i at time t, $S_i(t)$ is the activation value for neuron i at time t and T is the threshold level, a parameter that yields functions of different slopes. As T approaches zero the transfer function takes the value 1 if the activity level $S_i(t)$ is positive and zero otherwise. The sigmoid transfer function produces a continuous value in the [0,1] range.

2.2. Training and testing an ANN

Once the architecture and the activation and transfer functions are chosen, the weights of the connections have to be determined so that the network can provide the desired output. This is done through the process of network training that has two phases, learning and testing. In the so called supervised learning the network is provided with a set of inputs for which the outputs are known and are used as targets in the process of learning. The network processes the inputs and calculates corresponding outputs which are then compared with the known targets. The difference in the values of an output and its corresponding target is the error. If the percentage error of an output is within a prespecified range, called the training tolerance, the output is said to be good, otherwise it is called bad. If all outputs are good then the network is considered 100 percent or fully trained. If there are bad outputs then a change must be made in the weights of some of the connections. Determining the weights to be changed is in general a difficult problem. The operations used to perform this adjustment are known as learning rules. Currently the most popular form of learning system is called the backpropagation. The learning rules of backpropagation aim at minimizing the sum of the squared errors of the network by changing the values of the weights and biases of the network in the direction of the steepest decent with respect to this sum. In other words if we define the total error of the system E to be $E=\Sigma_{ip}(t_{ip}-y_{ip})^2$ where t_{ip} is the target output for the output neuron i corresponding to the input/output pair p and y_{ip} indicates the actual output for that neuron on that pair p, then provided that the output functions are differentiable, the change in the weight w_{ji} corresponding to pattern p is $\Delta_p w_{ji} = - \eta(\partial E/\partial y_{ip})\cdot(\partial y_{ip}/\partial w_{ji})$ where η is a constant between 0 and 1, called the learning rate. This process is repeated in many iterations or runs until the number of good outputs reaches a high percentage, not necessarily 100 percent. The usefulness of a neural network is determined not by how well it has been trained but by how well it tests, because it is possible that a fully trained network has learned to memorize data and lost the ability to generalize. During the testing phase the network is provided with a different set of input/target-output data that it has not seen before. The inputs are processed by the trained network and its performance is measured by the percentage of good outputs it calculates. If this percentage is not satisfactory the training of the network must continue, otherwise we have a useable neural network.

3. METHOD AND DATA

Our task is to predict the yield of the Canadian 90-day Treasury Bills (TBs) one month ahead using information from the current economic data, as for example the level of economic activity (GDP), inflation, liquidity (e.g. the rate of growth of M_2, which includes currency outside banks, demand and saving deposits of chartered banks), the change in the value of the country's currency (exchange rate) etc. Of course predicting interest rates is a difficult problem and it is a challenge to identify the major predictors. The aim of this paper is not to derive sophisticated models but rather to asses the effectiveness and easiness of use of ANN models in predicting interest rates by comparing the performance of ANN models with that of simple multivariate linear regression models.

To accomplish the aim of the paper we tried various approaches for the construction of a model to predict interest rates using the following method. Starting from a selected a set of economic indices which we judged to be useful in predicting the interest rates, we obtained two types of models to predict the yield of the Canadian 90-day Treasury Bills, a simple multivariate linear regression model and an artificial neural network model. The various approaches differed in the set of the initially selected predictors.

To obtain the multivariate linear regression model we used the computer package MINITAB, and we arrived at the best model derived by this method by following the standard procedure of eliminating predictors that were not statistically significant (value of $p>0.05$). The aptness of the models were tested by verifying that the standardized residuals are independent, normally distributed and have constant variance.

To derive the artificial neural network model we used the computer package BRAINMAKER PROFESSIONAL. All networks had one hidden layer, used activation and transfer functions (1) and (2) given above and were trained using backpropagation, with training tolerance and testing tolerance equal to 0.10. We run several trials by varying the number of neurons in the hidden layer, the learning rate and the number of iterations used to train the network.

We obtained data from January 1979 until December 1994 from the Bank of Canada (Bank of Canada Review) on a monthly basis. The data from January 1979 until December 1993, a total of 180 months, were used to construct and test the models. Then the models were compared by their performance (in terms of the average percentage error) in predicting the interest rates for each month of 1994, one month ahead at a time.

To derive the models to predict the interest rates we used the following two approaches.

Approach 1. We have started with sixteen predictors. From the Canadian indices we used: the exchange rate of the Canadian dollar with the US dollar (exra), the consumer price index in Canada (CPI), GDP in Canada (GDP), the unemployment rate (ura), money supply M_1 (ms1), money supply M_2 (ms2), money supply M_3 (ms3), the changes in M_1, M_2, M_3 (cms1, cms2, cms3), the level of residential construction in terms of the number of units started (Resco), the Toronto Stock Exchange index (TSE), and the Canadian International Reserves (Intre). We also used the following US indices: the real rate of inflation of US (inUS), the index of inflation in US (indUS), and the Dow Jones industrial average index (Dow).

Approach 2. Here we take the view that if the interest rate is considered to be purely the price of money and not a policy variable, then it should be determined, like any other commodity in the free market, by the interaction of supply and demand factors. Thus for this approach we start with eight predictors related to the supply and demand sides of the money market. These predictors are: GDP, M_1, M_2, change in M_1, change in M_2, TSE, Implicit Price Index (Imprin) and the Capacity Utilization Rate (Caput). The money supply indices and their changes are obviously indicators on the supply side, while GDP, the Capacity Utilization Rate and the inflation rate as measured through the Implicit Price Index represent the demand side. The data available for the Implicit Price Index and the Capacity Utilization Rate were on a quarterly basis, so we adjusted for monthly values by linear interpolation and extrapolation. This approach was also repeated by lagging the variables for one and two months. This means that the value of interest rates of this month is estimated based on the values of the predictors one or two months ago respectively.

Below we report on the best models derived from the above described approaches and the corresponding results.

4. THE REGRESSION MODELS

For all the models we have followed the procedures of the statistical software (MINITAB) to eliminated highly correlated variables and insignificant variables (with p-value > 0.05). We stopped when all the predictors had p-values less than 5% and the overall p-value of the regression was zero.

4.1 Model R1

We have started with the 16 predictors as described in approach 1 in Section 3. The resulting model is shown in Table 1.

Table 1. The model R1

Intb90=25.3-0.0488 CPI-29.5 indUS+0.000104 GDP-0.609ura - 0.155 cms1-0.000204 ms1-0.0227 Resco-0.00102 TSE

Predictor	Coefficient	Stdev	t-ratio	p
Constant	25.322	3.516	7.20	0.000
CPI	-0.048777	0.004264	-11.44	0.000
indUS	-29.451	3.837	-7.68	0.000
GDP	0.00010394	0.00001614	6.44	0.000
ura	-0.6095	0.1111	-5.49	0.000
cms1	-0.15502	0.06009	-2.58	0.011
ms1	-0.00020394	0.00004240	-4.81	0.000
Resco	-0.022746	0.003988	-5.70	0.000
TSE	-0.0010230	0.0004424	-2.31	0.022

s=1.319 R-sq=84.4% Rsq(adj)=83.7% p=0.000

As the table shows, the predictors of this model are: CPI, inflation index in U.S., GDP, unemployment rate, M_1, change in M_1, residential construction and TSE. While the fitting is high the signs of some coefficients does not agree with economic theory, for example the sign for CPI that implies that interest rates increase when inflation decreases. The predictions obtained from this model are shown in Table 2.

The check mark (✓) means that the trend is correctly predicted. The absolute percentage error is calculated as follows:

$$E\ (\%) = |\text{Predicted Value} - \text{Actual Value}|\ /\ \text{Actual Value}$$

Table 2. Prediction of interest rates using model R1

Intb90=25.3-0.0488 CPI-29.5 indUS+0.000104 GDP-0.609ura - 0.155 cms1-0.000204 ms1-0.0227 Resco-0.00102 TSE					
Month	Trend	Predicted Value	Actual Value	Absolute Error	Percentage Error %
1		6.831	3.63	3.63	100
2	✓	8.281	3.85	4.431	115.09
3		8.021	5.38	2.641	49.09
4	✓	8.423	5.81	2.613	44.97
5		7.048	6.33	0.718	11.34
6	✓	7.272	6.67	0.602	9.03
7	✓	7.026	5.79	1.236	21.35
8	✓	6.555	5.35	1.205	22.52
9		7.257	5.29	1.967	37.18
10		6.806	5.37	1.436	26.74
11		6.529	5.78	0.749	12.96
12		5.996	7.18	1.184	16.49
Mean				1.86767	38.9

4.2 Model R2

Starting with the eight predictors as described in approach 2, and using the same procedure as in Model R1 we obtained the second regression model as shown in Table 3.

Table 3. Model R2

Intb90=9.37+0.000042 GDP-0.514 cms1+0.874 cms2-0.000375 ms1 +0.360 Imprin-0.0965 Caput				
Predictor	Coefficient	Stdev	t-ratio	p
Constant	9.369	3.099	3.02	0.003
GDP	0.00004242	0.00000937	4.53	0.000
cms1	-0.5140	0.1138	-4.51	0.000
cms2	0.8738	0.3146	2.78	0.006
ms1	-0.00037515	0.00006074	-6.18	0.000
Imprin	0.35995	0.06904	5.21	0.000
Caput	-0.09648	0.02722	-3.54	0.001

s=2.126 R-sq=59.1% Rsq(adj)=57.7% p=0.000

Again we observe that there are two inappropriate signs for cms2 (the change in M_2) and Caput (the Capacity Utilization Rate). The predictions obtained from this model are shown in Table 4.

Table 4. Prediction of interest rates using model R2

Intb90=9.37+0.000042 GDP-0.514 cms1+0.874 cms2-0.000375 ms1 +0.360 Imprin-0.0965 Caput					
Month	Trend	Predicted Value	Actual Value	Absolute Error	Percentage Error %
1	✓	3.027	3.63	0.603	16.61
2		2.45	3.85	1.4	36.36
3	✓	2.99	5.38	2.39	44.42
4		2.12	5.81	3.69	63.51
5		0.842	6.33	5.488	86.7

Table 4. Prediction of interest rates using model R2 (continued)

Month	Trend	Predicted Value	Actual Value	Absolute Error	Percentage Error %
6	✓	3.71	6.67	2.96	44.38
7	✓	2.59	5.79	3.2	55.27
8		2.74	5.35	2.61	48.79
9		4.46	5.29	0.83	15.69
10		4.115	5.37	1.255	23.37
11		3.134	5.78	2.646	45.78
12		0.876	7.18	6.304	87.8
Mean				2.78133	47.39

4.3 Model R3

This model was obtained by starting with the eight predictors as described in approach 2, with the difference that the predictors were lagged. We made various trials by lagging the predictors by one and two months following the same procedure as in Model R1. The best model derived from these attempts has the independent variables lagged by two months and is shown in Table 5.

Table 5. Model R3

Intb90 = 0.78 + 0.000047 lag2_GDP - 0.000403 lag2_ms1 + 0.426 lag2_Imprin				
Predictor	Coefficient	Stdev	t-ratio	p
Constant	0.776	2.883	0.27	0.788
lag2_GDP	0.00004659	0.00000957	4.87	0.000
lag2_ms1	-0.00040274	0.00006506	-6.19	0.000
lag2_Impri	0.42610	0.06931	6.15	0.000

s=2.301 R-sq=51.8% Rsq(adj)=50.9% p=0.000

The signs of the coefficients of this model are in agreement with economic theory. The predictions obtained from this model are shown in Table 6.

Table 6. Prediction of interest rates using model R3

Intb90 = 0.78 + 0.000047 lag2_GDP - 0.000403 lag2_ms1 + 0.426 lag2_Imprin					
Month	Trend	Predicted Value	Actual Value	Absolute Error	Percentage Error %
1		4.082	3.63	0.452	12.45
2		3.18	3.85	0.67	17.4
3	✓	3.70	5.38	1.68	31.23
4		2.91	5.81	2.9	49.91
5	✓	3.271	6.33	3.059	48.33
6	✓	3.59	6.67	3.08	46.18
7	✓	2.83	5.79	2.96	51.12
8		3.32	5.35	2.03	37.94
9	✓	2.798	5.29	2.492	47.11
10	✓	3.38	5.37	1.99	37.06
11	✓	4.146	5.78	1.634	28.27
12		4.026	7.18	3.154	43.93
Mean				2.17508	37.58

5. THE ANN MODELS

We used the two approached described is Section 3 to derive ANN models that would correspond to the regression models obtained from the two approaches. One difference is that the ANN models do not eliminate independent variables but rather use all the predictors fed to them as inputs. The software we have used is the BrainMaker Professional version 3.0 (by California Scientific Software). The transfer function used for all models was the sigmoid. The networks were trained with backpropagation with training tolerance and testing tolerance equal to 0.10. The number of neurons in the input layer was equal to the number of the input variables increased by one to account for a bias variable. The output layer had one neuron since the desired output had a single value. The actual output was the that of the one month ahead forecast value of the change in the yield of the Canadian 90-day Treasury Bills. The use of differences in numbers is recommended because neural networks that identify trends respond much better to changes in the values of input and output variables than to precise numeric values.

All models had a single hidden layer. It has been shown [4, 6] that a network with a single hidden layer having an adequate number of neurons can represent efficiently any functional relationship between input and output variables. Such a network also needs much less time to train. In addition a network with too many hidden layers or neurons may result in overfitting and thus loose the capacity to generalize when presented with new data. For all approaches we have constructed and tested two types of models differing in the number of neurons in the hidden layer. The number of hidden neurons were determined according to the following rules. In the first type the number of hidden neurons was equal to the number of inputs, if the number of inputs was bigger than 10, and 10 otherwise. In the second type the number of hidden neurons was equal to (the number of inputs + the number of outputs)/2, an empirical formula that has been found to work well according to the BRAIMAKER manual [1]. In all cases the models of type 2 were found to outperform the models of type 1, therefore we report below only models of type 2.

We experimented with learning rates of 0.5 and 1 and with training runs of 500 and 1000. In all cases a learning rate of 1 yielded a better model. The same holds for training runs of 1000.

As previously said the data from January 1979 until December 1993, a total of 180 months, were used to train and test the models. Out of this 10% was reserved for testing and the rest was used for training the network. The input data was tested for any identifiable cycles. No cycles were found.

Below we report on three of the best models derived from the above described approaches and the corresponding results. They were all trained with 1000 runs, used learning rate of 1 and had a learning and testing tolerance of 0.10.

5.1 Model N1

This is a three-layer neural network with 17 neurons in the input layer, 9 neurons in the hidden layer and 1 neuron in the output layer. The inputs used were the 16 predictors described in approach 1 in Section 3. The predictions obtained from this model are shown in Table 7.

Table 7. Prediction of interest rates using model N1

Month	Predicted Change	Trend	Predicted Value	Actual Value	Absolute Error	Percentage Error %
1	-.0201	✓	3.835	3.63	0.205	5.65
2	0.0302	✓	3.662	3.85	0.188	4.88
3	0.0016	✓	3.851	5.38	1.529	28.42
4	-0.1831		5.204	5.81	0.606	10.43
5	-0.0722		5.745	6.33	0.585	9.24
6	-0.131		6.204	6.67	0.466	6.99
7	-0.084	✓	6.588	5.79	0.798	13.78
8	-0.0134	✓	5.78	5.35	0.43	8.04
9	-0.0722	✓	5.287	5.29	0.003	0.06
10	-0.0974		5.193	5.37	0.177	3.3
11	-0.0789		5.295	5.78	0.485	8.39
12	-0.2402		5.548	7.18	1.632	22.73
Mean					0.592	10.16

5.2 Model N2

This is a three-layer neural network with 9 neurons in the input layer, 5 neurons in the hidden layer and 1 neuron in the output layer. The inputs used were the 8 predictors described in approach 2 in Section 3. The predictions obtained from this model are shown in Table 8.

Table 8. Prediction of interest rates using model N2

Month	Predicted Change	Trend	Predicted Value	Actual Value	Absolute Error	Percentage Error %
1	.1595		4.015	3.63	0.385	10.61
2	.1175	✓	3.749	3.85	0.101	2.62
3	.0637	✓	3.913	5.38	1.467	27.27
4	.2032	✓	5.591	5.81	0.219	3.77
5	.2099	✓	6.027	6.33	0.303	4.79
6	.3577	✓	6.692	6.67	0.022	0.33
7	.4148		7.086	5.79	1.296	22.38
8	.3510		6.145	5.35	0.795	14.86
9	.4030		5.757	5.29	0.467	8.83
10	.3946	✓	5.685	5.37	0.315	5.87
11	.3493	✓	5.723	5.78	0.057	0.99
12	.4232	✓	6.212	7.18	0.968	13.48
Mean					0.53292	9.65

5.3 Model N3

This is a three-layer neural network with 4 neurons in the input layer, 3 neurons in the hidden layer and 1 neuron in the output layer. The inputs used were the 3 predictors lag2_GDP, lag2_ms1, lag2_Imprin, which were the predictors used in Model R3 in Section 4, representing the values of the Canadian GDP, money supply M1 and Implicit Price Index, all lagged by two months. The predictions obtained from this model are shown in Table 9.

Table 9. Prediction of interest rates using model N3.

Month	Predicted Change	Trend	Predicted Value	Actual Value	Absolute Error	Percentage Error %
1	.1595		4.0155	3.63	0.3855	10.62
2	.1595		3.7915	3.85	0.0585	1.52
3	.1645	✓	4.0145	5.38	1.3655	25.38
4	.1628	✓	5.5508	5.81	0.2592	4.46
5	.1628	✓	5.9838	6.33	0.3462	5.47
6	.1612	✓	6.4962	6.67	0.1738	2.61
7	.1612		6.8332	5.79	1.0432	18.02
8	.1612	✓	5.9552	5.35	0.6052	11.31
9	.1578	✓	5.5118	5.29	0.2218	4.19
10	.1578		5.4488	5.37	0.0788	1.47
11	.1578	✓	5.5318	5.78	0.2482	4.29
12	.1612	✓	5.9502	7.18	1.2298	17.13
Mean					0.5	8.87

6. CONCLUSION AND REMARKS

As it is easily observed the ANN models outperformed the simple linear multivariate regression models in all cases by a wide margin. The mean percentage error for the forecasts of the ANN models was around 10% as compared to 40% for the regression models. The ANN models were also better able to predict the trend in interest rates.

The ANN models have been criticized as black boxes which even if can be trained to recognize and predict patterns they give us no insight for the relationship of the output variables to individual input variables. However even for regression models the relationships derived may be inaccurate and great care has to be exerted in this direction. In addition for linear regression to be valid we must exercise great care that the variables used satisfy the underlying statistical assumptions while no such effort is needed for ANN models. ANN models can handle any functional relationship between input output variables linear or not while the linear regression models would suffer in the case of nonlinear relationships. A great amount of effort would have to be exerted to try to adjust the variables with nonlinear relationship in linear regression models while no additional effort is needed for ANN models.

It could be possible to improve the performance of the ANN models by specifying tighter tolerances for training and testing. However this may have the drawback that the networks would take longer to train and they may loose some of the capacity to generalize.

REFERENCES

[1] California Scientific Software. *BrainMaker User's Guide and Reference Manual.* 6th ed. 1993.

[2] Hammerstrom D. Neural networks at work. *IEEE Spectr.* June 1993, 26-32.

[3] Hawley DD, Johnson JD, Raina D. Artificial neural systems: A new tool for financial decision making. *Financial Analyst Journal,* Nov-Dec. 1990, 63-72.

[4] Hornik K. Approximation capabilities of multilayer feedforward networks. *Neural Networks,* 1991, 4(2):251-257.

[5] Ledbetter W, Cox J. Are OR techniques being used? *Industrial Engineering,* 1977, 9(2):19-21.

[6] Lippman RP. An introduction to computing with neural networks. *IEEE ASSP Magazine,* 1987, 4(2):4-22.

[7] Marquez L, Hill T, Worthley R, Remus W. Neural network models as an alternative to regression. In *Neural Networks in Finance and Investing,* Trippi RR, Turban E, eds. Chicago: Irwin, 1996.

[8] Rumelhart DE, Widrow B, Lehr MA. The basic ideas in neural networks. *Commun. ACM,* 1994, 37(3):87-92.

[9] Schwartz EI. Where neural networks are already at work: Putting AI to work in the markets. *Bus. Week,* Nov. 2, 1992, 136-137.

[10] Schwartz EI, Treece JB. Smart programs go to work: How applied-intelligence software makes decisions for the real world. *Bus. Week,* Mar. 2, 1992, 97-105.

[11] Swanson NR, White H. A model selection approach to assessing the information in the term structure using linear models and artificial neural networks. In *Neural Networks in Finance and Investing,* Trippi RR, Turban E, eds. Chicago: Irwin, 1996.

[12] Trippi RR, Turban E, eds. *Neural Networks in Finance and Investing.* Chicago: Irwin, 1996.

[13] Widrow B, Rumelhart DE, Lehr MA. Neural networks: applications in industry, business and science. *Commun. ACM,* 1994, 37(3):93-105.

[14] Zadehi F. *Intelligent Systems for Business: Expert Systems with Neural Networks.* Belmond: Wadsworth Publishing, 1993.

V. MULTICRITERIA ANALYSIS IN COUNTRY RISK EVALUATION

ASSESSING COUNTRY RISK USING MULTICRITERIA ANALYSIS

Michael Doumpos[1], Constantin Zopounidis[1], Thomas Anastassiou[2]

[1] Technical University of Crete
Department of Production Engineering and Management
Decision Support Systems Laboratory
University Campus, 73100 Chania, Greece

[2] Athens University of Economics
Department of Business Administration
76 Patission Str.
Athens 10434, Greece

Abstract: Country risk assessment is a decision problem which has gained an increasing interest both from the macroeconomic and the microeconomic point of view, mainly during the last two decades. Banks and international lending institutions are interested in developing effective country risk models to determine the creditworthiness of countries. This paper presents the contribution of multicriteria analysis in country risk assessment. Initially, the studies of Mondt and Despontin (1986) and Oral et al. (1992) proposing a multiobjective programming approach and a generalized logit model respectively, are discussed. Then, focusing on the preference disaggregation approach, three multicriteria methods are applied in the case study of Tang and Espinal (1989). The obtained results are very encouraging, proving that multicriteria analysis methods could be used as an alternative tool to statistical approaches in analyzing the preferences of the decision makers (managers of banks and lending institutions) in assessing country risk.

Keywords: Multicriteria analysis, Country risk, Preference disaggregation.

1. Introduction

During the 1970s and the 1980s, the world economy has experienced a severe recession, mainly due to the two oil crises in 1973 and 1979. As a consequence, the external debts of most of the countries have increased tremendously and the problem of assessing country risk has attracted the attention of banks, governments and international institutions as well. Although the world economy is slowly starting to upturn, the impacts of the recession are still evident for many countries.

Several attempts, mainly by banks, have been made to establish efficient procedures for estimating country risk. These procedures were initially based on devising checklist systems which proved to be insufficient due to the difficulty in selecting the economic indicators and determining their relative importance (Saini and Bates, 1984).

Then, more sophisticated multivariate statistical techniques were proposed. Saini and Bates (1984) reviewed the applications of discriminant analysis, principal components analysis, and logit analysis in country risk assessment. Mumpower et al. (1987) applied factor, cluster and regression analysis to study the degree of political risk in 49 countries, while Cosset and Roy (1988;1989) studied the application of regression trees as an alternative to regression analysis for country risk assessment. Although these statistical approaches have been widely applied in the past for country risk assessment, their practical applications are restricted by significant limitations. Saini and Bates (1984) remark that "no institution lending money to developing countries is placing exclusive reliance on a statistical model to guide its actions". To justify their remark, Saini and Bates (1984) reported five possible drawbacks of the statistical techniques and the related studies which have been conducted in the past:

(i) The definition of the dependent variable: the classification of the countries in the rescheduling and the non-rescheduling ones is not always a realistic approach since it overlooks voluntary and non-voluntary reschedulings, as well as other substitutions for formal reschedulings.
(ii) The reliance on debt information which is incomplete at least as far as it concerns the long term case.
(iii) The statistical restrictions, such as the reduction of the original data, the determination of the importance of the explanatory variables, the difficulty in interpreting the obtained results, etc.
(iv) The exclusion of important social and political factors from the analysis, the assumption of stable statistical relationships across countries, and the overlooking of the dynamic nature of the world economy.
(v) The poor predictability of the statistical models, since statistically significant variables were found to be inadequate in making accurate predictions.

To overcome these limitations and difficulties, new methodological approaches have to be introduced in the assessment of country risk. Amongst them, multicriteria decision aid methods (MCDA) constitute a significant tool which can be used as an alternative to statistical techniques. MCDA methods are free of the aforementioned restrictive statistical assumptions, they incorporate the preferences of the decision maker (managers of banks and international institutions) into the analysis of country risk, they are capable of handling qualitative social and political factors, and they are easily updated taking into account the dynamic nature of the world economy.

This paper presents the application of three multicriteria analysis methods in the assessment of country risk, based on the preference disaggregation approach of MCDA (Zopounidis, 1997). More specifically, the UTASTAR method (UTilités Additives, Siskos and Yannacopoulos, 1985), the UTADIS method (UTilités Additives DIScriminantes, Devaud et al., 1980; Jacquet-Lagrèze and Siskos,

1982; Jacquet-Lagrèze, 1995) and a variant of the UTADIS method (UTADIS I, cf. Zopounidis and Doumpos, 1997c) are applied in a case study derived from the study of Tang and Espinal (1989), to analyze the preferences of the managers of two lending institutions. The UTASTAR method is used to develop a model which ranks a sample of 30 countries according to their creditworthiness, while the UTADIS and the UTADIS I methods are used to classify the countries in classes of risk.

This paper is divided in 4 sections. Initially, in section 2 a brief overview of the applications of MCDA approaches in country risk assessment is presented. Section 3, focuses on the three proposed preference disaggregation approaches (the UTASTAR, UTADIS and UTADIS I methods) and their application in the aforementioned case study. Finally, in section 4 the concluding remarks as well as some future research directions are discussed.

2. Multicriteria analysis in the assessment of country risk: an overview

The flexibility of MCDA methods, their adaptability to the preferences of the decision makers and to the dynamic environment of decisions related to country risk, as well as to the subjective nature of such decisions (Chevalier and Hirsch, 1981), has already attracted the interest of many researchers in developing more reliable and sophisticated models for country risk assessment.

Mondt and Despontin (1986) used the perturbation method to assess country risk. The perturbation method, a variant of the well known STEM method (Banayoun et al., 1971), is based on the multiobjective programming approach of MCDA, and it is used to determine the proportion of each country in the portfolio of a bank. The method proceeds in two steps: (i) initially, a first compromise solution is determined, and (ii) through an interactive phase, some perturbations are performed to determine a feasible and acceptable portfolio. The aim of this formulation is to maximize the return of the portfolio (measured through an interest criterion, e.g. the fixed international basic interest rate) and minimize the corresponding risk. This approach was applied in a sample of 10 countries (9 European countries and the United States) evaluated along 5 criteria, the inflation risk, the exchange risk, the political risk, the social risk and the growth risk. The first compromise solution obtained in the first step of the method is presented in Table 1 (the proportion of each country in the portfolio, and the contribution of each country to the total risk measured using a scale from 0-10 with higher values corresponding to higher contribution to risk). The same table also presents the risk level of the compromise portfolio on each one of the 5 criteria, measured through a scale from 0 to 10 with higher values corresponding to higher risk level on each specific criterion. The return of this portfolio is 10.741%.

Based on this initial compromise solution several perturbations can be performed interactively with the decision maker to achieve a flexible and acceptable portfolio (cf. Mondt and Despontin, 1986). The two authors in their application performed five perturbations. For illustrative purposes one of these perturbations is presented in Table 2. This perturbation involves the increase of the part of Germany in the portfolio obtained by the first compromise solution. In the new solution, the return of the portfolio is slightly decreased compared to the first compromise solution, to 10.644%.

Table 1: The first compromise solution (Source: Mondt and Despontin, 1986)

Countries	Portfolio composition	Contribution to risk	Criteria	Risk level
Germany	20.2%	1	Inflation risk	3
Finland	1.3%	2	Exchange risk	4
The Netherlands	6.9%	2	Political risk	2
Great Britain	9.8%	2	Social risk	3
United States	34.3%	3	Growth risk	3
Denmark	1.7%	2		
France	9.0%	2		
Italy	14.4%	9		
Greece	0.4%	3		
Spain	1.9%	3		

Table 2: Solution obtained after the perturbation (Source: Mondt and Despontin, 1986)

Countries	Portfolio composition	Contribution to risk	Criteria	Risk level
Germany	23.8%	2	Inflation risk	3
Finland	1.3%	2	Exchange risk	4
The Netherlands	6.9%	2	Political risk	3
Great Britain	9.8%	2	Social risk	3
United States	29.5%	2	Growth risk	3
Denmark	1.7%	2		
France	9.1%	2		
Italy	15.5%	10		
Greece	0.4%	3		
Spain	1.9%	3		

This approach provides flexibility to the decision maker who through perturbation of "what-if" analysis can derive a wide spectrum of solutions to select the one which is more consistent with his/her preferences. However, this approach studies the country risk problem from the portfolio construction point view. Consequently, although it provides the contribution of each country to the risk of the whole portfolio, it does not provide an overall country risk rating (ranking of the countries) according to the creditworthiness of the countries.

Oral et al. (1992) proposed a generalized logit model for the evaluation of country risk. The generalized logit model is of the following form:

$$r_i = \frac{e^{\left(u_0+\sum u_{ij}\right)}}{1+e^{\left(u_0+\sum u_{ij}\right)}}$$

where r_i is the risk rating of country i, u_0 is a constant and $u_{ij}=\beta_{ij}x_{ij}$ (x_{ij} is the score of a country i with respect to criterion j). Unlike the statistical logistic regression the parameters β_{ij} of the proposed generalized logit model are estimated through a mathematical programming formulation which is able to consider the impacts of countries being in different geographical regions or even countries with different political and economic characteristics.

This approach was applied in a sample of 70 countries for two different years, 1982 and 1987. The country risk indicators used in this application included 8 economic-political factors: reserves to imports ratio, net foreign debt to exports ratio, GNP per capita, current account balance to GNP ratio, investment to GNP ratio, export variability, export growth rate, and political instability. Furthermore, the geographical location of each country was considered as an additional criterion. The results obtained by the proposed generalized logit model were compared with those obtained by logit analysis and regression tree analysis using four different criteria: (i) coefficient of simple correlation, (ii) Spearman's correlation coefficient, (iii) Kindallstan's correlation coefficient, and (iv) the mean value of absolute deviation. The comparison of the three methods depicted the superiority of the generalized logit model to both, the logit and the regression tree analysis. A cross validation stage was also performed to test the predictability of the three methods. Again the results obtained by the generalized logit model were superior to those obtained by the two other statistical methods. As far as the importance of country risk indicators is concerned, the three different models provided similar results which were almost stable for both years, 1982 and 1987. According to the generalized logit model the most important economic-political indicators for the year 1982 were found to be the net foreign debt/exports, the GNP per capita, and the investment/GNP. Furthermore, it was found that developed countries and countries geographically located in Southeast Asia are evaluated as countries of low risk, while countries geographically located in Central America are evaluated as countries of high risk. The same factors were also found to be important for 1987.

Although the proposed generalized logit model constitutes an advantageous alternative to the classical statistical approaches, the mathematical programming formulation used to estimate the parameters of the model is a rather complicated one.

3. Preference disaggregation analysis in the assessment of country risk

Generally, four different approaches can be distinguished in MCDA (Zopounidis, 1997): (i) the outranking relations, (ii) the multiattribute utility theory, (iii) the multiobjective programming, and (iv) the preference disaggregation, on which this paper focuses. The preference disaggregation approach refers to the analysis (disaggregation) of the global preferences of the decision maker to deduce the relative importance of the evaluation criteria, using ordinal regression techniques based mainly on linear programming formulations. In the case of country risk assessment, the decision maker (i.e. the manager of a bank or a lending institution) expresses indirectly his preferences in a form of an a priori ranking or classification of the countries (global preferences). Using a preference disaggregation approach the aim is to derive the relative importance of the evaluation criteria (i.e. economic, social, and political factors, etc.) and develop the corresponding preference model which is as consistent as possible with the global preferences and the decision policy of the decision maker.

Cosset et al. (1992) applied a preference disaggregation methodology in the evaluation of country risk, based on the MINORA multicriteria decision support system (Siskos et al., 1993) which implements the UTASTAR method. The UTASTAR method, a variant of the UTA method (Jacquet-Lagrèze and Siskos, 1982), performs an ordinal regression based on the preference disaggregation approach of MCDA. Given a preordering of a set of alternatives (i.e. countries) defined by the decision maker, the aim of the UTASTAR method is to estimate a set of additive utility functions which are as consistent as possible with the decision maker's preferences. The additive utility function has the following form:

$$U(g) = \sum_{i=1}^{n} u_i(g_i)$$

where $g = (g_1, g_2, \ldots, g_n)$ is the vector of a country's performance on n evaluation criteria and $u_i(g_i)$ is the marginal utility of criterion g_i representing its relative importance in the ranking model. The estimation of the marginal utilities is achieved through the following linear programming formulation:

Minimize $F = \sum_{a \in A} \left\{ \sigma^+(a) + \sigma^-(a) \right\}$

s.t.

$u[g(a)] - u[g(\beta)] + \sigma^+(a) - \sigma^-(a) - \sigma^+(\beta) + \sigma^-(\beta) \geq \delta$ if a is preferred to β

$u[g(a)] - u[g(\beta)] + \sigma^+(a) - \sigma^-(a) - \sigma^+(\beta) + \sigma^-(\beta) = 0$ if a is indifferent to β

$$\sum_i \sum_j w_{ij} = 1$$

$$w_{ij} \geq 0,\ \sigma^+(a) \geq 0,\ \sigma^-(a) \geq 0,\ u_i(g_i^*) = \sum_{k=1}^{a_i - 1} w_{ik} \quad \forall a \in A,\ \forall\ i,\ j$$

where A is the set of reference countries used to develop the additive utility model, $u[g(a)]$ is the global utility of a country $a \in A$, σ^+ and σ^- are two error functions, δ is a threshold used to ensure the strict preference of a country a over a country β, a_i is the number of subintervals $[g_i^j, g_i^{j+1}]$ into which the range of values of criterion g_i is divided, and w_{ij} is the difference $u_i(g_i^{j+1}) - u_i(g_i^j)$ of the marginal utilities between two successive values g_i^j and g_i^{j+1} of criterion i ($w_{ij} \geq 0$). In a second stage the method proceeds in a post-optimality analysis to identify other optimal or near optimal solutions which could better represent the preferences of the decision maker. A detailed description of the UTASTAR method can be found in Siskos and Yannacopoulos (1985).

The MINORA system was applied in a sample of 76 countries (for the year 1986) to develop a ranking model of the countries according to their creditworthiness. Using a reference set of 22 countries an additive utility model was interactively developed which consistently represented the preferences of an expert. The evaluation criteria used in this application and their corresponding weights are presented in Table 3.

Table 3: Weights of evaluation criteria (Source: Cosset et al., 1992)

Criterion	Weight
GNP per capita	42.2%
Propensity to invest	20.1%
Net foreign debt to exports	2.6%
Reserves to imports ratio	0.6%
Current account balance on GNP	18%
Export growth rate	4%
Expert variability	11.2%
Political risk	1.3%

This additive utility model was extrapolated to the rest of the countries. The correlation between the global utilities of the countries with their actual country risk rating was very satisfactory ($r=0.856$). European countries (e.g. Norway, Switzerland, Denmark, West Germany, France, United Kingdom, etc.), United States, Canada and Japan were found to be the best countries according to their creditworthiness, followed by countries such as Singapore, Austria, Rumania, Portugal, Greece, New Zealand, Mexico, Sri Lanka, Egypt, etc. Finally, countries such as Nigeria, Argentina, Bolivia and Zambia were found to be the most risky ones.

3.1. Case study

The preference disaggregation approach was applied in the assessment of country risk using a case study by Tang and Espinal (1989). The aim of this application is twofold: (i) to develop a model for the ranking of the countries from the less risky to the most risky ones, and (ii) to develop a model to classify the countries in predefined homogeneous classes according to their risk. This section is devoted to the description of the case study, and to the development of a ranking model for country risk assessment.

The application involves a sample of 30 countries from different geographical regions. The assessment of their country risk is based on 14 economic indicators involving the external repayment capability, the liquidity, the per capita income and population increases, and the purchasing power risk (Table 4).

Table 4: Economic indicators used in country risk assessment (Source: Tang and Espinal, 1989)

	External repayment capability
X_1	Recent increases of foreign exchange earnings
X_2	Foreign exchange earnings' instability
X_3	Foreign exchange earnings' concentration index
X_4	Current account imbalance as percentage of gross external revenues (GER) during recent periods
X_5	Current account imbalance as percentage of GER increases during recent period
X_6	Imbalance between external debit and credit interest as percentage of GER during recent periods
X_7	Imbalance between external debit and credit interest plus imbalance between profits due and receivable on foreign equity investments as percentage of GER
	Liquidity
X_8	Gross international reserves as percentage of gross external expenditures
X_9	Debt service as a percentage of GER
X_{10}	Interest earned on international assets as percentage of interest due on external debts
X_{11}	Credit balances with banks as a percentage of the amount due to banks reporting to the Bank of International Settlements
	Per capita income and population increases
X_{12}	Per capita income increases
X_{13}	Population increases
	Purchasing power risk
X_{14}	Purchasing power change risk

Tang and Espinal (1989) in their study, using the Delphi method and with the cooperation of experts from international lending institutions determined that the external repayment capability is the most important indicator, both in the short as well as in the long term, followed by liquidity, per capita income and population increases, and finally purchasing power risk. Based on the criteria's weights, determined through the Delphi method, Tang and Espinal (1989) developed two multiattribute country risk models for the long/medium term and for the short term. Comparing the results of these two models with those obtained by the methodologies of two financial institutions A and B, it is evident that the long/medium and short term models can hardly represent the decision policy of the two institutions. More specifically, according to Kendall's τ rank correlation coefficient the only rankings that depict satisfactory consistency, are those obtained by the long/medium and the short term model (τ=0.8391), whereas the rankings defined by the two institutions and the rankings according to the two developed country risk models depict significant differences (the Kendall's τ rank correlation coefficient varies between 0.6045 and 0.6643, see Table 5).

Table 5: Kendall's τ rank correlation coefficient between institution's rankings and the models of Tang and Espinal (1989)

	Long/Medium term model	Short term model	Instit. A	Instit. B
Long/Medium term model	1	0.8391	0.6643	0.6045
Short term model	0.8391	1	0.6505	0.6551
Institution A	0.6643	0.6505	1	0.6367
Institution B	0.6045	0.6551	0.6367	1

In this application the aim is to examine in what extend the UTASTAR method can consistently represent the decision policy (ranking) of the two institutions. Therefore, the UTASTAR method was applied to develop two additive utility models (country risk models) which could explain the preferences of the managers of the two institutions. Table 6 presents the rankings provided by the two institutions (original rankings), the global utilities of each country according to the two developed country risk models and the corresponding estimated rankings in parentheses. The developed country risk models represent very satisfactory the decision policy of the two institutions. More specifically, as far as the country risk model for institution A is concerned, there is only one inconsistency involving Israel (Kendall's τ=0.972). According to the country risk rating of institution A, Israel is ranked in the 20th place, below countries such as Thailand, Indonesia, Venezuela, Mexico, Egypt and Brazil. On the other hand, the estimated country risk model ranks Israel in the 14th place, above the aforementioned countries, which seems to be a more realistic estimation. In the case of institution B, the developed additive utility model is fully consistent with the decision policy of this institution.

Table 6: Ranking according to the country risk models for institutions A and B

Countries	Institution A		Institution B	
	Original ranking	Global utility	Original ranking	Global utility
U.S.A	1	0.613 (1)	12	0.430 (12)
Switzerland	2	0.611 (2)	1	0.694 (1)
Japan	3	0.554 (3)	2	0.441 (2)
Germany	4	0.489 (4)	3	0.547 (3)
U.K	5	0.476 (5)	9	0.448 (9)
Canada	6	0.473 (6)	13	0.426 (13)

Table 6: Ranking according to the country risk models for institutions A and B (continued)

Countries	Institution A		Institution B	
	Original ranking	Global utility	Original ranking	Global utility
Australia	7	0.471 (7)	8	0.451 (8)
France	8	0.469 (8)	4	0.477 (4)
Belgium	9	0.467 (9)	6	0.458 (6)
Italy	10	0.464 (10)	5	0.473 (5)
Malaysia	11	0.462 (11)	14	0.423 (14)
Spain	12	0.460 (12)	10	0.444 (10)
South Korea	13	0.458 (13)	16	0.409 (16)
Thailand	14	0.415 (15)	17	0.357 (17)
Indonesia	15	0.413 (16)	18	0.354 (18)
Venezuela	16	0.411 (17)	7	0.454 (7)
Mexico	17	0.409 (18)	22	0.339 (22)
Egypt	18	0.407 (19)	25	0.329 (25)
Brazil	19	0.404 (20)	23	0.336 (23)
Israel	20	0.431 (14)	15	0.420 (15)
Sri Lanka	21	0.400 (21)	11	0.436 (11)
Turkey	22	0.398 (22)	21	0.344 (21)
Chile	22	0.395 (23)	20	0.347 (20)
Kenya	24	0.393 (24)	19	0.351 (19)
Argentina	25	0.391 (25)	28	0.319 (28)
Philippines	26	0.389 (26)	29	0.316 (29)
Jamaica	27	0.270 (27)	30	0.312 (30)
Costa Rica	28	0.267 (28)	24	0.333 (24)
Honduras	29	0.243 (29)	26	0.326 (26)
Zambia	30	0.182 (30)	27	0.322 (27)

The weights of the evaluation criteria in the two additive utility models are presented in Table 7. From this table it is clear that there are significant differences between the weights in the two models corresponding to Institutions A and B. This is not surprising since it has already been observed that there were also significant differences between the two rankings, and therefore to the decision policies, of the two institutions. This fact complies with the general finding that in preference modelling each decision maker has his own decision policy.

In both country risk models the external repayment capability is the dominant factor, followed by liquidity, per capita income and population increases, and purchasing power risk. In their study, Tang and Espinal have also concluded the same results. However, as far as the specific country risk indicators are concerned there are differences among the two developed models. In the country risk model for institution A, the most important criteria are the current account imbalance as percentage of GER increases during recent period (X_5), the purchasing power change risk (X_{14}), the interest earned on international assets as percentage of interest due on external debts (X_{10}), and the recent increases of foreign exchange earnings (X_1). On the other hand in the country risk model for institution B the most important criteria are the gross international reserves as percentage of gross external expenditures (X_8), the per capita income increases (X_{12}), and the foreign exchange earnings' instability (X_2).

Table 7: Weights of evaluation criteria for the two country risk models developed for institutions A and B

Criteria	Country risk model for institution A	Country risk model for institution B
X_1	10.044%	4.471%
X_2	0.012%	10.128%
X_3	3.844%	3.899%
X_4	8.991%	9.398%
X_5	16.823%	6.183%
X_6	6.037%	7.467%
X_7	9.629%	5.446%
External repayment capability	*55.380%*	*46.992%*
X_8	3.998%	15.474%
X_9	0.000%	4.340%
X_{10}	12.716%	1.964%
X_{11}	0.000%	1.415%
Liquidity	*17.714%*	*23.193%*
X_{12}	8.767%	13.120%
X_{13}	3.921%	7.956%
Per capita income and population increases	*12.688%*	*21.076%*
X_{14}	14.219%	8.739%
Purchasing power risk	*14.219%*	*8.739%*

3.2. Application of the UTADIS and UTADIS I methods

The problem of assessing country risk has been studied in the international literature both as a ranking problem, as well as a classification problem (cf. Saini and Bates, 1984). The latter involves the classification of a set of countries into two or more predefined homogeneous classes (i.e. countries facing debt service problems or not, rescheduling countries or not, etc.). The UTADIS method, a variant of the UTA method, is well adapted in the study of classification problems. The basic difference between the two methods is that instead of comparing each alternative (country) with the others so that a predefined ranking can be reproduced as consistently as possible through an additive utility model, the UTADIS method performs comparisons between the alternatives and the thresholds (utility thresholds) which are used to distinguish the classes so that the alternatives can be classified in their original class with the minimum misclassification error. In this case the estimation of the additive utility model and the utility thresholds is achieved through the following linear programming formulation.

$$\text{Minimize } F = \sum_{a \in C_1} \sigma^+(a) + \ldots + \sum_{a \in C_k} \left[\sigma^+(a) + \sigma^-(a)\right] + \ldots + \sum_{a \in C_Q} \sigma^-(a)$$

s.t.

$$u[g(a)] - u_1 + \sigma^+(a) \geq 0 \qquad \forall a \in C_1$$

$$\left.\begin{aligned} u[g(a)] - u_{k-1} - \sigma^-(a) &\leq -\delta \\ u[g(a)] - u_k + \sigma^+(a) &\geq 0 \end{aligned}\right\} \quad \forall a \in C_k$$

$$u[g(a)] - u_{Q-1} - \sigma^-(a) \leq -\delta \qquad \forall a \in C_Q$$

$$\sum_{i=1}^{m} \sum_{j=1}^{a_i-1} w_{ij} = 1$$

$$u_{k-1} - u_k \geq s \qquad k = 2, 3, \ldots, Q-1$$

$$w_{ij} \geq 0, \ \sigma^+(a) \geq 0, \ \sigma^-(a) \geq 0,$$

where $C_1, C_2, \ldots, C_Q$ are the Q ordered predefined classes (C_1 the best, C_Q the worst), $u_1, u_2, \ldots, u_{Q-1}$ are the corresponding utility thresholds which distinguish the classes (i.e. the utility threshold u_k distinguishes the classes C_k and C_{k+1}, $\forall k \leq Q-1$), σ^+ and σ^- are two misclassification error functions, s is a threshold used to ensure that $u_{k-1} > u_k$ ($s>0$), and δ is a threshold used to ensure that $u[g(a)] < u_{k-1}$, $\forall a \in C_k$, $2 \leq k \leq Q-1$ ($\delta \geq 0$). The w_{ij} and a_i have the same meaning as in the UTASTAR method. A detailed description of the method can be found in Devaud et al. (1980) and Zopounidis and Doumpos (1997a).

Zopounidis and Doumpos (1997c) proposed a variant of the UTADIS method (referred as UTADIS I) which apart of minimizing the misclassification errors also accommodates the objective of maximizing the distances (variation) of the global utility of an alternative (country) from the utility thresholds. This is achieved through the following linear program:

$$\text{Minimize } F = p_1 \sum_a \left[\sigma^+(a) + \sigma^-(a)\right] - p_2 \sum_a \left[d^+(a) + d^-(a)\right]$$

s.t.

$$u[g(a)] - u_1 + \sigma^+(a) - d^+(a) = 0 \qquad \forall a \in C_1$$

$$\left.\begin{aligned} u[g(a)] - u_{k-1} - \sigma^-(a) + d^-(a) &= -\delta \\ u[g(a)] - u_k + \sigma^+(a) - d^+(a) &= 0 \end{aligned}\right\} \quad \forall a \in C_k$$

$$u[g(a)] - u_{Q-1} - \sigma^-(a) + d^-(a) = -\delta \qquad \forall a \in C_Q$$

$$\sum_{i=1}^{m} \sum_{j=1}^{a_i-1} w_{ij} = 1$$

$u_{k-1} - u_k \geq s \qquad k = 2, 3, \ldots, Q-1$

$w_{ij} \geq 0, \sigma^+(a) \geq 0, \sigma^-(a) \geq 0, d^+(a) \geq 0, d^-(a) \geq 0$

where d^+ and d^- are the distances between the global utilities and the utility thresholds, and p_1 and p_2 are weighting parameters of the two objectives of minimizing the misclassification errors and maximizing the distances.

Both the UTADIS method and its variant (UTADIS I) are implemented in the FINCLAS multicriteria decision support system (Zopounidis and Doumpos, 1997b) which is specifically designed for the study of financial classification problems. Although the present form of the FINCLAS system focuses on the assessment of corporate failure risk, it can be easily adapted to a wide range of financial classification problems such as company acquisition, venture capital investments, portfolio selection and management and country risk assessment among others.

According to the country risk rating provided by Institutions A and B (scores), a subjective classification of the countries in three classes was defined (of course, any other classification scheme could be easily adopted to estimate country risk): (i) the first class including creditworthy countries (class C_1), (ii) the second class including countries which are in an intermediate situation as far as their creditworthiness is concerned (class C_2), and (iii) the third class including the most risky countries (class C_3). This trichotomic approach provides much more flexibility compared to the classical dichotomous approach, since using an intermediate (uncertain) class the strict assignment of a country as a risky or a non-risky one is overcame (see Roy and Moscarola, 1977; Zopounidis, 1987; Ferhat and Jaszkiewicz, 1996 for some classification methodologies based on the trichotomic approach). Tables 8 and 9 present the classification results obtained by the application of the UTADIS and UTADIS I methods in the case study of Tang and Espinal (1989), for both institutions A and B.

Table 8: Classification results for institution A

		UTADIS		UTADIS I	
Countries	Original Class	Global Utility	Estimated Class	Global Utility	Estimated Class
Switzerland	1	0.8586	1	0.8900	1
Japan	1	0.7914	1	0.9786	1
Germany	1	0.7287	1	0.9245	1
United Kingdom	1	0.6925	1	0.8769	1
France	1	0.6862	1	0.8870	1
Belgium	1	0.6820	1	0.8632	1
South Korea	1	0.6802	1	0.5424	1
Italy	1	0.6726	1	0.9092	1
Malaysia	1	0.6413	1	0.4918	1
Spain	1	0.6124	1	0.8061	1
United States	1	0.6062	1	0.5381	1

Table 8: Classification results for institution A (continued)

		UTADIS		UTADIS I	
Countries	Original Class	Global Utility	Estimated Class	Global Utility	Estimated Class
Indonesia	1	0.5939	1	0.4292	1
Canada	1	0.5591	1	0.4464	1
Australia	1	0.5582	1	0.3895	1
Thailand	1	0.5582	1	0.3895	1
Utility threshold u_1		0.5582		0.3895	
Israel	2	0.5572	2	0.3885	2
Sri Lanka	2	0.5562	2	0.3885	2
Venezuela	2	0.4463	2	0.3885	2
Egypt	2	0.4387	2	0.2316	2
Chile	2	0.4372	2	0.3040	2
Kenya	2	0.4321	2	0.2834	2
Philippines	2	0.3824	2	0.2182	2
Turkey	2	0.3389	2	0.1395	2
Brazil	2	0.3369	2	0.2239	2
Mexico	2	0.3224	2	0.1524	2
Argentina	2	0.3209	2	0.1852	2
Utility threshold u_2		0.3082		0.1395	
Jamaica	3	0.3072	3	0.1385	3
Costa Rica	3	0.3057	3	0.1385	3
Zambia	3	0.3039	3	0.1385	3
Honduras	3	0.2042	3	0.1343	3

Table 9: Classification results for institution B

		UTADIS		UTADIS I	
Countries	Original Class	Global Utility	Estimated Class	Global Utility	Estimated Class
Switzerland	1	0.9061	1	0.9985	1
Germany	1	0.8762	1	0.9965	1
Japan	1	0.8070	1	0.9952	1
United Kingdom	1	0.8025	1	0.9934	1
Belgium	1	0.7899	1	0.9544	1
Italy	1	0.7855	1	0.9100	1
France	1	0.7762	1	0.9112	1
United States	1	0.7089	1	0.9932	1
Venezuela	1	0.6637	1	0.7731	1
Malaysia	1	0.6223	1	0.5747	1
Spain	1	0.6161	1	0.6535	1
Canada	1	0.5909	1	0.6396	1

Table 9: Classification results for institution B (continued)

Countries	Original Class	UTADIS Global Utility	UTADIS Estimated Class	UTADIS I Global Utility	UTADIS I Estimated Class
Australia	1	0.5836	1	0.4659	1
Sri Lanka	1	0.5816	1	0.5128	1
Utility threshold u_1		0.5785		0.4659	
Israel	2	0.5283	2	0.4010	2
Indonesia	2	0.5288	2	0.4649	2
South Korea	2	0.4673	2	0.4649	2
Kenya	2	0.4494	2	0.2188	2
Turkey	2	0.4030	2	0.2159	2
Thailand	2	0.3953	2	0.2868	2
Chile	2	0.3448	2	0.2159	2
Utility threshold u_2		0.3285		0.2159	
Egypt	3	0.3275	3	0.2149	3
Jamaica	3	0.3222	3	0.2149	3
Costa Rica	3	0.3120	3	0.0095	3
Zambia	3	0.2763	3	0.1505	3
Argentina	3	0.2603	3	0.2149	3
Philippines	3	0.2415	3	0.0102	3
Mexico	3	0.2413	3	0.0301	3
Brazil	3	0.2258	3	0.1861	3
Honduras	3	0.1673	3	0.0059	3

According to the obtained results, both methods (UTADIS and UTADIS I) are able to develop a classification model which correctly classifies all countries to their original class for both institutions A and B. Furthermore, through the global utilities of the countries the competitive level between the countries of the same class can be examined to determine which ones are the most/less risky ones. The obtained results depict some similarity: Switzerland, Germany, Japan, and United Kingdom are almost in any case the best four countries according to their creditworthiness, whereas Honduras is the most risk country of all. Furthermore, an interesting additional feature of the country risk models developed by the UTADIS I method, is that the global utilities of most of the developed countries (Japan, Germany, Switzerland, United Kingdom, France, Belgium, Italy, etc.) differ significantly from the global utilities of other countries.

Table 10 presents the weights of the evaluation criteria for all the classification country risk models developed for the two institutions by the UTADIS and the UTADIS I methods. The weights of the evaluation criteria in the country risk models developed though the UTADIS and the UTADIS I methods differ compared to those obtained through the UTASTAR method. This is mainly caused by the different approach and aim of these methods. The UTASTAR method aims at developing a ranking country risk, whereas the UTADIS and the

UTADIS I methods aim at developing a classification country risk model which could correctly classify the countries into their original class.

Table 10: Weights of evaluation criteria in the classification country risk models

Criteria	Country risk model for institution A		Country risk model for institution B	
	UTADIS	UTADIS I	UTADIS	UTADIS I
X_1	1.430%	0.000%	2.880%	0.000%
X_2	5.460%	0.000%	6.460%	0.255%
X_3	7.700%	11.973%	0.110%	0.475%
X_4	14.160%	16.670%	13.280%	31.822%
X_5	5.890%	0.000%	8.980%	0.094%
X_6	20.930%	11.874%	24.780%	28.367%
X_7	2.750%	0.140%	6.780%	17.331%
External repayment capability	*58.320%*	*40.657%*	*63.270%*	*78.344%*
X_8	8.780%	7.809%	10.760%	0.727%
X_9	7.570%	8.704%	7.820%	0.000%
X_{10}	0.830%	0.000%	0.360%	0.000%
X_{11}	0.000%	0.000%	0.000%	0.000%
Liquidity	*17.180%*	*16.513%*	*18.940%*	*0.727%*
X_{12}	10.290%	7.312%	0.740%	0.151%
X_{13}	7.130%	34.700%	15.720%	20.770%
Per capita income & population increases	*17.420%*	*42.012%*	*16.460%*	*20.921%*
X_{14}	7.070%	0.818%	1.330%	0.008%
Purchasing power risk	*7.070%*	*0.818%*	*1.330%*	*0.008%*

4. Concluding remarks and future directions

This paper focused on the application of the preference disaggregation approach of MCDA in the study of country risk. The country risk problem in this application was studied as a ranking as well as a classification problem. In both cases the obtained results are very satisfactory since the obtained country risk models are consistent with the preferences and the decision policy of the managers of two leading institutions. The use of the three methods (UTASTAR, UTADIS and UTADIS I), illustrated their ability in deriving flexible decision models taking into account the preferences of the decision makers. The decision maker plays a significant role in the decision process by interacting with the methods to take decisions in real time. Furthermore, these methods are free of restrictive statistical assumptions, they are able of incorporating in the decision process qualitative social and political factors, and they can be easily adapted to the changes in the decision environment. Moreover, the Delphi method that Tang and

Espinal (1989) proposed as a tool for country risk assessment, involves a rather complicated and time consuming process of gathering expert's opinions through questionnaires and a series of interviews (Lindstone and Turoff, 1975). The proposed three methods need significantly less information involving only the determination of a preordering of the countries according to their creditworthiness, or an a priori classification of a reference set of countries in classes of risk. This information (preordering or classification) can be easily obtained based on past decisions that the decision maker has already taken.

The scope of this case study was to investigate the potentials of the application of preference disaggregation approaches in the assessment of country risk. In the future, these approaches could be applied in new data, taking also into account social and political factors which were not considered in this application, so that this research could also be of practical interest.

Based on this approach a multicriteria decision support system (MCDSS), such as the FINCLAS system (FINancial CLASsification, cf. Zopounidis and Doumpos, 1997b) could be developed to provide real time support in the study of decision problems related to country risk assessment. Using the three powerful disaggregation methods presented in this paper, and based on economic, social and political indicators, the FINCLAS system could provide integrated support to analysts in the study of country risk, either by ranking the countries according to their creditworthiness, or by classifying them into classes of risk.

References

Benayoun, R., De Montgolfier, J., Tergny, J. and Larichev, O. (1971), "Linear programming with multiple objective functions: Step method (STEM)", *Mathematical Programming* 1, 3, 366-375.

Chevalier, A. and Hirsch, G. (1981), "The assessment of political risk in the investment decision", *Journal of the Operational Research Society* 32, 7,599-610.

Cosset, J.C. and Roy, J. (1988), "Expert judgments of political riskiness: An alternative approach", *Document de Travail* 88-12, Université Laval, Québec, Canada.

Cosset, J.C. and Roy, J. (1989), "The determinants of country risk ratings", *Document de Travail* 89-43, Université Laval, Québec, Canada.

Cosset, J.C., Siskos, Y. and Zopounidis, C. (1992), "Evaluating country risk: A decision support approach", *Global Finance Journal* 3, 1, 79-95.

Ferhat, A.B. and Jaszkiewicz, A. (1996), "Trichotomy based procedure for multiple criteria choice problems", *Cahier du LAMSADE*, no 143, Univesité de Paris-Dauphine.

Jacquet-Lagrèze, E. (1995), "An application of the UTA discriminant model for the evaluation of R & D projects", in: P.M. Pardalos, Y. Siskos, C. Zopounidis (eds.), *Advances in Multicriteria Analysis*, Kluwer Academic Publishers, Dordrecht, 203-211.

Jacquet-Lagrèze, E. and Siskos, Y. (1982), "Assessing a set of additive utility functions for multicriteria decision making: The UTA method", *European Journal of Operational Research* 10, 151-164.

Lindstone, H.A. and Turoff, M. (1975), *The Delphi Method: Techniques and Applications*, Addison-Wesley, Reading, MA.

Mondt, K. and Despontin, M. (September 1986), "Evaluation of country risk using multicriteria analysis", *Technical Report*, Vrije Universiteit Brussel.

Mumpower, J.L., Livingston, S. and Lee, T.J. (1987), "Expert judgments of political riskiness", *Journal of Forecasting* 6, 51-65.

Oral, M., Kettani, O., Cosset, J.C. and Daouas, M. (1992), "An estimation model for country risk rating", *International Journal of Forecasting* 8, 583-593.

Roy, B., and Moscarola, J. (1977), "Procédure automatique d'examem de dossiers fondée sur une segmentation trichotomique en présence de critères multiples", *RAIRO Recherche Operationnele* 11, 2, 145-173.
Saini, K.G. and Bates, Ph.S. (1984), "A survey of the quantitative approaches to country risk analysis", *Journal of Banking and Finance* 8, 341-356.
Siskos, Y. and Yannacopoulos, D. (1985), "UTASTAR: An ordinal regression method for building additive value functions", *Investigação Operacional* 5, 1, 39-53.
Siskos, Y., Spiridakos, A. and Yannacopoulos, D. (1993), "MINORA: A multicriteria decision aiding system for discrete alternatives", *Journal of Information Science and Technology* 2, 2, 136-149.
Tang, J.C.S. and Espinal, C.G. (1989), "A model to assess country risk", *OMEGA: International Journal of Management Science* 17, 4, 363-367.
Zopounidis, C. (1987), "A multicriteria decision-making methodology for the evaluation of the risk of failure and an application", *Foundations of Control Engineering* 12, 1, 45-67.
Zopounidis, C. (1997), "Multicriteria decision aid in financial management", in: J. Barcelo (ed.), *Proceedings of EURO XV-INFORMS XXXIV Joint International Meeting* (in press).
Zopounidis, C., and Doumpos, M. (1997a), "Preference disaggregation methodology in segmentation problems: The case of financial distress", in: C. Zopounidis (ed.), *New Operational Approaches for Financial Modelling*, Springer-Verlag, Berlin-Heidelberg (in press).
Zopounidis, C. and Doumpos, M. (1997b), "FINCLAS: A multicriteria decision support system for financial classification problems", in: C. Zopounidis (ed.), *New Operational Tools in the Management of Financial Risks*, Kluwer Academic Publishers, Dordrecht (in press).
Zopounidis, C. and Doumpos, M. (1997c), "A multicriteria sorting methodology for financial classification problems", *Working Paper* 97-03, Technical University of Crete, Chania, Greece.

Author Index

Printed in the United States
81717LV00002B/60

9 780792 380559